Interfacial Design
and Chemical Sensing

ACS SYMPOSIUM SERIES **561**

Interfacial Design and Chemical Sensing

Thomas E. Mallouk, EDITOR
Pennsylvania State University

D. Jed Harrison, EDITOR
University of Alberta

Developed from a symposium sponsored
by the Division of Colloid and Surface Chemistry
at the 206th National Meeting
of the American Chemical Society,
Chicago, Illinois,
August 22–27, 1993

American Chemical Society, Washington, DC 1994

Library of Congress Cataloging-in-Publication Data

Interfacial design and chemical sensing / Thomas E. Mallouk, editor, D. Jed Harrison, editor.

p.   cm.—(ACS symposium series, ISSN 0097–6156; 561)

"Developed from a symposium sponsored by the Division of Colloid and Surface Chemistry at the 206th National Meeting of the American Chemical Society, Chicago, Illinois, August 22–27, 1993."

Includes bibliographical references and indexes.

ISBN 0–8412–2931–7

1. Separation (Technology)—Congresses.   2. Chemical detectors—Congresses.

I. Mallouk, Thomas Edward   II. Harrison, Daniel Jed   III. American Chemical Society.   Division of Colloid and Surface Chemistry. IV. American Chemical Society.   Meeting (206th: 1993: Chicago, Ill.) V. Series.

TP156.S45I56   1994
660'.2842—dc20   94–19957
                                                                      CIP

Foreword

THE ACS SYMPOSIUM SERIES was first published in 1974 to provide a mechanism for publishing symposia quickly in book form. The purpose of this series is to publish comprehensive books developed from symposia, which are usually "snapshots in time" of the current research being done on a topic, plus some review material on the topic. For this reason, it is necessary that the papers be published as quickly as possible.

Before a symposium-based book is put under contract, the proposed table of contents is reviewed for appropriateness to the topic and for comprehensiveness of the collection. Some papers are excluded at this point, and others are added to round out the scope of the volume. In addition, a draft of each paper is peer-reviewed prior to final acceptance or rejection. This anonymous review process is supervised by the organizer(s) of the symposium, who become the editor(s) of the book. The authors then revise their papers according to the recommendations of both the reviewers and the editors, prepare camera-ready copy, and submit the final papers to the editors, who check that all necessary revisions have been made.

As a rule, only original research papers and original review papers are included in the volumes. Verbatim reproductions of previously published papers are not accepted.

M. Joan Comstock
Series Editor

Contents

STRUCTURALLY TAILORED INTERFACES

CHEMICAL SENSOR DESIGNS

INDEXES

x

Preface

RESEARCH IN THE AREA OF CHEMICAL SENSING has grown dramatically in both academia and industry during the past decade. The demand for new sensors that operate with high specificity and sensitivity, as well as low cost, has been driven by an increasing need for environmental monitoring, by modernization of industrial process control systems, and by the desire for improved biomedical technology. The design of these sensors and of interfaces that bind analytes with molecular specificity is a problem that crosses traditional lines of analytical, biological, inorganic, organic, and physical chemistry. Within the past few years, parallel discoveries have been made in these different disciplines that are now beginning to have broad impact on the science of chemical sensing. New paradigms for the synthesis and characterization of designed interfaces and for their integration into practical devices have emerged from this work.

The purpose of this volume is to bring together some of the best new research on chemically sensitive interfaces from different relevant disciplines. Twenty-six chapters from leading scientists describe new advances in the synthesis of materials, molecular recognition in analyte–receptor systems, the preparation and characterization of surface thin films, new strategies for chemical signal transduction, and the microfabrication and miniaturization of sensor devices. An overview chapter gives a brief historical review of chemical sensors, as well as a general discussion of rational interfacial design, techniques for the characterization of surface structures, and device technology. It is the hope of the editors and authors of this volume that the reader will be introduced, through brief accounts of recent fundamental research, to the full scope of problems and techniques currently associated with the advancing field of chemical sensors.

Generous support by the Division of Colloid and Surface Chemistry of the American Chemical Society and by the Petroleum Research Fund has helped to make the symposium and the writing of this book possible. Their contributions are gratefully acknowledged.

THOMAS E. MALLOUK
Department of Chemistry
Pennsylvania State University
University Park, PA 16802

D. JED HARRISON
Department of Chemistry
University of Alberta
Edmonton, Alberta T6G 2G2
Canada

March 28, 1994

Chapter 1

Chemically Sensitive Interfaces

D. Jed Harrison[1] and Thomas E. Mallouk[2]

[1]Department of Chemistry, University of Alberta, Edmonton,
Alberta T6G 2G2, Canada
[2]Department of Chemistry, Pennsylvania State University, University
Park, PA 16802

This chapter provides a brief overview of the design,
characterization, and function of chemically sensitive interfaces.
Chemical sensing is an interdisciplinary problem involving the
synthesis of new materials, the characterization of materials and
interfaces, the design of new methods of signal transduction, and
the development of techniques for microfabrication and
miniaturization. This chapter summarizes current research in each
of these areas, referring to contributions in the literature and to
specific chapters in this book in which recent advances are
described.

The present level of interest in chemical sensors is driven by the perceived
need for improved methods of monitoring chemical concentrations in a
variety of fields. The principal applications are in industrial process
control, environmental monitoring and remediation, the automotive
industry, clinical diagnostics and biomedical research. The eventual
benefits of having better, or a greater variety of sensors include increased
efficiency and productivity in the manufacturing and energy sectors,
reduced pollution, and improved health care (possibly at lower cost). The
performance and benefits associated with sensors that are presently
available certainly indicate that these are realistically anticipated
outcomes of the extensive effort now being devoted to chemical sensor
development[1-5].

The selectivity of most chemical sensors arises from selective reactions
occurring at interfaces or from differences in the affinity of two contacting
phases for the molecule or ion of interest (the analyte). From this
perspective the study of sensors is an interfacial science, and research on
the chemistry of interfaces plays a significant role in sensor development.

Historical perspective. It is useful to begin this overview chapter by
considering the definition of a chemical sensor. In fact, there are many

0097–6156/94/0561–0001$08.00/0
© 1994 American Chemical Society

definitions available, and we offer here yet another, in which the sensor is described in terms of its function. A sensor is generally required to be a relatively small device, which may be regarded as portable, but a key criterion is that it be able to determine the concentration of one component selectively, in the presence of many others, without significant interference. It is generally assumed that the sensor will function with little or no prior work-up of the sample.

The most well known and well utilized chemical sensor is the pH-selective glass electrode. This sensor is based on electrochemical principles, and is relatively robust, highly selective, and useful for a broad range of samples. Its development is interesting, in that the pH sensitivity of glass was recognized early on, and the sensor was used for many years before the mechanism of its function was understood. In the 1960's and early 70's other sensors based on electrochemical reactions became more commonplace. These included ion-selective electrodes such as the fluoride electrode, based on LaF_3, the silver chalcogenide and halide family of crystalline membranes, and polymer/liquid membranes incorporating a selective ligand for ions of interest, such as K^+ and Ca^{2+}. The enzyme electrodes, which incorporate enzyme reactions that yield electrochemically detected products were also introduced at that time, and were preceded by the now classical Clark and Severinghaus electrodes. These sensors have withstood the tests of time, harsh sample conditions, and economics and are used extensively in a variety of industries. Other sensing schemes based on electrolytic cells were also developed, and are used commercially in sensors for H_2S, Cl_2 and, more recently, in breathalyzers for alcohol. The O_2 sensor, based on a redox cell formed in a ceramic that is conducting at high temperature, is now the property of most automobile owners, owing to the need to control emissions from auto exhaust.

Despite the success of various sensors the need for further improvement in selectivity, durability, detection limits, cost of manufacture and an expanded range of detectable analytes is significant. Gas phase sensing in particular requires improved selectivity; ion sensing would benefit from improved detection limits, and biomedicine demands sensors for many more biochemical analytes than can be readily determined at present.

The sensors described above do not represent the complete range of those available, particularly for gas phase sensors. However, they do represent a significant number of the types of selectivity and transduction schemes that have been used commercially. A key point to make is that many of these sensors were developed through trial and error, and most took advantage of materials that were already available or that could be easily adapted to the application. It seems likely that further development in selective interfaces for chemical sensors will require novel materials, microfabrication strategies, and/or signal transduction techniques.

Rational design of interfaces.   Chemical sensors are inherently multicomponent systems, which contain a responsive "material" - something that interacts chemically with an analyte of interest - interfaced to a measurement device.   Other components, such as chemical, biochemical, or electrochemical relays, membrane barriers or microcolumns that separate out interferences, structural support materials, optical excitation sources, electronic measurement devices, and signal processing software form part of the integrated system that comprises a functioning sensor.   The design of even the simplest such devices is an interdisciplinary task involving elements of interfacial molecular recognition, separations, and signal transduction.   As the repertoire of synthetic tools for preparing complex, supramolecular chemical systems expands[6,7], one can now think of designing, at both the molecular and the macroscopic levels, interfacial assemblies that carry out several of these functions simultaneously.

Two rather different recent examples that illustrate the idea of "integrated" design are a glucose sensor based on internally wired enzymes[8], and a micromachined silicon-based sensor for potassium, phosphate, and other analytes (de Rooij).   In the former, glucose oxidase is modified by covalently binding a pentaamine ruthenium (II) complex at internal sites, in order to electrochemically link the internal flavin group with external redox mediators.   The modified enzymes are encapsulated in a biocompatible hydrogel, an epoxy resin that contains side chains binding both proteins and electroactive relays.   Two other enzymes (lactate oxidase and horseradish peroxidase) are also incorporated into the resin, in order to eliminate the response from ascorbate, urate, and other reducing interferents found in physiological fluids. This multifunctional assembly, coated onto a carbon electrode, acts as an amperometric sensor for glucose in the 1-100 mM range.   In the latter example, microfabrication techniques are used to create piezoelectric pumps on a silicon wafer. Ion-sensitive field effect transistors (ISFETs) detect K^+ ions as the analyte solution (a few microliters) is impelled by an injector micropump into a flow loop and over the ISFET gate region.   In more complex designs, two auxiliary micropumps mix ascorbic acid and molybdate into the analyte stream in order to carry out a multi-step colorimetric analysis for phosphate.

Organized monolayer and multilayer assemblies provide examples of systems that can be engineered at the molecular level, yet perform dual functions, such as molecular recognition and signal amplification, simultaneously.   For example, biomimetic "channel" membranes composed of Langmuir-Blodgett films of polycyclic amines, calixarenes, or cyclodextrins, are selectively opened or closed by analyte ions (Odashima, Umezawa).   The host layers respond selectively to strongly bound analytes, such as ATP^{4-}, and collectively amplify the signal by opening channels to electroactive markers.   In cases where the monolayer is composed of two different components, it may combine the functions of analyte binding (or gating) and surface passivation. Alkanethiols and related molecules adsorb on gold electrode surfaces as dense monolayers that block non-specific responses from electroactive markers (Majda). When these monolayer-forming thiols are adsorbed together with "template" molecules, the latter form defects of molecular dimensions that pass electroactive markers of appropriate size and shape (Crooks). In other cases alkanethiol monolayers can be patterned photolithographically, or by means of a "rubber stamp" technique[9]. The monolayers themselves can then act as templates for the deposition of large-scale structures such as diffraction gratings and microelectronic circuits[10, 11].

Materials for chemically sensitive interfaces. Because the problem of chemical sensing is a broad one, involving analytes that range from single atoms or diatomics to large, complex biomolecules, it follows that an increasing range of materials are finding applications in this area. Finding a material that is responsive to and selective for the analyte of interest is often the "weak link" in the design of a sensor which, thanks to newly developed techniques for signal transduction, might be expected to be magnificently sensitive if the right material could be found. For example, surface acoustic wave (SAW) devices can detect the adsorption of sub-nanogram amounts of material on their surfaces[12]; however, their response is unselective unless there is an overlayer that binds the analyte of interest. The combination of various chemically sensitive materials - metals, metal oxides, polymers, and monolayer-forming organic compounds - with SAW signal transduction has resulted in extremely sensitive devices with improved selectivity (A. J. Ricco, R. A. McGill). Multicomponent sensors containing several SAW devices acting in concert, each with a chemically different overlayer, are described in the chapter by Ricco and coworkers. In this case a pattern recognition strategy is used to maximize the selectivity of the overall sensor response. Zeolite films have also been used to impart selectivity to SAW and related quartz crystal microbalance (QCM) devices (Bein).

An attractive alternative to the use of "off the shelf" materials is the synthesis of new ones, designed for specific binding or reactivity with analytes of interest. Useful materials of this type range from single receptor molecules and simple supramolecular systems (Czarnik) to complex or extended structures such as ion channels (Fyles) or zeolites that are designed for a specific application in molecular recognition or catalysis (Davis). Synthetic principles are now beginning to emerge for the preparation of materials such as non-silicate molecular sieves and coordination solids[13], which have predictable bond connectivity and microscopic channel networks[14, 15, 16]. Despite advances in the rational synthesis of these materials, chance and the prepared mind are still indispensable tools in their "design." Coordination solids containing molecule-sized voids may be interfaced to QCM devices, to form the basis of selective sensors for ammonia and other small molecules (Mallouk).

A new generation of polymeric, biological, thin film, and solid materials is also now finding applications in novel sensor designs. Electropolymerized porphyrins (Bachas) and metal polypyridyl complexes (Abruña) are active materials in sensitive fiber-optic and amperometric detectors for biochemically interesting small molecules such as nitrite and nitric oxide. Using avidin-biotin coupling it is now possible to prepare complex interfacial assemblies in which binding of larger biomolecules, such as proteins, oligonucleotides, and enzyme substrates, can be monitored. Enzymatic amplification techniques have been used to create extremely sensitive QCM-based biosensors of this type[17]. Thin film superlattice assemblies prepared by epitaxial growth of phthalocyanine and fullerene layers provide an interesting and potentially broad opportunity to design new conductimetric sensors for gas-phase analytes (Armstrong). Langmuir-Blodgett and metal phosphonate multilayer structures also suggest interesting new opportunities for structural design (Talham). Finally, new solid materials, such as luminescent porous silicon and high T_c superconductors have interesting interfacial properties that can be modulated photochemically, thermally, or by adsorption of analyte molecules. These materials are now beginning to be explored as active elements in chemical sensors (Bocarsly, McDevitt).

Characterization of interfaces. The characterization of multicomponent interfacial assemblies, in which structural details on the length scale of angstroms or nanometers determine device function, can be a formidable task. Often in these cases a complete picture of the structure and chemistry of the interface must come from the application of several complementary techniques, in much the same way as a complete description of an elephant requires complementary data from several blind men.

Tools for characterization of interfaces at the macroscopic level, for example electrochemistry, vibrational spectroscopy, and electron and ion spectroscopies, have been complemented recently by techniques that now allow one to probe local structure in considerable detail. Scanning tunneling microscopy (STM) and scanning force microscopy (AFM), apart from providing maps of surface topology and electronic structure, can be modified to give chemical information on the scale of molecular dimensions. For example, friction force microscopy (FFM), a modification of AFM, has been used to map frictional force, which is related to intermolecular packing forces and molecular dynamics, in monolayer and multilayer thin films[18]. Scanning electrochemical microscopy (SECM), an electrochemical version of STM, gives information about the conductivity of substrate materials as well as providing a local probe of corrosion and other electrochemical phenomena[19]. In favorable cases, scanning probe microscopies can be used to monitor directly the construction of interfacial structures, for example the formation of metal overlayers on self-assembling monolayers (McCarley), and the epitaxial nucleation and growth of organic crystals on organic substrate materials (Ward).

The availability of synchrotron radiation sources has stimulated the development of interfacial characterization techniques based on scattering of x-rays and photoelectrons. For example, near-edge x-ray absorption fine structure (NEXAFS) analysis of surface monolayer films[20] gives information on molecular orientation that is complementary to that usually obtained by vibrational spectroscopy[21]. X-ray diffraction from interfaces, in particular x-ray standing wave techniques, now appear to be very promising as tools for providing vertical profiles of well-ordered surface structures at angstrom resolution[22].

Finally it should be noted that ultrahigh vacuum techniques such as photoelectron and ion spectroscopies are being applied in creative ways to the characterization of complex interfaces. In this area examples are too numerous to cite in this brief overview, and the reader is referred to the chapter by Pomerantz and coworkers for the application of two of these techniques, electron energy loss spectroscopy (EELS) and secondary ion mass spectrometry (SIMS), to the study of organic monolayer films.

Strategies for signal transduction. Chemists have historically relied on relatively few designs for signal transduction, principally potentiometric and amperometric electrochemical sensors, and have concentrated their abilities on the materials and chemical systems at the front end of the device, although this trend is now changing rapidly. New strategies for converting chemical signals into electronic ones have created new opportunities for the synthetic chemist to make and implement functional materials for chemical sensors. At the same time, as microfabrication techniques push closer to molecular dimensions, chemistry becomes more important in the fabrication process, and the distinction between the chemically responsive material and the transduction end of the device blurs.

Electrochemical methods of transduction from a chemical recognition event to an electronic signal continue to be among the most heavily utilized. In potentiometric measurements there is no net flow of current and the signal is a voltage difference. Typically, the response of such sensors obeys the Nernst equation (1);

$$E = E_c + RT/zF \ln(a) \qquad (1)$$

where E_c is a constant dependent on the sensors design, z is the charge of the ionic species measured, and a is the activity of the analyte, while R,T and F have their usual meanings[23]. This sensing scheme allows the determination of ions that are not redox-active, and so can be quite general. It does suffer from the logarithmic relationship between analyte concentration and signal, which results in lower precision than with amperometrically based sensors. On the other hand, potentiometric sensors frequently exhibit a linear dynamic range of 4 to 8 orders of magnitude. Because these sensors are very sensitive to the change in charge at an interface they may be well suited for use with monolayer films[24]. Conventional ion-selective electrodes can also be coupled to enzyme or immunosensor reaction schemes to detect the products of biochemical reactions[25].

Voltammetric sensors usually involve control of the electrode potential and measurement of the current flow at the applied voltage[25, 26]. The analyte concentration is usually linearly proportional to the current, which can lead to good precision. Current measurements are inherently less sensitive to noise than potentiometric measurements, and this often means better detection limits and improved precision relative to potentiometric measurements. Amperometric detection is a subset of voltammetry, in which the potential is held constant. Amperometric detection usually provides the best detection limits, due to low background signal, although pulsed potential methods can also lead to improved signal to noise performance. Use of some form of voltammetry has become the dominant electrochemical detection scheme for sensors developed more recently, presumably due to these advantages. The products of various enzyme reactions can be detected amperometrically, and selective polymers or other semipermeable coatings that allow only certain redox species to reach the electrode are also easily compatible with voltammetric detection. One difficulty is that most electrode surfaces are easily deactivated (poisoned) by electrode reaction products, or adsorption of proteins, sulfur containing compounds and other chemicals. This means the sample solutions must be carefully controlled, or protective coatings must be used for the sensor when it is in a harsh environment [27].

Thermocouples and thermopiles have been used as transduction devices for chemical sensors[3]. These are usually based on a measurement of the heat of reaction and have been coupled to specific enzyme reactions for selectivity. Sensitivity and selectivity of these devices has yet to reach truly satisfactory levels, but there may be specific applications in which they are practical. They do offer the advantage of reduced susceptibility to fouling, compared to electrochemical detection methods.

Changes in mass arising from complexation or absorption reactions provide another means of signal transduction. The mass changes are usually very small, but the use of oscillating, piezoelectric devices for which the frequency of oscillation is sensitive to the mass of the device presents a way of measuring sub-nanogram mass changes[12]. Application of a voltage to a piezoelectric material such as quartz sets up an oscillation. The oscillation may involve the bulk material, as in the quartz crystal microbalance (QCM)[28], or a surface wave, such as in the surface acoustic wave (SAW) device[12]. The QCM can be used for gas or liquid phase determinations, while the SAW device, which is much more sensitive owing to its ten-fold higher frequency of oscillation, is generally believed to function best in the gas phase. Other oscillation modes also exist, but are not as thoroughly explored at present as the two mentioned here. Originally, researchers hoped that these sensors would respond only to changes in mass, but it is now recognized that there is also substantial sensitivity to changes in the visco-elastic properties of the selective coating materials, as well as to their conductivity, and to the efficiency of coupling the acoustic energy across the interface between the transducer and the coating[29]. Nevertheless, these devices can be highly sensitive and such effects can be controlled, accounted for, or even taken advantage of to improve sensitivity or selectivity[30]. These devices have allowed materials that might not have provided a good signal with other transduction methods to be utilized for sensing applications. The transducers themselves are not highly selective, so that considerable development of selective membranes and coatings has been required.

Optical methods of detection were not as common in the earlier years of sensor development as they have become recently. Optical fibers have attracted considerable attention as a means of developing remote sensing techniques[31]. The simplest schemes involve direct measurements of the absorbance or fluorescence of a sample solution. Wavelength ranges from the near infrared to the ultraviolet have been used with these fibers. More complicated sensors require the use of a surface coating on the tip of the fiber where it is exposed to solution. The coatings provide chemical selectivity and generate an optical signal as a result of binding or other chemical reaction. The thermodynamics of such sensors have been the subject of controversy recently[32], but valid relationships between signal and concentration have been experimentally verified. Fluorescence detection can be extremely sensitive, and so offers detection limits that can exceed those of electrochemical methods[33]. However, these systems again rely almost entirely on the chemistry of the selective coatings for their performance, so that generalized comparisons to other transduction methods are not that helpful. The optical detection methods do offer some advantages over other methods though, in that optical methods of analysis are so common, so they are easily incorporated into existing analytical protocols and measurement systems.

In physical sensors, such as pressure sensors and accelerometers, capacitance measurements are a common means of signal transduction. Such methods can also be used with chemical sensors[34], however, they have not received extensive attention to date. For sensors applied to the liquid phase, this may be due to the complexities associated with the impedance of the double layer that forms in solution at a charged interface. This tends to make the sensor sensitive to changes in ionic strength, unless the time constant of the selective coating that is relevant to the analytical measurement is considerably different from that of the double layer. Further, many chemical reactions associated with a coating may cause a relatively small change in capacitance compared to changes induced in other parameters. Some thin film coatings, though, could perhaps be very well suited to capacitive transduction schemes, since their dielectric constants or thickness may change significantly upon absorption of an analyte.

Microfabrication and miniaturization. Integrated circuit (IC) technology is the combination of photolithographic techniques used to pattern a surface with various thin film deposition, oxide growth, and silicon doping methods for the fabrication of microelectronic circuits. A detailed discussion of these methods is outside the scope of this chapter, and the interested reader is referred to literature sources[2-4]. This technology allows the batch fabrication of hundreds of integrated electronic circuits on wafers of silicon. Various aspects of the technology can also be used to fabricate sensors, and have been applied to a variety of chemical sensors. The power of the technology is tremendous, and it has attracted considerable attention from sensor researchers for this reason.

Since the introduction of the ion-selective field effect transistor (ISFET) by Bergveld[35], and its substantial characterization by Janata and colleagues[36], there has been a tremendous interest in the use of integrated circuit (IC) fabrication technology for chemical sensors. At the same time, sensors for many physical parameters, such as pressure, temperature, and electric or magnetic fields, have been successfully developed using IC technology. Commercialization of chemical sensors fabricated using IC techniques has developed more slowly, however, there are now at least 3 different pH sensitive ISFETs on the market in North America.

The potential benefits of using IC techniques to fabricate sensors have often been presented[36], but will be briefly discussed here. One advantage often cited is the ability to integrate both the sensing element and signal conditioning electronics on a single structure. This can improve signal to noise performance, may serve to linearize the sensor's output, and can greatly ease the engineer's task of incorporating the sensor into a larger system such as in an automobile. However, these advantages are apparently not as significant in chemical analysis as they have been with physical sensors, at least given the present state-of-the-art. A more significant advantage is the ability to easily fabricate devices with small size, which can be used to reduce sample and reagent volumes, or to make a more portable or less obtrusive sensor.

The ability to make many sensors for the same or different chemicals is another potential advantage, but relatively few concepts presented in the literature have really utilized this. A stimulating exception is the multichannel microphysiometer developed by Molecular Devices, which consists of a system with multiple microwells on a silicon substrate, into which biological cells are immobilized[37]. Passage of agonists or antagonists over the cells changes their rate of metabolic function, which is evidenced by a change in the rate the cells acidify the solution in the microwell. A pH sensor integrated into the bottom of each well measures the rate of pH change, providing evidence of changes in the metabolic rate of the cells. Eight sensor/microwell combinations are integrated onto a device, making parallel toxicology studies easy to perform, and it is clearly possible to produce many more sensors in parallel. While alternative technologies for such analyses exist, only IC technology could realistically provide a means to introduce a pH sensor into the bottom of each microwell. In contrast, the need for single stand-alone pH ISFETs designed to replace the glass pH electrode in typical laboratory situations is much less obvious, unless the pH sensitive layer on the ISFET has better performance than does glass.

Another interesting advantage of IC technology is that it allows for the use of materials with interesting properties for sensing that might not otherwise function with conventional transducers, or fabrication of transducers with improved performance. For example, highly resistive materials can be coated on the gate of an ISFET or between interdigitated electrodes spaced a few micrometers apart and used for sensing. When used in conventional designs their high resistance would lead to low sensitivity or poor signal to noise, if a signal could be detected at all. The impedance matching of the ISFET gate, or the short separation of contact points with the interdigitated electrodes can overcome these problems. Similarly, small changes in capacitance, associated with absorption of an analyte into a polymer coating can be detected more easily with these devices than with conventional transducers.

The surface acoustic wave device (SAW) is an example of a transducer that is batch fabricated using IC technology and provides improved performance. SAW devices operate at much higher frequencies than the quartz crystal oscillator (or microbalance, QCM), and this results in improved detection limits[12, 29, 30]. This can make measurements of absorption into films coating the SAW device possible, under circumstances where the QCM is insufficiently sensitive. On the other hand the QCM can be used in aqueous systems, while the SAW device is essentially restricted to gas phase measurements. Here too, IC techniques have provided means to fabricate thin membranes that can be made to oscillate at frequencies similar to the SAW device, but in a mode that is not over-damped in aqueous solutions. Nevertheless, regardless of the specific oscillator involved, it is the coating films and interfaces that provide the chemical specificity required of the sensor.

In electronics, the ability to make large numbers of chips has driven the cost of very complex circuits to extremely low levels. It had been hoped that the same would be true of IC-fabricated chemical sensors. This has not yet really been the case. A large part of this stems from the complexity of packaging an integrated chemical sensor, since the sensing element *must* be exposed to the gas or solution test phase, but the delicate electronics and other components of the silicon substrate must be well isolated. Techniques to overcome this problem have been developing[38, 39]. However, chemical sensing applications are likely to be unique enough that no one solution is viable, so that packaging will probably always represent a significant cost of a sensor and limit the advantage that IC technology offers in lowering fabrication cost. This is a problem not addressed by researchers focusing on designing highly selective interfaces, and while it is apparently less exciting, it is a highly important key to the overall development of successful integrated sensors.

The ability to fabricate large numbers of sensors does offer another advantage, which is related to the reduced cost per fabricated, but unpackaged sensor. When an array of sensors is prepared with somewhat differing selectivities it is possible to use multi-variate analysis techniques, or pattern recognition, to analyze mixtures[40]. This can offer advantages in sensor development, since the design of highly selective sensing elements can be extremely costly and time consuming. Pattern recognition methods are still developing and can have difficulties with quantitative analysis in complex mixtures, or when a sensor array sees a sample composition it was not "trained" for. Nevertheless, the judicious use of partially selective sensing elements in an array, combined with pattern recognition remains a promising field.

In the past decade a technology known as micromachining has developed out of IC technology[41]. It uses the same techniques, but instead of fabricating electronic circuits on a chip, three dimensional physical structures on a micrometer scale can be made. Intriguing devices such as 100 micrometer electrostatic motors, tweezers, bridges and oscillating beams, 10 nm filters, micropumps and valves have all been made in silicon and other IC compatible materials.

Micromachining methods are now being applied to the fabrication of integrated chemical analysis systems on a chip[42]. A system approach to sensor design offers advantages compared to fabricating stand-alone sensors on a chip. This is because a variety of elements can be incorporated together to maximize performance and lifetime of the sensor. Manz, de Rooij, and their colleagues have used this approach to make flow injection systems on a chip, using micromachined pumps and valves for fluid control and optical detection for sensing, as discussed earlier[43, 44]. In this way complete assays can be performed on a chip, with masking reagents or color forming reagents added in sequence, so that the device functions like a bench top system, yet has the size, speed and selectivity of a sensor. There is considerable promise to this approach, as it can be used to integrate an optimal sample treatment with a sensor/detector on chip. Such integration could reduce the demand for extreme selectivity and durability in harsh conditions, by allowing for pretreatment of the samples before they reach the sensor itself, thus improving the ultimate performance and lifetime.

Outlook. The increasing need for practical sensors of various types is helping to drive scientific development in a variety of areas. The purpose of the Symposium upon which this book was based, and of the collection of chapters in this book, is to present the best new science in this area to scientists and engineers interested in various aspects of chemical sensing, and to see if such a collection of fundamental research can stimulate the development of practical devices. New developments in materials design, construction of "molecule precise" interfacial assemblies, microfabrication, and signal transduction techniques are expected to lead to a generation of faster, more sensitive, and more accurate devices for a variety of applications.

One of the constant challenges in this area is the integration of molecular (chemical) and engineering approaches to the design of better sensors. IC technology leads to devices with micrometer scale dimensions. In contrast, many of the approaches to designing selective interfaces discussed in subsequent chapters involve nanometer or molecular dimensions. In this sense they do not take direct advantage of IC techniques. Rather, if the sensitive interfaces are designed in a manner that is compatible with IC fabrication, then the new materials or interfaces that are developed will be able to make use of integration technologies. The reverse is certainly true, that is, sensors fabricated using IC technology can clearly benefit from further development of chemically selective interfaces, as the range of selective materials presently available appears to limit their development. Microfabrication of sensor arrays combined with further development of selective interfaces represents a synergistic combination of research areas that will undoubtedly prove fruitful.

Literature Cited

(1) *Chemical Sensors*, Edmonds, T.E., Ed., Blackie, London, 1988.
(2) Janata, J. *Principles of Chemical Sensors,* Plenum Press, New York, **1989**.
(3) *Solid State Chemical Sensors*, Janata, J.; Huber, R.J., Ed. Academic Press, New York, 1985.
(4) Bergveld, P.; Sibbald, A. *Comprehensive Analytical Chemistry, Vol XXIII, Analytical and Biomedical Applications of Ion-Selective Field-Effect Transistors.* Svehla, G. Ed. Elsevier, New York, 1988.
(5) *Biosensors and Chemical Sensors*, Edelman, P.G.; Wang, J., Ed. ACS Symp. Ser. 487, 1992.
(6) Lehn, J. M. *Angew. Chem. Int. Ed. Engl.* **1990**, *29*, 1304.
(7) Bard, A. J., *Integrated Chemical Systems*, Cornell University Press: Ithaca, NY (1994).

(8)   Maidan, R.; Heller, A. *J. Am. Chem. Soc.* **1991**, *113*, 9003.
(9)   Kumar, A.; Whitesides, G. M. *Science* **1994**, *263*, 60.
(10) Dulcey, C. S.; Georger, J. H., Jr.; Krauthamer, V.; Stenger, D. A.; Fare,
      T. L.; Calvert, J. M. *Science* **1991**, *252*, 551.
(11) Tarlov, M. J.; Burgess, D. R. F., Jr.; Gillen, G. *J. Am. Chem. Soc.* **1993**,
      115, 5305.
(12) Ballantine, D. S., Jr.; Wohltjen, H. *Anal. Chem.* **1989**, *61*, 704A.
(13) Stein, A.; Keller, S. W.; Mallouk, T. E. *Science* **1993**, *259*, 1558.
(14) Robson, R.; Abrahams, B. F.; Batten, S. R.; Gable, R. W.; Hoskins, B.
      F.; Liu, J. in *Supramolecular Architecture, ACS Symp. Ser. 499*,
      Bein, T. Ed., American Chemical Society, Washington, DC, 1992, pp.
      256-273.
(15) Haushalter, R. C.; Mundi, L. A. *Chem. Mater.* **1992**, *4*, 31.
(16) Bowes, C. L.; Ozin, G. A. *Mater. Res. Soc. Symp. Proc.* **1993**, *286*
      (Nanophase and Nanocomposite Materials), 93.
(17) Ebersole, R. C.; Miller, J. A.; Moran, J. R.; Ward, M. D. *J. Am. Chem.
      Soc.* **1990**, *112*, 3239.
(18) Overney, R.M. et. al. *Nature* **1992**, *359*, 133.
(19) Bard, A. J.; Denault, G.; Lee, C.; Mandler, D., Wipf, D. O., *Acc. Chem.
      Res.* **1990**, *23*, 357.
(20) Outka, D.A.; Stohr, J.; Rabe, J.P.; Swalen, J.D. *J. Phys. Chem.* **1988**,
      88, 4076.
(21) (a) Allara, D. L.; Nuzzo, R. G. *Langmuir* **1985**, *1*, 45; (b) Allara, D. L.;
      Nuzzo, R. G. *Ibid.* **1985**, *1*, 52.
(22) Abruña, H. D.; Bommarito, G. M.; Acevedo, D.; *Science* **1990**, *250*,
      69.
(23) Cammann, K. *Working with Ion-Selective Electrodes: Chemical
      Laboratory Practice* Springer-Verlag, Berlin, **1979**.
(24) Aizawa, M.; Kato, S.; Suzuki, S. *J. Memb. Sci.* **1977**, *2*, 125.
(25) Guilbault, G.C. *Analytical uses of Immobilized Enzymes* Marcel
      Dekker, New York, **1984**.
(26) Heineman, W.R.; Halsall, H.B. *Anal. Chem.* **1985**, *57*, 1321A.
(27) Moussy, F.; Harrison, D.J.; O'Brien, D.W.; Rajotte, R.V. *Anal. Chem.*
      1993, *65*, 2072.
(28) Ward, M.D.; Buttry, D.A. *Science* **1990**, *249*, 1000.
(29) Martin, S.J.; Granstaff, V.E.; Frye, G.C.; *Anal. Chem.* **1991**, *63*, 2272.
(30) Ricco, A.J.; Martin, S.J. *Thin Solid Films* **1992**, *206*, 94.
(31) Dessy, R.E. *Anal. Chem.* **1989**, *61*, 1079A.
(32) Janata, J. *Anal. Chem.* **1992**, *64*, 921A.
(33) Seitz, W.R. *CRC Crit. Rev. Anal. Chem.* **1988**, *19*, 135.
(34) Bataillard, P.; Gardies, F.; Jaffrezic-Renault, N.; Martelet, C.; Colin, B.;
      Mandrad, B. *Anal. Chem.* **1988**, *60*, 2374.
(35) Bergveld, P. *IEEE Trans. Biomed. Eng.,* **1972**, *BME-19*, 342.
(36) Janata, J.; Huber, R.J. *Ion Sel. Rev.* **1979**, *1*, 31.
(37) Parce, W.J. et al. *Science* **1989**, *246*, 181.
(38) Goldberg, H.D.; Liu, D.P.; Hower, R.W.; Poplawski, M.E.; Brown
      Tech.Dig. IEEE Solid-State Sens. Actuat. Workshop Hilton Head
      Island, S.C., June 22-25, **1992**, pp 140.

(39) Bowman, L.; Meindl, J.D. *IEEE Trans. Biomed. Eng.***1986**, *BME-33*, 248.
(40) Ballantine, D.S.; Rose, S.L.; Grate, J.W.; Woltjen, H. *Anal.Chem.* **1986**, *58*, 3058.
(41) Wise, K.D.; Najafi, K. *Science* **1991**, *254*, 1335.
(42) Harrison, D.J.; Fluri, K.; Seiler, K.; Fan, Z.; Effenhauser, C.S.; Manz, A. *Science* **1993**, *261*, 895.
(43) van der Schoot, B.H.; Jeanneret, S.; van den Berg, A.; de Rooij, N.F. *Sens. Actuat. B,* **1992**, 57.
(44) Effenhauser, C.S.; Manz, A.; Widmer, H.M. *Anal. Chem.* **1993**, *65*, 2637.

RECEIVED March 25, 1994

New Materials

Chapter 2

Design of Thin Films with Nanometer Porosity for Molecular Recognition

T. Bein[1] and Y. Yan[2]

Department of Chemistry, Purdue University, West Lafayette, IN 47907

Chemical sensors based on selective sorption in microporous oxide films are described. Several strategies for the formation of thin films with well-defined sub-nanometer pores are presented. These include the attachment of molecular sieve crystals and organically modified clays on gold substrates. Zeolite crystals are attached to gold electrodes on quartz crystal microbalances *via* molecular anchoring groups, and combined with glass overlayers. In situ nitrogen and organic vapor adsorption studies on the films show that substantial zeolitic microporosity is accessible such that highly selective sensors based on molecular size could be designed. Uptake of molecules small enough to enter the zeolite pores can be one hundred times greater than that of molecules with kinetic diameters greater than the zeolite pores. The selective adsorption can be tailored by variation of the zeolite composition, pore sizes, and other parameters. The selectivity of these stable films was further controlled by ion exchange in the film, effectively incorporating "gate" functions into the layers. The structure and sorptive behavior of the above films will be compared with microporous films based on organically modified clays.

A chemical sensor[1] is a device that can monitor concentrations of gases or liquids *continuously on site* by converting a chemical interaction (mostly on surfaces) into an electronic or optical response. Chemical sensors will play an increasingly important role in environmental monitoring (ground water, plant effluents, waste disposal sites, car exhaust), automated manufacturing processes, and medical monitoring. The ability to continually measure important parameters in these and other applications will be the key to greater efficiency, time- and energy-savings, and environmental safety and health in the industry of the coming decades. For example, it will be much more efficient to place *in-situ* sensors in the periphery of a waste disposal site to monitor

[1]Current address: YTC America, Inc., 550 Via Alondra, Camarillo, CA 93012

0097–6156/94/0561–0016$08.00/0
© 1994 American Chemical Society

toxic effluents and ground water quality instead of collecting soil samples that have to be analyzed individually in a central laboratory.

A number of physical devices with chemical sensitivity have been developed previously, including the quartz crystal microbalance (QCM) and other acoustic wave devices, semiconductor gas sensors, and various chemically sensitive field effect transistors. However, based on their intrinsic detection principles, most of the known solid state chemical sensors are not <u>selective</u>, i.e., they respond to more than one or a few chemical species. There is an urgent demand for new families of selective, microscope sensors that can eventually be integrated into microelectronic circuits. We have embarked on a program aimed at the design of *conceptually new microporous thin films with molecular recognition capabilities*. On the surface of chemical sensors, these membranes will serve as "molecular sieves" that control access of selected target molecules to the sensor surface.

Piezoelectric acoustic wave device such as the quartz crystal microbalance[2] offer many attractive features as vapor phase chemical sensors: small size, ruggedness, electronic output, sensitivity, and adaptability to a wide variety of vapor phase analytical problems.[3] The QCM is based on a piezoelectric quartz substrate coated with keyhole pattern electrodes on opposite surfaces on the crystal. Mass changes Δm (in g) per face of the QCM cause proportional shifts Δf of the fundamental resonance frequency F, according to

$$\Delta f = -2.3 \times 10^{-6} F^2 \, \Delta m / A$$

where A is the electrode surface area (in cm^2). It follows that the 9 or 5 MHz QCM is capable of detection of mass changes at the nanogram level.[4,5,6]

Zeolite molecular sieves are crystalline, porous inorganic solids, typically aluminosilicates, with channel diameters ranging from 0.3 to 1.2 nm and beyond, and crystal sizes typically between 0.5 and 5 μm.[7,8] If species are selectively adsorbed into the well-defined channel systems of zeolites, the sensor response can be made selective for those species, while larger molecules are only adsorbed on the outer surface of the sensor membrane. Furthermore, many zeolites show ion exchange capability which introduces numerous possibilities for intrapore modifications.

Experimental

Porosity characterization of the thin films.

Nitrogen sorption isotherms. These were obtained at liquid nitrogen temperature in a computer-adjusted mass flow controller system (Unit Instruments Inc.). Nitrogen partial pressures in helium were adjusted over the range of 0-0.95. QCMs coated with zeolite and clay thin films were usually pretreated at 200°C in helium for 2 h. The zeolites were dehydrated under these conditions, as shown by stabilization of the QCM frequency and from related FTIR experiments of similar films. Data acquisition and analysis was performed with DAS-16 analog-digital I/O boards (Keithley MetraByte Co.). The frequency changes of the coated QCMs upon nitrogen sorption/desorption and the nitrogen partial pressures were monitored at intervals of one second. **Dynamic vapor sorption kinetics and isotherms.** Different vapor concentrations were generated with gravimetrically calibrated vapor diffusion tubes at 25°C under constant flow of carrier gas (15 ml/min helium). Vapor concentrations were adjusted by dilution with a second computer controlled helium flow (0-200 ml/min). Sorption measurements were carried out similar to those described above for nitrogen. Equilibrium was usually assumed and the next partial pressure was adjusted when the frequency change of the QCM was less than 1 Hz in 90 seconds. The close coincidence of adsorption and desorption branches of many isotherms shows that the measurements are often close to true equilibrium.

Results and Discussion

Zeolite Single Layers Attached through Molecular Coupling Agents. Our recent development of molecular sieve-based composite films has introduced a novel means for tailoring vapor/surface interactions. [9-15] Our initial approach was focused on zeolite/glass microcomposites derived from sol/gel suspensions that were coated on surface acoustic wave devices. However, even though zeolite/glass composites have favorable stability and show zeolite microporosity, the glass matrix can introduce some additional nonselective porosity and/or obstruct part of the zeolitic porosity.

An alternative means of attachment is a monolayer of a reactive coupling agent that can bind the zeolite crystals to the sensor surface. The coupling agent 3-mercaptopropyl-trimethoxysilane, $HS(CH_2)_3Si(OCH_3)_3$, was used as a bifunctional molecular precursor for anchoring zeolite crystals to the gold electrode. [11,15] Gold shows strong specific interactions with thiol groups that permit formation of self-assembled monolayers in the presence of many other functional groups. [16-18] Figure 1 summarizes the idealized process of the formation of a zeolite-coated QCM. A related system consisting of SnO_2/cationic disilane/zeolite arrangements has been reported for the self-assembly of redox chains on electrodes. [19]

Adsorption of many organic vapors, including small alcohols, chlorinated hydrocarbons, benzene, toluene, as well as water was demonstrated in these films. Dynamic sorption isotherms of organic vapors and nitrogen as well as the transient sorption behavior of organic vapor pulses were studied to characterize the zeolite-coated QCMs. The regular micropores (0.3-0.75 nm) of the QCM-attached zeolite crystals were found to efficiently control molecular access into the coating. Selectivity of the frequency response in excess of 100:1 towards molecules of different size and/or shape (molecular sieving) could be demonstrated.

The thin films can equilibrate and desorb vapors within a few seconds to minutes, often at room temperature, while the bulk materials need substantial heating (ca. 200-300°C) to remove the absorbed vapors (the last ppm desorption steps of water from polar zeolite films at r.t. can take 30-50 min). Thus the sensor response occurs at a satisfactory time scale. The kinetics of vapor desorption from the zeolite layers are strongly dependent on the adsorbate/zeolite combination, thus providing an additional capability for molecular recognition.

Ethanol sensor. As indicated above, zeolites discriminate molecules not only based on size but also by the strength of surface interactions (heat of adsorption). We demonstrate that silicalite-based films on QCMs show selective response towards ethanol in competition with water and larger organics such as 2,2,3-trimethylpentane.[12] The preparation of the microporous layer involved two steps: First, silicalite crystals (about 3 μm diameter) were chemically anchored to the QCM gold electrodes via a thiol-organosilane coupling layer. Silicalite is a crystalline molecular sieve of approximate composition SiO_2 with a pore system of zig-zag channels along A (free cross-section ca. 5.1 x 5.5 Å), linked by straight channels along B (5.3 x 5.6 Å).[20] The silicalite crystals bonded to the QCM electrodes were then further coated with an amorphous, porous silica layer, prepared via sol-gel processing from $Si(OEt)_4$. The films have the following typical adsorption characteristics: Saturation of the micropores at 20°C (derived from α-plots) was achieved with 0.12 μg/(μg film) of ethanol, corresponding to a microporosity of 0.15 cm³/g.

The selective vapor responses of the QCM coated with the silicalite composite layer are illustrated by the sorption of pure vapors of ethanol (ca. 4.3 Å diameter), 2,2,3-trimethylpentane (iso-octane, ca. 6.2 Å) and water (2.65 Å) in the range of 0-2000 parts per million (ppm) in moles (Figure 2). The sorption of ethanol (top curve) shows the largest and fairly linear response as a function of vapor concentration. In contrast, the response of the sensor to iso-octane (bottom line) exhibits an almost negligible change with increasing vapor concentration. Exclusion of 2,2-dimethylbutane from the microporous coating was also observed. The results show that novel sensing layers with highly effective molecular sieving functions and very low external surface areas can be designed. Sorption of the small water molecules (middle line) shows also a small response. Although water sorption slightly increases with increasing concentration, it already approaches a rectilinear isotherm at these low concentrations. This behavior differs drastically from the sorption of ethanol.

The nature of the hydrophobic sensing layer and the preferential adsorption of ethanol over water on a pre-loaded (7% water, which appears to block strongly adsorbing defect sites) QCM can be understood by evaluating the isosteric heats of absorption. [21] Water sorption on a polar surface is through specific interactions with cations or hydroxyls. When such energetic sites (cations) are "blocked" by molecules presorbed in the silicalite layer, the isosteric heat of adsorption of water over the rang of 8-14 ng/μg in the sensing layer is about 6.6 kcal mol^{-1}, substantially below that of the heat of liquefaction of water (10.5 kcal mol^{-1} at 25°C) and similar to that reported for pure silicalite. Thus it is energetically unfavorable for water to condense into such a system. In contrast, for ethanol an isosteric heat of 11 kcal mol^{-1} is obtained over a wide sorption range of 20-80 ng/μg. This value is *greater* than the heat of liquefaction of ethanol (10.1 kcal mol^{-1} at 25°C). The favorable sorption of ethanol vs. water also illustrates that pronounced micropore filling by ethanol can occur at low concentrations. Some initial competitive measurements of ethanol with other vapors confirm the high selectivity discussed above.[12]

Gate functions. An intriguing possibility exists to manipulate adsorption into the zeolite-based films by appropriate gate functions. If molecular entry into the zeolite pore volume is controlled on the crystal surface or inside the crystal, one can "switch" the selectivity from one molecular size to another. We demonstrate this concept with the example of zeolite A, the well-known material used to dry solvents.[22] The eight-ring openings of zeolite A into its large cage have a diameter of only about 4.5 Å. When this zeolite is exchanged with large, singly-charged cations, the free opening

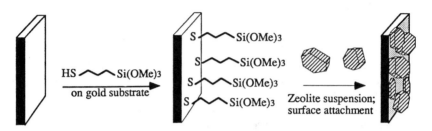

Figure 1. Anchoring process of zeolite crystals on a gold surface *via* a thiol-alkoxysilane layer.

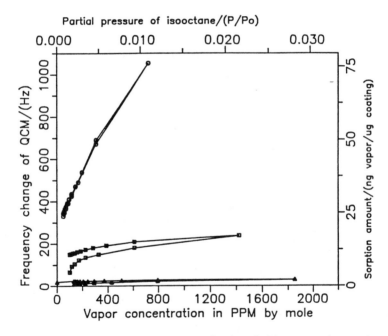

Figure 2. Sorption isotherms at 20°C of ethanol (o), water (squares) and isooctane (Δ) on a QCM coated with silicalite/silica composite. From ref. 12. The QCM frequency shifts in this and the following plots were multiplied with (-1).

is narrowed and the molecular selectivity changes. Thus, zeolite 5A contains Ca(II), 4A contains Na(I), and 3A is the potassium form. The numbers indicate the approximate effective pore openings in Å.

Piezoelectric devices (QCMs) were coated with the Ca-form of zeolite A, using a bifunctional thiol-monolayer for attachment as described above. This film shows a Type I nitrogen sorption isotherm indicating microporosity accessible to nitrogen, as expected for this zeolite. In order to switch the film to a different selectivity, the entire QCM was submerged in dilute Na(I), K(I), or Rb(I) solutions. Ion exchange could be demonstrated by the weight change of the film. Nitrogen sorption isotherms are extremely sensitive to this treatment: the Na-form is already almost completely blocked (sorption is extremely slow), while no nitrogen sorbs in the K(I) and Rb(I) forms. Accordingly, the sorption isotherms corresponding to the latter status of the film are Type II with only external sorption. These changes in pore opening become even more obvious when organic vapors are sorbed in the different versions of the films. For example, the Ca-form sorbs water (2.7 Å), ethanol, and n-hexane, while the Na-form already shuts down for hexane (4.3 Å), and sorbs ethanol (4.3 Å) only weakly. A film highly selective for only water is formed when it is exchanged with potassium ions (Figure 3). Ethanol, hexane, and other larger hydrocarbons are effectively excluded from this film (Figure 4). These dramatic changes show the powerful control of sorption into such films by simply exchanging the charge-balancing cations of the zeolite structure.

Immobilization of Organoclays on Piezoelectric Devices. Organoclays (layered silicate materials ion-exchanged with alkyl ammonium cations) are interesting complements to zeolite sorbents because their sorption selectivities can be controlled to a large extent by the choice of layer charge density, and structure and chain length of the alkylammonium ions.[23-27] Different combinations of these factors will produce very different interlayer environments, containing polar mineral surfaces, hydrophobic organic chains, and paraffin-like layers. The organoclays can offer preferential sorption of nonpolar and flat molecules (Figure 5).

The host discussed in the following was hectorite of approximate composition $Na_{0.3}(Li_{0.3}Al_{0.01}Mg_{2.67})Si_4O_{10}(OH)_2$, with cation exchange capacity of 8 meg/10 g, and charge density of *ca.* one monovalent cation per 100 Å^2. Hectorite is a layered material of the smecitite family, having two tetrahedral silicate sheets fused to an edge-sharing octahedral sheet containing Mg and Li. Intercalation was achieved by exchanging 3 times with three-fold excess (based on cation exchange capacity) with tetramethylammonium (TMA) chloride, tetrapropylammonium (TPrA) bromide and tetrapentylammonium TPeA) bromide, respectively. Intercalation of long-chain organic cations (3-fold excess) was carried out 3 times in 30% acetonitrile/water solutions of trimethyloctadecylammonium (TrMOA) bromide, dimethyldioctadecyl-ammonium (DMDOA) bromide and methyltrioctadecylammonium (MTrOA) bromide, followed by washing in water until bromide-free. The coatings on the QCMs with 6 MHz resonance frequency were formed by dipping the crystals in the respective water or chloroform suspensions of the intercalated silicates. The films were heated at 200°C in nitrogen overnight.

The effect of intercalating various alkylammonium cations into the hectorite host is to increase the interlayer spacing of the Na-form (2.8 Å when partially hydrated) to 4.2, 4.9, and 5.5 Å, when TMA, TPrA, and TPeA ions are intercalated,

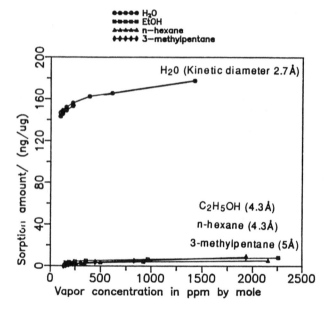

Figure 3. Sorption isotherms at 20°C of water (o), ethanol (squares), n-hexane (Δ) and 3-methylpentane (rhombus) on a QCM coated with potassium-exchanged zeolite A/silica composite.

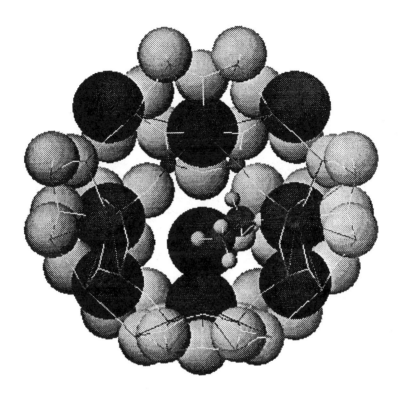

Figure 4. View into the eight ring of potassium-exchanged zeolite A. The blocked path of an ethanol molecule is shown.

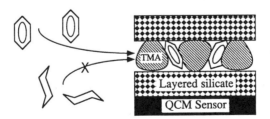

Figure 5. Schematic representation of selective benzene vs. cyclohexane sorption in a TMA-hectorite film on a QCM sensor. From ref. 14.

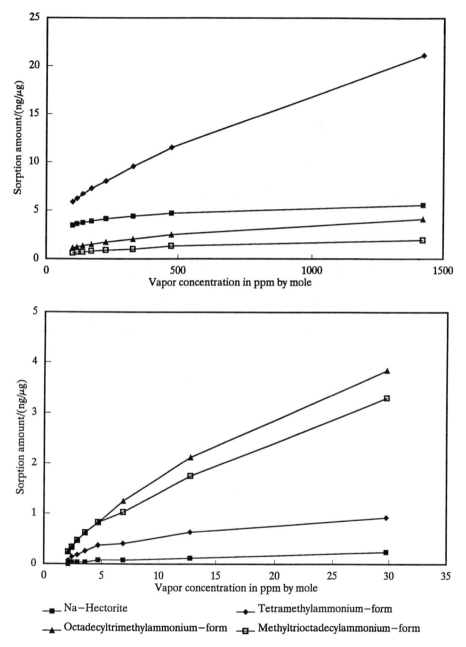

Figure 6. Vapor sorption isotherms at 25°C on Na and alkylammonium forms of hectorite. (A) water sorption (top). (B) toluene sorption (bottom).

respectively. These increased interlayer spacings are due to about one layer of hydrocarbon chains between the sheets of the clay mineral (9.6 Å thick). Intercalation of TrMOA results in an interlayer spacing of ll.1 Å, indicating 2-3 layers of hydrocarbon chains parallel to the clay sheets.[28] Finally, when the areal density of octadecyl chains is doubled (dimethyldioctadecylammonium, DMDOA, 21.1 Å) or tripled (methyltrioctadecylammonium, MTrOA, 24.5 Å), they are forced to orient almost normal to the clay sheets, in a paraffinic arrangement similar to Langmuir films.

Nitrogen sorption in the Na- and TMA-forms follows Type I isotherms and clearly shows microporosity for nitrogen with 1.2 and 2.0 wt% of nitrogen in the pores (derived from t-plots with the TPrA film as reference). The latter films show also significant external surface area. In contrast, all other organoclays show Type II isotherms and no microporosity. Vapor sorption isotherms at low partial pressures provide detailed information on the selective sorption behavior. Water sorption in the organoclays most closely tracks that of nitrogen (Figure 6A). The interlayer cavities produced on intercalation of TMA strongly sorb water at 25°C. Less water is sorbed in the Na-form and even less in the TrMOA and MTrOA forms. This behavior can be associated with the hydrophobic nature of the interlayer phase in the two latter organoclays.

A drastic reversal of the selectivities is observed when hydrocarbons are sorbed in the organoclays. Cyclohexane sorption is now most favored in the TrMOA form, and in the following sequence in the other sorbent films: MTrOA > TMA > Na-form. As shown in Figure 6B, toluene (and benzene) are absorbed even more strongly in the organoclays. We associate this increase over the sorption of more saturated six-rings with the compact, flat shape of the aromatics that should allow the molecules to slip into the gaps between the hydrocarbon chains of both the trimolecular horizontal TrMOA structure and the paraffinic more vertical MTrOA structure (relative to the clay sheets). Benzene shows also surprisingly high sorption in the TMA form. The benzene molecule can just barely be accommodated in this form when in a tilted upright orientation, as discussed by Barrer.[24] Liquid-like phases with different densities might occur at high loadings but are considered unlikely at the low loadings present in our measurements. Another driving force resulting in the preferred sorption of aromatics could be their higher polarizability. This could play a role in the interactions with organoclays containing TMA. but much less so in the long hydrocarbon (C_{18}) chain systems. There is less sorption of toluene than benzene which could be due to less efficient packing. In summary, the interplay of size exclusion from rigid pores and sorption (partitioning) into the organic phases can cause unique selectivities in the organoclays that complement the molecular sieving of porous framework hosts such as zeolites.

This overview shows the rich variety of sorption behavior of microporous solids that can be combined with piezoelectric devices to create selective sensors. Future work will explore additional parameters such as acid base reactions and coordination chemistry, as well as issues including chemical interferences under real world conditions.

Acknowledgments. The authors appreciate funding for different aspects of this work from the National Science Foundation, and from the U.S. Department of Energy (New

Mexico WERC Program). We thank Drs. Kelly Brown, C. Jeffrey Brinker, and Gregroy C. Frye for their contributions to this program.

Literature Cited

1. *Biosensors and Chemical Sensors*, Edelman, P.G.; Wang, J., Editors, ACS Symp. Ser. 487, ACS: Washington, D.C., **1992**.
2. Ward, M.D.; Buttry, D.A., *Science*, **1990**, *249*, 1000.
3. Guilbault, G.G.; Kristoff, J.; Owen, D. *Anal. Chem.* **1985**, *57*, 1754.
4. King, W.H. *Anal. Chem.*, **1964**, *36*, 1735.
5. Hlavay, J.; Guilbault, G.G. *Anal. Chem.*, **1977**, *49*, 1892.
6. Ballantine, D.S.; Wohltjen, H. *Anal. Chem.*, **1989**, *61*, 704A.
7. Breck, D.W. *Zeolite Molecular Sieves*, Krieger: Malabar, Florida, **1984**.
8. Barrer, R.M. *Hydrothermal Chemistry of Zeolites*, Academic Press: London, **1982**.
9. Bein, T.; Brown, K.; Frye, C.G.; Brinker, C.J. *J. Am. Chem. Soc.*, **1989**, *111*, 7640.
10. Bein, T.; Brown, K.; Brinker, C.J., *Zeolites: Facts, Figures, Future*, in P.A. Jacobs, R. A. Van Santen (Eds.), *Stud. Surf. Sci. Catal.* 49, Elsevier: Amsterdam, **1989**, pp 887.
11. Yan, Y.; Bein, T. *J. Phys. Chem.* **1992**, *96*, 9387.
12. Yan, Y.; Bein, T. *Chem. Mater.* **1992**, *4*, 975.
13. Bein, T.; Brown, K; Frye, C.G.; Brinker, C.J., U.S. Patent No. 5,151,110, Sept. 29, **1992**.
14. Yan, Y.; Bein, T.; *Chem. Mater.* **1993**, *5*, 905.
15. Yan, Y.; Bein, T.; *Microporous Materials*, **1993**, *1*, 401.
16. Bain, C.D.; Troughton, E.B.; Tao, Y.T.; Evall, J.; Whitesides, G. M.; Nuzzo, R. G. *J. Am. Chem. Soc.* **1989**, *111*, 321.
17. Wasserman, S.R.; Biebuyck, H.; Whitesides, G.M. *J. Mater*, Res., **1989**, *4*, 886.
18. Porter, M.D.; Bright, T.B.; Allara, D.L.; Chidsey, C.E.D. *J. Am. Chem. Soc.*, **1987**, *109*, 3559.
19. Li, Z.; Lai, C.; Mallouk, T.E. *Inorg. Chem.*, **1989**, *28*, 178.
20. Flanigen, E.M.; Bennett, J.M.; Grose, R.W.; Cihen, J.P.; Patton, R.L.; Kirchner, R.M.; Smith, J.V., *Nature*, **1978**, *271*, 512.
21. The isosteric heats were obtained from the sorption isotherms in the range of 0-40°C by the relation: $Q_{st} = RT_1T_2\ln(P_1/P_2)_a/(T_1-T_2)$, where P_1 and P_2 are the equilibrium concentrations for the same sorption amount in the sensing layer, a, at the temperatures T_1 and T_2, respectively.
22. Yan, Y.; Bein, T., to be published.
23. Barrer, R.M. Phil. *Trans. R. Soc.* Lond. **1984**, *A 311*, 333.
24. Barrer, R.M.; Perry, G.S. *J. Chem. Soc.* **1961**, 850.
25. Lee, J.-F.; Mortland, M.M.; Boyd, S.A.; Chiou, C.T. *J. Chem. Soc.*; Faraday Trans, 1, **1989**, *85*, 2953.
26. Lee, J.-F.; Mortland, M.M.; Chiou, C.T.; Kile, D.E.; Boyd, S.A. *Clays Clay Min.* **1990**, *38*, 113.
27. Lao, H.; Latieule, S.; Detellier, C. *Chem. Mater.* **1991**, *3*, 1009.
28. Lagaly, G. *Solid State Ionics* **1986**, *22*, 43.

RECEIVED April 20, 1994

Chapter 3

Zeolite Synthesis: Can It Be Designed?

Mark E. Davis

Chemical Engineering, California Institute of Technology,
Pasadena, CA 91125

The possibility of designing zeolites by their controlled synthesis is
discussed. The feasibility of creating pore architectures through the
design of organic structure-directing agents is described for high-
silica zeolites. The first example of such a design is presented. A
strategy for building pure-silica zeolites from molecular precursors
is outlined. It is concluded that the total design of zeolite structures
through their designed synthesis is not currently obtainable but
information providing "pieces to this puzzle" are quickly being
assembled.

Zeolites and zeolite-like molecular sieves are crystalline oxides that have high
surface-to-volume ratios and are able to recognize, discriminate and organize
molecules with differences that can be less than 1 Å. The close connection between
the atomic structure and the aforementioned macroscopic properties of these
materials has led to uses in most fields of molecular recognition. For example,
zeolites and zeolite-like molecular sieves can reveal marvelous molecular
recognition specificity and sensitivity that can be applied to catalysis, separations
technology and chemical sensing (1,2). Additionally, they can serve as hosts to
organize guest atoms and molecules that endow the composite materials with
optoelectric and electrochemical properties (3,4). Because of the high level of
control of the atomic structure that is necessary to create high-performance sensors
(whether they be chemical, electrical or optical sensing) with zeolites or zeolite-like
molecular sieves, one would like to design and synthesize these materials with
desired architectures and properties. Although this lofty goal has to date remained
elusive, advances have been made to allow one to now seriously consider the
possibility of designing molecular sieves. In this paper I will discuss (i) the design
of pore architectures of molecular sieves through the use of organic structure-
directing agents and (ii) the feasibility of synthesizing molecular sieves by "Lego
chemistry" (5); that is, the assembly of molecular precursors into a supramolecular
solid.

Zeolite and Zeolite-Like Molecular Sieve Synthesis

Crystallization is the result of self-assembly with molecular recognition events
occurring throughout the entire nucleation and crystal growth processes. Typically,
non-molecular reactants such as alumina and silica are combined with alkali-metal

0097–6156/94/0561–0027$08.00/0

hydroxides and water in closed, reaction environments to crystallize zeolites. These batch reaction mixtures are heated to temperatures of 50-200°C for specified periods of time to allow the crystallization process to occur (from now on called hydrothermal syntheses). A short review of this topic is available (6).

Several factors complicate the design of zeolite synthesis. First, the molecular mechanisms of self-assembly occurring during zeolite synthesis are unknown. Second, since one cannot draw an analogy to organic synthesis, the vast information of organic transformations provide little insight into zeolite synthesis. For example, with organic syntheses, numerous reaction steps involving different reagents and chemistries with separations in between can be utilized to create a final product by design. Such is not the case with zeolite syntheses. Separation of useful and unwanted products is typically difficult. Thus, one must design a synthesis medium that in one step spontaneously self-assembles into only one product with the correct architecture and atomic ordering. In view of this daunting task, one might expect that zeolite syntheses could never be designed.

To date, the complete design of a zeolite synthesis has not occurred. However, great strides are currently being made in attacking this problem. Here, I will cover two aspects of this ongoing effort. First, I will discuss the feasibility of designing pore architectures through the use of organic structures-directing agents. Second, I will explore the possibility of creating zeolites through "Lego chemistry".

Designing Pore Architectures with Structure-Directing Agents

In addition to the reactants described above, organic molecules can be added into zeolite synthesis mixtures and they can ultimately end-up residing within the product crystals. This phenomenon has led to much work on elucidating the roles of the organic species in zeolite crystallization. If there is a close connection between the size and shape of the organic molecule and the space it occupies in the zeolite crystal (structure-direction), then the design of pore architectures (organic removed by decomposition to form pore space) may be feasible by designing the organic. What is structure-direction? The concept of structure-direction is that the organic molecules organize the inorganic oxide species around themselves to provide the building units for the formation of the crystalline structures. This effect has often been called templating. However, the correspondence between the organic and inorganic is normally not sufficiently high to call this effect templating, especially when it is compared to its use in a biological context (6). Since not all organic species structure-direct the synthesis of zeolites and certain zeolites and molecular sieves can be synthesized from numerous different organic molecules, the concept of structure-direction has been questioned by many. Can structure-direction be proven and used successfully for designing pore architectures of zeolites? There is not a clear answer to this question as yet. However, the concept of structure-direction can be illustrated by the work of Liebau (7) and Gies and Marler (8). Silica, water and organic molecules were combined to crystallize a wide variety of porosils (general term to denote xM · SiO_2 where M is a guest molecule and x can vary). Porosils include clathrasils (crystals with cagelike voids occupied by guest species with openings too small to expel the guest species without decomposition (8)) and pure-silica zeolites, e.g., ZSM-48, ZSM-22, ZSM-5, ZSM-11, ferrierite and ZSM-12. Aqueous solutions containing silica and organic species were sealed in silica tubes and heated to 150-200°C for up to 6 months in order to prepare these materials. In general, the size of the clathrating silica cage was found to correlate with the size of the guest. Also, the symmetries of the cages were higher than that of the guest species, presumably due to the rotational freedom of the guests. When linear diamines were used, zeolites with one-dimensional channels were formed. For example, diaminopropane yielded ZSM-48. Although this molecule still contains an axis of rotation, the elongated dimension gives rise to a shape appropriate for channel formation (as opposed to cages). Finally, triethyl-, tripropyl- and tributylamine were used to crystallize ZSM-5 while tetrabutylammonium was used to synthesize ZSM-11. Both ZSM-5 and ZSM-11 have intersecting channels. In these cases, the three-dimensional geometry of the aqueous tri- and tetraalkylammonium species are not masked by hydrophobic hydration. Gies and Marler conclude that since there are no ionic interactions between the guest molecules and the silicate structures, the close guest-host geometrical fit most likely reflects an optimized configuration for van der Waals interactions (8). This work leads to a fairly convincing argument for structure-direction in the synthesis of pure-silica zeolites. The reasons for structure-direction with silica are that the surface of a forming silica crystal will be hydrophobic and will prefer to interact with an organic species rather than be exposed to the aqueous environment. Hydrothermal syntheses utilizing pure-silica and water-soluble organic species, e.g., amines, quaternary ammonium ions, are likely to reveal structure-direction due to the favorable energetics of organic-silicate interactions that will spontaneously lead to self-assembly. Thus, for pure-silica syntheses, the possibility of designing pore architectures through the design of organic structure-directing agents appears high. Other factors that support the premise that pure-silica zeolites can be designed are: (i) the silicon-oxygen-silicon angle is very flexible (9) and (ii) there is a lack of variability in the enthalpy of formation for pure-silica zeolites (10). Quantum mechanical calculations have revealed that the Si—O—Si angle can vary greatly (~ 120°–180°) with little difference in energy. Thus, numerous atomic arrangements of tetrahedral silicon that utilize bridging oxygen are feasible. Additionally, Petrovic *et al.* (10) recently showed that the enthalpy of formation for several pure-silica zeolites is essentially the same. These two factors support the hypothesis that for pure-silica zeolites, numerous structures are energetically feasible and obtainable through structure-direction by design.

The results from Liebau (7) and Gies and Marler (8) suggest that organic molecules that are water-soluble and stable to the synthesis conditions should structure-direct via hydrophobic interactions with silica. However, this is not always the case. There are many examples where organics fail to produce zeolites from pure-silica reaction mixtures. However, most of these investigations have not employed long crystallization times; Gies and Marler reported times as long as 6 months. Thus, the kinetics of the crystallization may be important. Goepper *et al.* showed that pure-silica ZSM-12 can be synthesized in 84 days at 150°C in the absence of alkali-metal ions(11). If alkali-metal ions are added, e.g., K^+/SiO_2 or $Na^+/SiO_2 = 0.3$, ZSM-12 is crystallized in 7 days at 150°C. Although the alkali-metal ions do not ultimately reside in the ZSM–12 crystals, Goepper *et al.* suggest that their role in nucleation and crystal growth is to increase the rates of these processes. We have examined this effect for several pure-silica zeolites and present here the results for ZSM-12 (pure-silica). The addition of alkali-metal ions increases the kinetics of nucleation and crystallization (Table I) that likewise affect the size of the crystals formed (Figure 1).

The information presented above suggests that the design of zeolite pore structures is feasible if pure-silica syntheses are used with water-soluble organics that do not decompose at synthesis conditions that may last for several months. For most practical cases, faster synthesis times and the addition of heteroatoms, e.g. Al^{3+}, are desired. Small amounts of alkali-metal cations and heteroatoms will most likely be acceptable. However, large amounts of alkali-metal ions and heteroatoms will alter the reaction chemistry sufficiently to override or modify the structure-directing effects of the organic species.

(A)

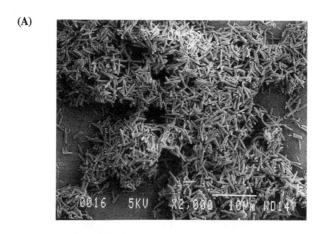

(B)

Figure 1. Scanning electron micrographs of ZSM-12. A. Synthesized with alkali-metal ions; B. Synthesized in the absence of alkali-metal ions.

Synthesis by Design

Recently, the first example of a zeolite whose pore system was formed by the purposeful design of its structure-directing agent was accomplished. Zones rationalized that a rigid polycyclic organocation of appropriate dimensions could be used to structure-direct the formation of a multidimensional pore system with pores in the size range of 5–8 Å (*12*). Figure 2A shows a fused two 5–ring diquaternary organocation that can structure-direct the formation of ZSM-12 (one-dimensional pore system). By design, Zones added a third fused ring to this organocation (Figure 2B) in order to disrupt the formation of channels and speculated that the zeolite formed via structure-direction with the three-ring cation would be multidimensional (Figure 3A) (*13*). Interestingly, Zones employed a high-silica, low alkali-metal synthesis medium that as outlined above is most appropriate for structure-direction to occur on reasonable time scales. The three-ring organocation structure-directed the synthesis of zeolite SSZ-26 (*12,13*). Very recently, we solved the structure of SSZ-26 (*14*) and the speculations of Zones and co-workers are correct. Figure 3 compares the speculation to the experimental finding. Although the atomic ordering necessary to form SSZ-26 was not designed or controlled, the pore structure was. These results show that in principle the design of pore architectures is possible through the design of organic structure-directing agents.

The view presented here is oversimplified for illustrative and length purposes. However, the premise that pore architectures should be designable via designed organic structure-directing agents is not.

Zeolites Via "Lego Chemistry"

Little is known about the mechanism of how the inorganic oxide framework forms. In light of this, we have taken an alternative approach to the synthesis of crystalline structures by design. We are attempting to synthesize pure-silica zeolites from molecular building units (*15*). This approach has been coined "Lego chemistry" (*5*). We are using spherosiloxanes as molecular building units due to the fact that their polycyclic ring structures are found in many zeolites. Figure 4 shows a schematic diagram of the O_h—$X_8Si_8O_{12}$ spherosiloxane. This spherosiloxane can be synthesized in multigram quantities for X=H (*16,17*) and synthetic methodologies exist for converting -X from -H to a wide variety of functional forms, e.g., -H to -Cl and -H to $-OCH_3$ (*18*). To date, we have synthesized multigram quantities of $X_8Si_8O_{12}$ where X=H, Cl and OCH_3 with structure verification by X-ray diffraction, infrared and Raman spectroscopies, mass spectroscopy and ^1H, ^{13}C and ^{29}Si NMR spectroscopy.

Corriu *et al.* (*19*) report the non-hydrolytic condensation of metal alkoxides and metal halides to form M–O–M' bridges with release of alkyl halides (RX):

$$M(OR)_n + M'X_{n'} = (RO)_{n-1}M - O - M'X_{n'-1} + RX$$

Here, we propose to use this type of condensation to produce microporous crystalline pure-silica materials by the reaction (Figure 5):

$$Cl_8Si_8O_{12} + (CH_3O)_8Si_8O_{12} \rightarrow SiO_2 + CH_3Cl$$

It is important to note that we do not employ water in these syntheses. Day *et al.* (*18*) synthesized a monolithic glass from the hydrolysis of $(CH_3O)_8Si_8O_{12}$:

$$(CH_3O)_8Si_8O_{12} + H_2O \rightarrow glass + H_2O + CH_3OH$$

Table I.   Synthesis of Pure-Silica ZSM-12 (From ref. 11).

Synthesis Composition	T, °C	t,d	Product
$0.06R \cdot 0.3 \ NaNO_3 \cdot SiO_2 \cdot 32 \ H_2O$	150	7	ZSM-12
$0.06R \cdot SiO_2 \cdot 32 \ H_2O$	150	84	ZSM-12
$0.06R \cdot 0.3 \ NaOH \cdot SiO_2 \cdot 32 \ H_2O$	150	7	ZSM-12
$0.06R \cdot 0.3 \ NH_4OH \cdot SiO_2 \cdot 32 \ H_2O$	150	7	Amorphous

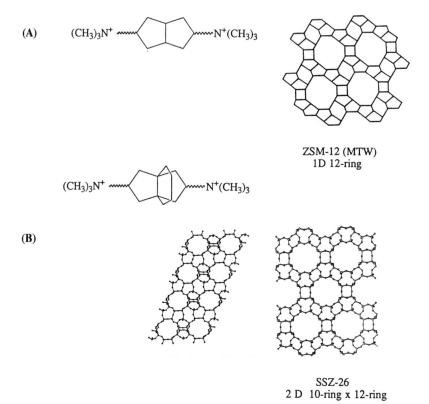

Figure 2.   Organic structure-directing agents and formed zeolites. A. Fused two 5-ring diquat and ZSM-12; B. Fused three-ring diquat and SSZ–26.

(A)

(B)

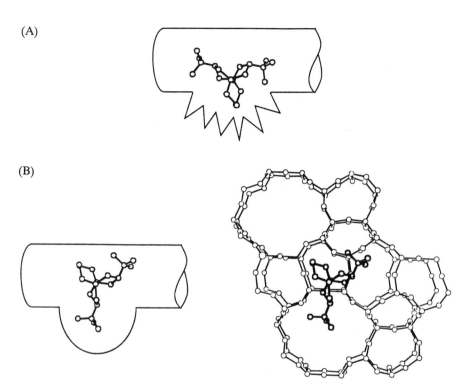

Figure 3. Diquaternary ammonium compound in zeolite pore system. A. Speculation of Zones and Santilli (13); B. Result from Lobo *et al.* (14).

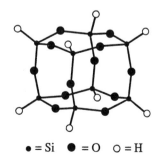

• = Si ● = O ○ = H

Figure 4. Schematic diagram of $H_8Si_8O_{12}$.

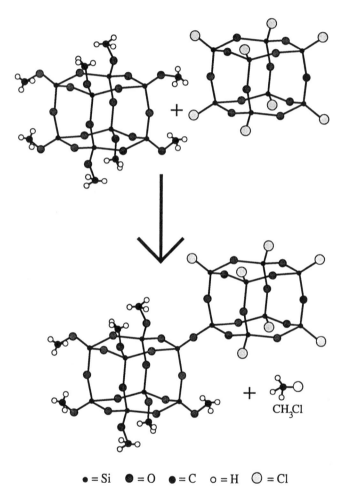

$\bullet = Si \quad \bullet = O \quad \bullet = C \quad o = H \quad \bigcirc = Cl$

Figure 5. Schematic diagram of the condensation reaction between $Cl_8Si_8O_{12}$ and $(CH_3O)_8Si_8O_{12}$.

Hydrolytic condensation pathways utilizing a single spherosiloxane reactant are most likely to allow random network formation. Thus, we avoid this type of synthetic process. We believe that the use of a binary component reaction scheme will offer control over the condensation processes not available in the single reagent, hydrolytic pathways. Also, Feher and co-workers have prepared the equivalent of spherosiloxane dimers (two double four-membered ring units). However, their synthesis procedures are not amenable to the formation of a zeolite via a "Lego" type synthesis (20).

A tremendous variety of structures can be theoretically obtained from spherosiloxane-like building units. In order to focus our initial efforts we have concentrated on the $X_8Si_8O_{12}$ unit with X = -Cl and -OCH$_3$ (denoted A and B, respectively). In the absence of water, A cannot react with A and B cannot react with B. Thus, only the reactions of A with B can occur. Additionally, with pure-silica, we expect that the formation of three fused 4–rings by the reaction of two –Cl groups on one A with two –OCH$_3$ group on one B will not occur because of the high strain on the middle ring. Thus, we expect that the most favorable scenario is that a molecule A will surround itself with 8 molecules of B. If such is the case, then the structure illustrated in Figure 6A will be formed.

(A)

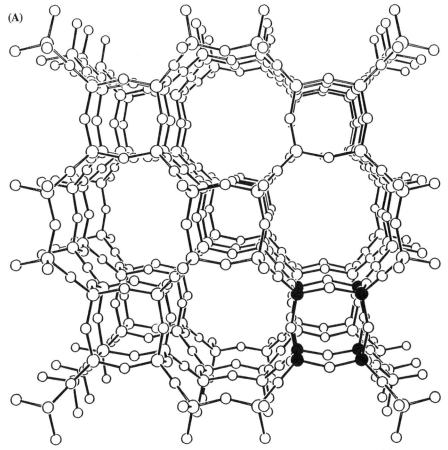

Figure 6. Schematic diagrams of structures that can be constructed from Si_8O_{12} building units. A. Structure containing only Si_8O_{12} units.

Continued on next page.

(B)

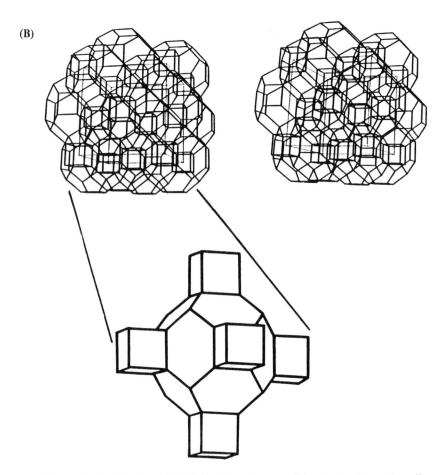

Figure 6.   Continued.   B. AlPO$_4$-16 or the pure-silica clathrasil octadecasil.

Moreover, by appropriate substitutions of the ligands on the spherosiloxane core, it is possible to produce an even wider range of structures. For example, if Cl$_8$Si$_8$O$_{12}$ is reacted with Si(OCH$_3$)$_4$, then:

$$Cl_8Si_8O_{12} + 8\ Si(OCH_3)_4 \rightarrow [(CH_3O)_3SiO]_8 - Si_8O_{12} + 8CH_3Cl$$

Denote the larger cluster as C. If C and A are now combined, then the structure shown in Figure 6B can be formed. This structure is that of AlPO$_4$-16 (21) and the pure-silica, clathrasil octadecasil (22). Although the structures shown in Figure 6 can be constructed from the building units described above, it is most likely that controlled combinations of A, B and C will be necessary to form these crystalline materials.

Can Zeolite Synthesis Be Designed?

At this point in time a completely designed synthesis of a zeolite has not occurred. However, the evidence presented here paints a bright picture for the future. The design of organic structure-directing agents and molecular building blocks is now a reality. The next step is to combine these elements with chemical compatibilities that lead to their self-assembly.

Literature Cited

1. Davis, M. E.; *Ind. Eng. Chem.Res.* **1991**, *30*, 1675-1683.
2. Ozin, G. A.; Kuperman, A.; Stein, A.; *Angew Chem.* **1989**, *101*, 373-390.
3. Stucky, G. O.; MacDougall, J. E.; *Science* **1990**, *247*, 669-678.
4. Rolison, D. R.; *Chem. Rev.* **1990**, *90*, 867-878.
5. Stein, A.; Keller, S. W.; Mallouk, T. E.; *Science* **1993**, *259*, 1558-1564.
6. Davis, M. E.; Lobo, R. F., *Chem. Mater.* **1992**, *4*, 756-768.
7. Liebau, F. In *Silicon Chemistry*; Editors: E. R. Corey, J. Y. Corey, P. P. Gaspar; Ellis Harwood Limited: Chichester, 1988, Chapter 29.
8. Gies, H.; Marler, B.; *Zeolites* **1992**, *12*, 42-49.
9. Newton, M. D.; Gibbs, G. V.; *Phys. Chem. Miner.* **1980**, *6*, 305-312.
10. Petrovic, F.; Navrotsky, A.; Davis, M. E.; Zones, S. I.; *Chem. Mater.*, in press.
11. Goepper, M.; Li, H. X.; Davis, M. E.; *J.C.S. Chem. Commun.* **1992**, 1665-1666.
12. Zones, S. I.; Olmstead, M. N.; Santilli, D. S.; *J. Am. Chem. Soc.* **1992**, *164*, 4195.
13. Zones, S. I.; Santilli, D. S. In *Proc. Ninth Inter. Zeolite Conf.*; Editors: R. von Ballmoos, J. B. Higgins; M. M. Treacy; Butterworth-Heinemann: Boston, 1993, pp.171-179.
14. Lobo, R. F.; Pan, M.; Chan, I.; Li, H. X.; Medrud, R. C.; Zones, S. I.; Crozier, P. A.; Davis, M. E.; *Science*, in press.
15. Nagel, J. F.; Davis, M. E., work in progress.
16. Agaskar, P. A.; *Inorg. Chem.* **1991**, *30*, 2707-2708.
17. Cagle, P. C.; Ph.D. Thesis, Univ. of Illinois, Champaign, 1992.
18. Day, V. W.; Klemperer, W. G.; Mainz, V. V.; Millar, D. M.; *J. Am. Chem. Soc.* **1985**, *107*, 8262-8264.
19. Corriu, R.; Leclerq, D.; Lefevre, P.; Mutin, P. H.; Vioux, A.; *Chem. Mater.* **1992**, *4*, 964-965.
20. Feher, F. J.; Weller, K. J.; *Organometallics* **1990**, *9*, 2638-2640.
21. Bennett, J. M.; Kirchner, R. M.; *Zeolites* **1991**, *11*, 502-506.
22. Caullet, P.; Guth, J. L.; Hazn, J.; Lamblin, J. N.; Gies, H.; *Eur. J. Solid State Inorg. Chem.* **1991**, *28*, 345-361.

RECEIVED March 25, 1994

Chapter 4

Design, Synthesis, and Characterization of Gated Ion Transporters

Gordon G. Cross, Thomas M. Fyles[1], Pedro J. Montoya-Pelaez,
Wilma F. van Straaten-Nijenhuis, and Xin Zhou

Department of Chemistry, University of Victoria, Box 3055, Victoria,
British Columbia V8W 3P6, Canada

Artificial channels for the transport of alkali metal cations through
bilayer membranes are readily prepared from a modular set of
subunits.   Structure-activity relationships and kinetic studies in
vesicles define the structural requirements of active materials. Active
transporters are symmetrical, and are apparently gated by random
reorientation or reorganization events in the membrane. Control of
ion transport in response to external stimuli such as light or
transmembrane potential could in principle be achieved by the
introduction of responsive functionality. Our approach to photo-gated
channels, channels switched on or off by light, involves an
intramolecular *trans* azo linked ammonium side arm to block the ion
channel in the dark state. Photo-isomerisation to the *cis* form would
alter the ammonium blocking position and open the channel for
transport. Progress in the synthesis of a suitable example is outlined.
Voltage-gated pores, switched by membrane potential, might be
derived from the aggregate pores formed by bolaform (two-headed)
amphiphiles.   Orientation of the dipoles of zwitterionic structures
would respond to the membrane potential. Preliminary studies of one
of these primitive voltage gates show partial control of ion transport.

Ion channels are everywhere in nature and are the heart of all the key life processes:
energy production, storage, and transformation, signal propagation and processing (1).
They act by creating, maintaining, and controlling chemical and potential gradients
across membranes. Natural ion channels are large protein aggregates, which span the
bilayer membrane. A burst of recent structural information from molecular biology
experiments (2) is beginning to probe the key amino acids, but a detailed molecular
scale view is still some time in the future.
    This is a pity because transport functions have clear implications for new
technology. As one example, the current thrust in biosensors seeks to transplant
natural transport systems into artificial sensors which could then exhibit the inherent
sensitivity of the biochemical apparatus. Can this be extended to other technological
goals such as molecular switches or ionic computers? Possibly, but the process will
be plagued with the problems of maintaining a "natural" environment in an artificial

[1]Corresponding author

device. An alternative strategy is to develop fully functional artificial systems that closely mimic natural systems. The mimics are likely to be smaller and more robust than natural transporters hence the transplanting step is likely to be "simple".

Following the tradition established for carrier transporters, design principles for ion channel mimics can be inferred from low molecular weight membrane spanning ionophores, such as gramicidin (3) or amphotericin (4). These sources suggest that artificial channels for the transport of ions across bilayer membranes could be designed according to the following principles: i - A channel would have a polar core surrounded by a non-polar exterior layer for simultaneous stabilization of an ion in transit and favorable interaction with membrane lipids; ii - A channel would have the overall length and shape to fit into a bilayer membrane approximately 40 Å thick.

Functional artificial ion channels have been reported which illustrate the general criteria. The most obvious course is to prepare oligopeptides with high helical content (5). Other reported systems are based on cyclodextrin (6), polymeric crown ethers (7), and "bouquet" shaped crown ether and cyclodextrin motifs (8). One of the most active systems is a simple tris-crown ether derivative reported by Gokel for the transport of sodium ions(9). All of these systems envisage a uni- or bi-molecular transmembrane structure, similar to the gramicidin structural paradigm.

An appealing alternative would exploit multicomponent aggregates, akin to an amphotericin pore (4), to achieve structures of the required size from modest molecular weight components. Simple functionalized amphiphiles apparently act using this mechanism to increase the permeability of synthetic (10), and natural phospholipid membranes (11). Specifically designed membrane disrupting agents also employ the same strategy (12). Fuhrhop has reported an intriguing system which can be reversibly switched, and which acts via aggregated structures in synthetic monolayer membranes (13). The best characterized system to date is an oligoethylene glycolic acid which forms single channels in planar bilayer membranes (14). Kobuke reports that this system is not voltage-gated, but does show the marked voltage and concentration dependence expected of an aggregate pore.

Our approach begins from the criterion ii (above) as it forces attention on the synthetic task. Effectively it defines a minimum molecular weight of 3500-4000 g/mol for a functional structure (cf. gramicidin dimer, 3740 g/mol). To build and optimize a suite of structures for structure-activity studies, or to optimize for a desired application, requires the targets to be assembled in large blocks - essentially a modular or "pre-fabricated" approach to the synthesis. In turn, this requires well behaved molecular components, and high yielding construction reactions. In addition, the components should provide a range of functionality to assist in the functioning of the ion channel mimic. This will include donor sites for interaction with a cation in transit, non-polar segments for interaction with membrane lipids, and polar groups to provide amphiphilic character.

Figure 1 illustrates our design proposal derived from these general synthetic and functional considerations. We assemble *unimolecular ion channels* (15) on a rigid framework ("core") at the bilayer mid-plane, by appending "wall" units of suitable size to extend to the bilayer faces. Polar head groups provide overall amphiphilic character. A molecular structure of one candidate is given below to illustrate the chemical identities of the "core" (18-crown-6), "wall" (macrocyclic tetraester), and "head groups" (mercapto-glucose). This structure can be assembled in two steps from a modular construction set consisting of crown ether polycarboxylic acids (2, 4, or 6 carboxylates, R,R- or R,S- isomers), macrocyclic tetraesters (22 to 36 membered rings, hydrocarbon and oligo-ether sides), and thiol nucleophiles (mercapto-glucose, mercapto-ethyl amine, mercapto-propanol, mercapto-acetic acid). A total of twenty-one examples have been prepared and characterized (16). The target structures have molecular weights ranging from 2300 to 4900 g/mol.

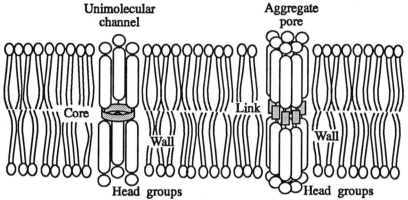

Figure 1: Schematic design proposals.

Structure of a unimolecular ion channel mimic.

Alternatively, simpler bola-amphiphiles could associate in the bilayer membrane as illustrated on the right of Figure 1 to form *aggregate pores* (17). Individual molecules are assembled from the same "wall" and "head group" components, linked at the bilayer mid plane by simple bi-functional linking reagents (tartaric acid, *m*-xylylene). A structure (below) illustrates one of fourteen examples prepared and characterized (18). The synthetic task is less demanding than in the case above as the candidate structures are all less than 2000 g/mol.

Structure of an aggregate pore-former.

The activity of both types of compounds has been assessed in bilayer vesicle membranes (15, 18-20) using a pH-stat technique to detect transporter activity. Vesicles are prepared in a pH 6.6 buffer, the external buffer is replaced by an unbuffered solution, and the external pH is controlled by a pH-stat titrimeter. Cation transport into the vesicle is accompanied by proton release, which is quantified by the pH-stat. Titrant volume added versus time can be treated to yield rate constants in all cases.

Survey experiments on all thirty-five compounds, and extensive investigations of the most active materials establish the following features (18-20):
i) Transporter activity is under structural control. The most active materials are sufficiently long to span the bilayer, have a balance of polar and non-polar functionality, and have neutral (unimolecular channels) or charged head groups (aggregate pores).
ii) Transport mechanisms have been established by a combination of kinetic techniques, and by comparison with known channel-forming and carrier compounds. Experiments include: apparent kinetic order in transporter, dependence of transport on cation concentration (saturable in some cases), cation selectivity among alkali metals, apparent activation energy, effects of additives and competing cations, and vesicle partitioning techniques.
iii) Among the unimolecular ion channels, transport mechanism is under structural control. Four or six wall units on a single core, and the ability to adopt a columnar shape are essential prerequisites for a channel mechanism. Fewer wall units, or alternate shapes result in an ion carrier mechanism.
iv) Active unimolecular ion channels are cation selective among the alkali metals, in some cases with marked non-Eisenman selectivity patterns (Na^+ >> Cs^+ > K^+, Rb^+ >> Li^+). In many cases, Li^+ is an inhibitor of Na^+ or K^+ transport. Octyl ammonium also inhibits alkali cation transport in some cases (19).
v) Aggregate pore formation is indicated by apparent kinetic orders greater than one. The aggregates are probably loosely structured, but some systems can discriminate between alkali metal cations and choline cations or sulfate anions.
vi) The transport process observed involves rate limiting initiation of active pores or channels. Ion translocation is rapid, and individual vesicles are equilibrated with the external solution during a single opening event. The openings are random, governed by thermal motions in the bilayer.
Closely related systems have been investigated by complementary techniques which clearly establish that channel-mimics of this type are incorporated into the bilayer membrane, albeit with a range of orientations ranging from fully membrane-spanning (as indicated in Figure 1), to U-shaped orientations with all head groups on one face of the bilayer.(8) Taken together, the experimental results are consistent with the static design proposals sketched in Figure 1. The dynamic situation must be more complex, and only a fraction of the transporters added will adopt the productive conformation drawn at any specified time.

Design of Gated Ion Transporters

The next step in the development of these artificial transport systems is to provide some means to control channel activity by external stimuli. Natural ion channels are controlled, or "gated", by stimuli such as the binding of small molecules (for example, neurotransmitters), or the sign and magnitude of the transmembrane potential gradient (voltage-gating, as in nerve impulse propagation) (1, 2, 21). Of course, artificial systems need not be constrained by biochemical paradigms: switching via light stimulus (photo-gated channel), or by chemical reaction (for example a redox-gated channel) are plausible alternatives. Nor does the gating need to be reversible. An irreversibly switched transporter might find application in drug delivery, or image immobilization.
In our exploration of gated ion transporters, we focused on two strategies: photo-gated unimolecular ion channels, and voltage-gated aggregate pores. In both cases we expect to exploit the modular assembly process outlined above to prepare examples for evaluation. The target structures would be derived from transporters of known activity and mechanism, by the addition of functionality which could respond to a switching stimulus.

For example, our design for a photo-gated ion channel, sketched in Figure 2, begins with the known ion channel compounds derived from the 18-crown-6 tetraacid and four suitable wall units. One of the wall units would be replaced by an azobenzene derivative terminated with a primary ammonium group. The switch portion of the molecule is akin to the "tail-biting" crown ethers described by Shinkai (22). Inspection of CPK models, and molecular mechanics modelling were used to decide which positional isomers would be required to give optimal binding of the ammonium group in the crown ether.

In the "dark" or resting state, the *trans* azo linkage would place the ammonium group in proximity to the crown ether for intramolecular complex formation. In this state transport through the channel would be inhibited, as has been observed for the parent channel and the intermolecular inhibitor, octyl ammonium (19).   Upon irradiation, photo-isomerization to the *cis* form would alter the intramolecular ion binding. This might involve complete exposure of the crown ether and complete channel opening, or might involve other reorganizations. In any event, the *cis* form is expected to provide ion translocation pathways absent in the *trans* form, thus is expected to accelerate ion transport.

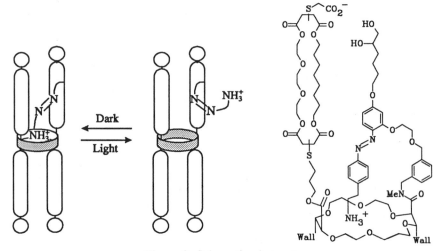

Figure 2: Schematic photogate.

A design proposal for a voltage-gated aggregate pore is sketched in Figure 3. The active subunit would be derived from a symmetrical precursor, by modification of head group charge to create a zwitterionic derivative. Molecular zwitterionic amphiphiles in a Langmuir-Blodgett film have recently been shown to act as "molecular rectifiers" (23).   In the pH range near neutral, suitable groups are carboxylate anions and ammonium cations. In the absence of a potential gradient, a random collection of orientations is expected. In the presence of a transmembrane potential, one of the dipolar orientations will be preferred, and the membrane potential would be expected to assist in the organization of aggregate pore structures. This would be voltage control of aggregation state, a type of voltage-gating. All dipoles are illustrated in a parallel alignment in the Figure, but a collection of parallel and anti-parallel orientations would be suitable as well. Depending on the ability of the head groups to penetrate the bilayer, pore formation might also be controlled by the sign of the transmembrane potential gradient. This would give rise to voltage-gating as observed with alamethicin and peptibol toxins (5,21). The structure illustrated below is one candidate for evaluation.

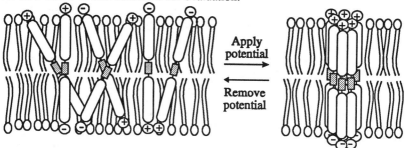

Figure 3: Schematic voltage gating mechanism.

Candidate voltage-gated pore-former.

Synthesis of a Photo-switch

The synthesis of the photo-gated ion channel proposed would use the chemistry of the modular construction set to link suitable wall and head group units to a crown ether derivative bearing the azobenzene side arm. The arm itself might be added in one step to the crown ether, but the preparation of mono-substituted derivatives of the tetraacid was not known at the outset of the synthesis. The regioselectivity of diazonium coupling to resorcinol derivatives was fairly clear from previous work (24). We chose to link the azo arm to a pre-formed crown ether derivative so that the two synthetic problems could be solved independently. The synthesis of the azobenzene derivative is outlined below. Diazotization of anilines requires acidic conditions (24), but the coupling to phenol and resorcinol requires basic conditions to provide activation by the phenoxide. Initially we used an acetamide protecting group for the primary amine, but at a late stage of the synthesis we were unable to cleave it without degradation of the other groups present.

As the synthesis was finally developed, a *tert*-butoxycarbonyl-protecting group (tBoc), was sufficiently acid stable to permit diazotization of the aniline. In model reactions coupling occurred predominantly *para* to the activating phenoxide, so resorcinol bearing the connecting arm to the crown ether, protected as the tetrahydropyranyl derivative (Thp), was coupled with the diazonium salt. The regiochemistry of the coupling was tedious to establish, as the NMR properties of the azobenzene nucleus provided ambiguous interpretations for most 1- and 2-D NMR techniques. Eventually a combination of ^{13}C - ^{13}C correlation spectroscopy, and synthesis of authentic 2-hydroxy-4-methoxyazobenzene for spectral comparison allowed the structure to be unambiguously assigned. Careful acidic hydrolysis removes the Thp group without cleaving the tBoc amine protecting group. The free phenolic hydroxyl group was elaborated to a six carbon side arm bearing a glycol as a polar group by use of one equivalent of base and the protected 6-bromo glycol.

We had initially anticipated preparing the mono-substituted crown ether derivative via a capped structure formed by reaction of the crown ether bis-anhydride with a suitable amino-alcohol. Previous work with bis-primary amines indicated that capping was a favorable process (25), but the secondary amino alcohol produced a complex mixture of products which resisted characterization. Moreover, a brute force approach using stoichiometric *p*-(N-methylaminomethyl)hydroxymethyl benzene and crown ether anhydride gave the required triacid (below; R = Me) in poor yield in mixture with bis-amides and apparently epimerized materials.

This disappointing result is unfortunately quite general. Secondary amines react with crown ether acid chlorides to give N-alkyl amides, but the slower reaction with crown ether anhydrides results in much poorer yields and significant competing epimerization via deprotonation-reprotonation at the carbon adjacent to the carboxylic acid. The monoamide-triacid (above; R = H) can be prepared from the corresponding primary amino-alcohol. The more reactive amine steers the dominant reaction towards amide formation without epimerization.

To complete the synthesis of the photo-switch component of the channel, Williamson ether synthesis using the primary alkoxide as the nucleophile and a suitable benzylic leaving group (Br, OTs) as the electrophile will be required. Our synthetic progress towards the final target will be reported subsequently.

Synthesis and characterization of a voltage-gated aggregate pore

The synthesis of the candidate voltage-gated pore-former follows directly from the reactions developed for the modular construction set. The synthetic route is outlined on the next page. Beginning with the macrocycle derived from maleic anhydride, 1,8-octanediol, and triethylene glycol, the symmetrical adduct with two mercapto acetate groups, and the dissymmetric adduct with one thiopropanol group, were prepared. Carboxylate alkylation of the di-acetate with the mesylate derivative of the alcohol proceeded to give a statistical mixture of products, from which the desired 1:1 ester was recovered by size-exclusion chromatography. The composition was established by a combination of NMR, analytical gel permeation, and FAB-mass spectroscopic techniques, and elemental analysis. The synthesis was completed by the Michael addition of N,N-dimethylaminoethane thiol to the remaining olefin. Again the structure was fully confirmed by spectroscopic and chromatographic techniques.

The vesicle assay previously used gives a good survey of transporter activity under a fixed set of conditions. However it is not suited to exploration of transport activity as a function of the sign and magnitude of the transmembrane potential. Bilayer conductance, or single channel recording techniques are required (21). In this technique, a lipid bilayer is formed across a small hole in a Teflon barrier, by direct application of the lipid in decane to the hole. Under favorable conditions, the lipid thins to a bilayer membrane which electrically isolates the two halves of the cell. Typically KCl is used as an electrolyte, and electrical contact is made via Ag/AgCl wires directly inserted into the solutions. A high impedance operational amplifier circuit (bilayer clamp) can then be used to apply a fixed transmembrane potential and to monitor the current which flows as a function of time. The two sides of the membrane are independently accessible, so different sequences of transporter addition, control of pH, and other variables are in principle possible.

Typical conductance recordings at two applied potentials (-80 and +80 mV) are illustrated in Figure 4, for the pore-former prepared above in 1 M KCl electrolyte at pH 2.0 (HCl). The bilayer was formed from a mixture of egg phosphatidyl choline : phosphatidic acid : cholesterol in an 8:1:1 mole ratio, and the transporter was added as a methanol solution to one side of the bilayer in the absence of a transmembrane potential. At each potential, channel openings are indicated by abrupt changes in the current, as the pore passes abruptly from a "closed" to an "open" state. The data presented is two "snapshots" from an experiment that lasted over two hours, involving several thousand pore openings at a range of applied potentials. The bilayer itself does not give rise to conductance steps or leakage currents in the absence of added pore-former. The conductance steps could indicate the formation of a water-filled aggregate, of simply a phase defect adjacent to added pore-former.

At each applied potential a histogram of the observed current step for each channel opening event was constructed. The maximum of the histogram distribution corresponds to the current carried by the most probable pore formed under the experimental conditions. The peak currents derived from histograms at various applied potentials can be combined to give the current-voltage response shown in Figure 5.   At pH 6.0 the response of the system is ohmic; the line drawn in Figure 5 was calculated from the least-squares fit to the data at pH 6.0. The data points given were derived at pH 2.0 (added HCl). For negative applied potentials the response is identical to that observed at pH 6.0. At positive transmembrane potentials, significant upward deviation to more highly conducting pores is evident.

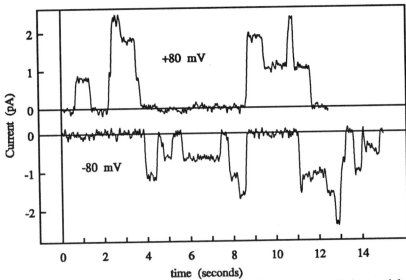

Figure 4: Transmembrane current as a function of time at two applied potentials.

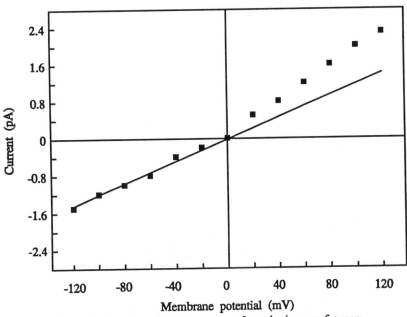

Figure 5: Current-voltage response of synthetic pore-former.

The results are somewhat at variance with those expected from the design proposal sketched in Figure 3. The lack of asymmetric current-voltage response at pH 6.0 indicates that the zwitterionic form of the pore-former does not discriminate between the orientations of the applied potential. Moreover, at pH 2.0 where asymmetrical responses are found, the carboxylic acid will be protonated, and only one of the head groups will be charged. The neutral head-group would be expected to penetrate the membrane more efficiently than the charged ammonium head-group which could lead to a preferred orientation of the pore-former. One likely possibility is that the ammonium head group would not penetrate the bilayer, but remain on the side where it was initially added. In this event, one sign of the applied potential would give a parallel alignment of pore-former dipoles which preserves the initial orientation. The other sign of the applied potential would require the charged head group to penetrate the bilayer. Alternatively, both head groups could penetrate the membrane but the pore would form with a parallel orientation of pore-former dipoles for one sign of the transmembrane potential, and an anti-parallel orientation of the dipoles for the other sign of the transmembrane potential.

Whatever the explanation, the data presented is preliminary evidence of voltage responsive behavior, probably governed by a preferred orientation of the pore-former dipoles in the membrane. Certainly it is not as marked as some naturally occurring materials, such as alamethicin (21). On the other hand, the compound is considerably simpler, and therefore easier to make and modify. Our continued explorations of this system will focus on enhancing the difference in hydrophilicity of the head groups, to enhance the differences in preferred orientation. We are confident that more convincing and useful voltage-gated materials will result.

Acknowledgments   The ongoing support of the Natural Sciences and Engineering Research Council of Canada is gratefully acknowledged.

Literature Cited
(1)   (a) Houslay, M.D.; Stanley, K.K. *Dynamics of Biological Membranes*; John Wiley and Sons: New York, 1982. (b) Robertson, R.N. *The Lively Membranes*; Cambridge University Press: Cambridge, 1983.
(2)   (a) Betz, H. *Biochemistry* **1990**, *29*, 3591.   (b) Stühmer, W. *Ann. Rev. Biophys. Biophys. Chem.* **1991**, *20*, 65.
(3)   (a) Jordan, P.C. *J. Phys. Chem.* **1987**, *91*, 6582. (b) Wallace, B.A.; Ravikumar, K. *Science* **1988**, *241*, 182. (c) Langs, D.A. *Science* **1988**, *241*, 188. (d) O'Connell, A.M.; Koeppe, R.E.; Andersen, O.S. *Science* **1990**, *250*, 1256.
(4)   (a) Hartsel, S.C.; Perkins, W.R.; McGarvey, G.J.; Cafiso, D.S. *Biochemistry* **1988**, *27*, 2656. (b) Bolard, J.; Legrand, P.; Heitz, F.; Cybulska, B. *Biochemistry* **1991**, *30*, 5707.
(5)   (a) Sansom, M.S.P. *Prog. Biophys. Molec. Biol.* **1991**, *55*, 139, and references therein. (b) Lear, J.D.; Wasserman, Z.R.; DeGrado, W.F. *Science* **1988** *240*, 1177. (c) Montal, M.; Montal, M.S.; Tomich, J.M. *Proc. Nat. Acad. Sci. U.S.A.* **1990** *87*, 6929.
(6)   Tabushi, I.; Kuroda, Y.; Yokota, K. *Tetrahedron Lett.* **1982**, *23*, 4601.
(7)   (a) Nolte, R.J.M.; van Beijnen, A.J.M.; Neevel, J.G.; Zwikker, J.W.; Verkley, A.J.; Drenth, W. *Isr. J. Chem.* **1984**, *24*, 297. (b) Neevel, J.G.; Nolte, R.J.M. *Tetrahedron Lett.* **1984**, *25*, 2263. (c) Kragten, U.F.; Roks, M.F.M.; Nolte, R.J.M. *J. Chem. Soc., Chem. Commun.* **1985**, 1275.
(8)   Pregel, M.J.; Jullien, L.; Lehn, J.M. *Angew. Chem. Int. Ed. Engl.* **1992**, *31*, 1637. (b) Jullien, L.; Lazrak, T.; Canceill, J.; Lacombe, L.; Lehn, J.-M. *J. Chem. Soc. Perkin Trans. 2* **1993**, 1011.

(9)   Nakano, A.; Xie, Q.; Mallen, J.V.; Echegoyen, L.; Gokel, G.W. *J. Am. Chem. Soc.* **1990**, *112*, 1287.
(10)   Kunitake, T. *Proc. N. Y. Acad. Sci.* **1985**, *471*, 70.
(11)   Menger, F.M.; Davis, D.S.; Perschetti, R.A.; Lee, J.J. *J. Am. Chem. Soc.* **1990**, *112*, 2451.
(12)   Nagawa, Y.; Regen, S.L. *J. Am. Chem. Soc.* **1992**, *114*, 1668.
(13)   (a) Fuhrhop, J.H.; Liman, U. *J. Am. Chem. Soc.* **1984**, *106*, 4643. (b) Fuhrhop, J.H.; Liman, U.; David, H.H. *Angew. Chem. Int. Ed. Engl.* **1985**, *24*, 339. (c) Fuhrhop, J.H.; Liman, U.; Koesling, V. *J. Am. Chem. Soc.* **1988**, *110*, 6840.
(14)   Kobuke, Y.; Ueda, K.; Sokabe, M. *J. Am. Chem. Soc.* **1992**, *114*, 7618.
(15)   Carmichael, V.E.; Dutton, P.J.; Fyles, T.M.; James, T.D.; Swan, J.A.; Zojaji, M. *J. Am. Chem. Soc.* **1989**, *111*, 767.
(16)   Fyles, T.M.; James, T.D.; Pryhitka, A.; Zojaji, M. *J. Org. Chem.* **1993**.
(17)   Fyles, T.M.; Kaye, K.C.; James, T.D.; Smiley, D.W.M. *Tetrahedron Lett.* **1990** *31*, 1233.
(18)   Fyles, T.M.; Kaye, K.C.; Pryhitka, A.; Tweddell, J.; Zojaji, M. *J. Supramol Chem.* **1994**.
(19)   Fyles, T.M.; James, T.D.; Kaye, K.C. *Can. J. Chem.* **1990**, *68*, 976.
(20)   Fyles, T.M.; James, T.D.; Kaye, K.C. *J. Am. Chem. Soc.* **1993**.
(21)   Hille, B. *Ionic Channels of Excitable Membranes*, Sinauer Assoc.: Plymouth, RI, 1984.
(22)   (a) Shinkai, S. In *Cation Binding by Macrocycles*; Gokel, G.W.; Inoue, Y., Eds.; Marcel Dekker: New York, NY, 1990; p 397. (b) Shinkai, S.; Manabe, O. *Top. Curr. Chem.* **1984**, *121*, 67.
(23)   Martin, A.S.; Sambles, J.R.; Ashwell, G.J. *Phys. Rev. Lett.* **1992**, *70*, 218.
(24)   (a) *Rodd's Chemistry of Carbon Compounds* Coffey, S. Ed.; Elsevier: Amsterdam, 1973; Vol. III, part C, p 136. (b) Machnova, O.; Sterbe, V.; Valter, K. *Coll. Czech. Chem. Commun.* **1972**, *37*, 1851.
(25)   Fyles, T.M.; Suresh, V.V.; Fronczek, F.R.; Gandour, R.D. *Tetrahedron Lett.* **1990**, *31*, 1101.

RECEIVED April 26, 1994

Chapter 5

Langmuir–Blodgett Monolayers as Templates for the Self-Assembly of Zirconium Organophosphonate Films

Houston Byrd, John K. Pike, Margaret L. Showalter, Scott Whipps, and Daniel R. Talham[1]

Department of Chemistry, University of Florida, Gainesville, FL 32611

Organized molecular assemblies are being explored for potential uses ranging from electronics and photonics applications to the active components in chemical sensors. These molecular assemblies are sometimes organized layer-by-layer, or even molecule-by-molecule. For one-layer-at-a-time depositions, the order of the active surface or "template" layer plays a critical role in organizing the assembly. In this article, we show how a Langmuir-Blodgett monolayer of octadecylphosphonic acid can be used as the template layer for monolayer and multilayer depositions of zirconium phosphonates. The highly organized and well characterized Langmuir-Blodgett template allows subsequent assembly steps to be quantified. The organization of molecules in the "self-assembled" layers reflects the order in the original template layer.

The ability to tailor the chemical and structural characteristics of surfaces at the molecular level is the key to many strategies for developing new electronic and optoelectronic devices, solar energy conversion media, functional coatings, and of course, chemical sensors. Many methodologies exist for chemically derivitizing surfaces, but procedures for controlling surface functionalities at the molecular level fall into two general categories, Langmuir-Blodgett (LB) and so-called "self-assembly" (SA) methods (1-3). In the LB technique (1-3), molecules (or polymers) are mechanically manipulated into an organized array on a water surface before transfer to a solid support. SA methods (3,4) rely on the affinity of chemical functionalities for specific chemical surfaces in order to bind and orient molecules. The SA of organic thiols on gold surfaces is a topical example (3).

Other SA methods take advantage of specific chemical interactions, but are not limited to a specific surface. An example here is the formation of Si-O linkages between chlorosilanes and surface hydroxyl functionalities (5). The OH group can be on an oxide surface, or it can be an alcohol from an organic surface-derivitizing agent. Several schemes for multilayer build-up take advantage of this Si-O linkage (5-7). Another approach to SA multilayers takes advantage of the strong affinity of phosphonate and phosphate groups for the Zr^{4+} ion. Mallouk and co-workers (8-12)

---

[1]Corresponding author

0097–6156/94/0561–0049$08.00/0

demonstrated that multilayers of α,ω-diphosphonic acids can be built-up one layer at a time by alternately adsorbing the α,ω-diphosphonic acid and Zr^{4+} ions from solution (Scheme 1).

Our interest in organic thin films is in using organized organic assemblies as templates for developing single layers of inorganic extended lattice systems (*13,14*). Inorganic layers formed in this way should provide an opportunity to study chemical, electronic, and magnetic interactions in the limits of two-dimensions. We have investigated routes for preparing single-layer analogs of the transition metal phosphonates (*13*). In the solid-state, these are mixed organic/inorganic layered solids where binding within layers is strong, and interlayer interactions are van der Waals in nature (*15*). The critical step in preparing the metal phosphonate layers should be the organization of the surface or "template" layer. A lesson from investigations of thiol adsorption onto gold is that the organization of the surface plays an important role in organizing the adsorbate (*3,16-18*). If inorganic layers are formed at an organic surface then the extent of structural coherence in the inorganic layer is expected to be limited by the size of organized domains in the organic template layer.

Several methods have been explored for preparing a phosphorylated surface for use in multilayer depositions (*8,9,19,20*). While most of these provide a high density of phosphonate binding sites, none are expected to provide the ordered array required to form an inorganic extended lattice. In an effort to produce an ordered array, we developed a Langmuir-Blodgett route to a phosphonic acid template layer (*13*). In this article, we compare this surface to a phosphonic acid template layer prepared by self-assembling long alkyl chain molecules. The LB template layer provides a more ordered surface, and we show that zirconium phosphonate bilayers can be produced at the LB template. In addition, the LB template can be used for the build-up of multilayers, using the method of Mallouk (*8-12*), and the organization of the LB template layer is reflected in the multilayer films that are produced. The results show that when developing functional assemblies the organization of the template layer plays an important role in controlling the architecture of the film.

The Template Layer.

Our approach to a "self-assembling" template layer is outlined in Scheme 2. In the first step, ω-hexadecenylbromide (*21*) is hydrosilylated with trichlorosilane and self-assembled to a silicon oxide surface to form a bromide terminated monolayer. Conversion of the bromide to the phosphonate, by reaction with triethylphosphite, is carried out on the assembled monolayer and followed by acid hydrolysis to the phosphonic acid. All of the conversions performed on the surface are followed by ATR-FTIR and XPS, and go to completion to the extent that can be determined by XPS (*22*). To form the zirconium phosphonate bilayer, the phosphorylated substrate is immersed into an aqueous solution of $ZrOCl_2$ to bind Zr^{4+} at the phosphonic acid sites. To form the complete bilayer, the zirconated surface is then rinsed with water and placed into a solution of octadecylphosphonic acid in order to self-assemble the capping layer according to Scheme 1.

The Langmuir-Blodgett template layer is prepared according to Scheme 3. A Langmuir monolayer of octadecylphosphonic acid on pure water is transferred to an OTS covered surface by dipping down though the air/water interface into a vial. The vial is removed from the trough, and $ZrOCl_2$ is added to the vial. After 30 minutes, the zirconated surface is rinsed with water and is then ready for binding the capping layer. If instead, zirconium ion is added directly to the LB subphase, zirconium binds to the

A) Single Layer

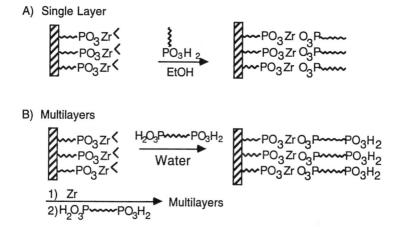

B) Multilayers

Scheme 1: Self-Assembly of Monolayer and Multilayered Films at the Zirconated Template Layer.

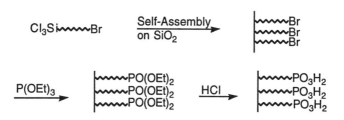

Scheme 2: Self-Assembly Method to a Phosphonic Acid Surface.

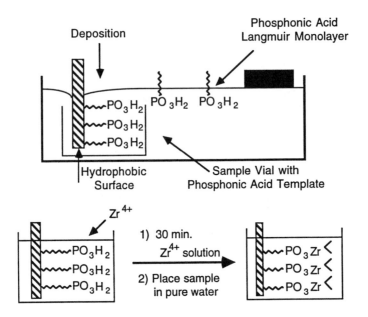

Scheme 3: Preparation of a Zirconated Langmuir-BlodgettTemplate Layer.

phosphonate monolayer, but the layer becomes too rigid to transfer using conventional techniques (23).
    The SA and LB template layers were characterized by ATR-FTIR (Figure 1) and XPS (Table 1). Three bands are seen in the IR between 2600 cm^{-1} and 3200 cm^{-1} that can be assigned (17,24) (from high energy to low energy) to the asymmetric methyl stretch, $v_a(CH_3)$, the asymmetric methylene stretch, $v_a(CH_2)$, and the symmetric methylene stretch, $v_s(CH_2)$. Information about the conformation and packing of the alkyl chains in the films can be obtained from the frequency and shape of these bands. For example, in solid octadecylphosphonic acid (13), where the hydrocarbon chains are close packed and all trans, $v_a(CH_2)$ appears at 2918 cm^{-1} with a fwhm of 20 cm^{-1}. As gauche bonds are introduced into the chains, $v_a(CH_2)$ moves to higher energy, and as the chains become more loosely packed, the fwhm increases. From the IR data shown in Figure 1, it is seen that $v_a(CH_2)$ for the LB template is at 2918 cm^{-1} with a fwhm of 20 cm^{-1}, characteristic of an all-trans and close-packed organization of the aliphatic chains. The same band for the SA template, however, is broader and shifted to 2922 cm^{-1}, indicating that the SA template is not as well organized. A similar result is seen when comparing the bilayers formed after self-assembling octadecylphosphonic acid at these template layers (Figure 1). The LB template provides a better organized surface, and this organization is carried over to binding of the second layer. This conclusion is backed up by the XPS analysis. Table 1 lists the P:Zr ratio of each of the zirconated template layers and the corresponding bilayers. The zirconated LB template has a P:Zr ratio of 1:1 which means that every phosphonic acid binds a zirconium ion from solution. This surface then assembles a complete capping layer resulting in a P:Zr ratio of 2:1 in the LB/LB bilayer (LB/LB designates a bilayer formed only by LB methods), which is consistent with the P:Zr content of the solid state zirconium phosphonates. The SA template is not completely zirconated, according to XPS, and the resulting SA/SA bilayers (SA/SA designates a bilayer formed only by SA methods) have a high P:Zr ratio. Combined, the IR and XPS analyses suggest that some surface sites in the more loosely packed template layer are inaccessible for binding species from solution, and that the disorder in the template is then transferred to subsequent layers. In contrast, the LB template layer provides a rigid, close-packed array of sites for binding subsequent layers.

Table 1. XPS Multiplex Data for Zirconium Phosphonate Layers

Film Type	Elements	Peak/Area[a]	Conc.(%)[b]
LB Template Layer	Zr	Zr 3d/1306	53.0
	P	P 2p/215	47.0
SA/SA Bilayer	Zr	Zr 3d/570	23.7
	P	P 2p/340	76.3
LB/SA Bilayer	Zr	Zr 3d/3795	31.0
	P	P 2p/1589	69.0
LB/LB Bilayer	Zr	Zr 3d/919	33.1
	P	P 2p/345	66.9

a. The units for Area are (counts-eV)/sec
b. The concentration is derived from the atomic and instrument sensitivity factors.

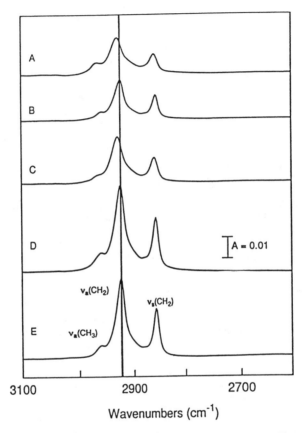

Wavenumbers (cm^{-1})

Figure 1. ATR-FTIR of the C-H stretch region for the A) Zirconated SA template layer, B) Zirconated LB template layer, C) bilayer formed by SA of octadecylphosphonic acid at A, D) bilayer formed by SA of octadecylphosphonic acid at B, E) bilayer formed by capping the LB template layer with another LB layer of octadecylphosphonic acid. The line at 2918 cm^{-1} represents the position of the $v_a(CH_2)$ band in crystalline octadecylphosphonic acid. Since there is no aliphatic CH$_3$ group in the SA template, the peak at 2962 cm^{-1} in A results from either unreacted ester groups or is a methylene stretch of the CH$_2$ adjacent to the phosphonate.

Zirconium Octadecylphosphonate Bilayers.

The LB template layer described above meets our requirements for an organized and close-packed template, and we have used this surface to build-up single layer and multilayer analogs of zirconium octadecylphosphonate. We can compare the structures obtained when completing the bilayer by capping the zirconated template layer with both an LB layer (LB/LB bilayer) and a self-assembled layer (this will be referred to as the LB/SA bilayer, because the capping layer is actually "self-assembled"). To complete the LB/LB bilayer (*13*), a substrate with the zirconated template is placed back into the LB trough containing pure H_2O, and a new octadecylphosphonic acid film is compressed. The octadecylphosphonic acid is then transferred to the substrate creating a Y-type zirconium octadecylphosphonate bilayer. The entire three-step deposition can be repeated to produce multilayers in this way. The LB/SA bilayer is prepared by simply placing the zirconated template into a solution of the octadecylphosphonic acid in 90% ethanol for one hour. ATR-FTIR (Figure 1) and XPS (Table 1) analyses show that both methods produce a complete bilayer. A P:Zr ratio of 2:1 is seen for both bilayers, indicating that each zirconium site in the template binds a phosphonic acid. The position, intensity, and shape of the C-H stretching bands are the same for both capping layers, as well as for the bilayers that are formed from each procedure. The all-trans and close-packing of the alkyl chains in the template layer are preserved in the bilayers.

There are differences, however, between the two bilayers. Contact angle measurements reveal a water-drop contact angle of 112° ± 1° on the surface of the LB/LB bilayer, and 110° ± 1° at the LB/SA bilayer (*25*). A more detailed investigation of the IR bands using polarized ATR-FTIR allows determination of the orientation of the alkyl chains in each layer. The tilt of the molecular axes from the surface normal can be determined from the dichroic ratio of the absorbance using s- and p-polarizations (*26,27*). The tilt angles determined in this way are illustrated in Figure 2. The LB and SA capping layers are oriented at tilt angles of 5° and 22°, respectively (*25*). Given the size of the phosphonate headgroup, a tilt of 22° is reasonable in order to achieve van der Waals packing of the alkyl chains. For the LB capping layer, the pressure used to align the molecules at the air-water interface orients the molecules nearly perpendicular to the surface, and this alignment is preserved when the film is transferred. The extremely strong zirconium-phosphonate bonding interaction does not allow the LB capping layer to relax after it is transferred to the zirconated template. The nearly perpendicular orientation of the LB capping layer is corroborated by ellipsometry and X-ray diffraction (*13*). It is interesting to note that the LB template layer has a tilt angle of 31°. It appears that the octadecylphosphonic acid layer relaxes under the water subphase after it is transferred to the OTS-covered substrate. Binding of zirconium ions then crosslinks the layer and holds it in place.

In order to compare the structures of the LB/LB and LB/SA bilayers to solid-state zirconium phosphonate structures, a model of zirconium octadecylphospohonate was generated using the SYBYL molecular modeling program (*13*). Using crystallographic coordinates for α-$Zr(HPO_4)_2 \cdot H_2O$ (*28*) to model Zr-O_3P binding, a bilayer was generated by grafting on a $(CH_2)_{17}CH_3$ chain in place of the phosphate OH. In the generated structure, the all-trans hydrocarbon chain lies at a tilt angle of 31.3° with respect to the zirconium ion plane which is a consequence of the phosphonate P-C bond orienting nearly perpendicular to the plane of metal ions. In the LB/SA bilayer, the alkyl groups orient in nearly this same way, and this bilayer closely resembles the structure seen in the solid-state analogs. However, the 5° tilt observed for the LB capping layer forces the Zr-O_3P bonding to differ from the bonding observed in α-$Zr(HPO_4)_2 \cdot H_2O$.

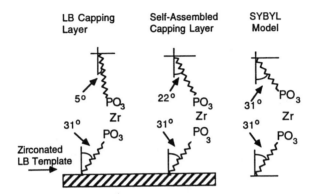

Figure 2. Tilt angles of the octadecylphosphonate molecules, determined by polarized ATR-FTIR, for the "LB/LB bilayer" and "LB/SA bilayer" formed at the zirconated LB template layer. For comparison, tilt angles from a SYBYL Model of zirconium octadecylphosphonate are shown.

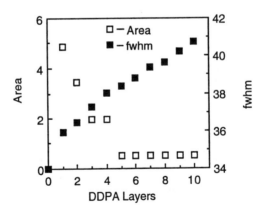

Figure 3. ATR-FTIR data for DDPA multilayers assembled at the LB template. Shaded boxes are the integrated area of the $v_a(CH_2)$ band for the complete film. Open boxes are the fwhm of the $v_a(CH_2)$ band for the outermost layer.

Multilayer Fabrication.

The LB template can also be used for depositing multilayers according to Mallouk's procedure (Scheme 1). By taking advantage of the well-characterized and highly organized nature of the LB template layer, each step of the multilayer deposition can be quantified (25).

The build-up of multilayers of 1,10-decanediyldiphosphonic acid was followed by ATR-FTIR, and Figure 3 shows the integrated intensity of the $v_a(CH_2)$ band as a function of the number of layers in the film. The integrated area per layer is much higher for the first few layers than for subsequent layers. After three or four layers, however, the increase in area with each additional layer becomes constant. It appears that a consequence of the high density of Zr^{4+} sites in the template layer is that once the flexible diphosphonic acid binds to the surface, it can bend over and bind another zirconium ion site within its reach. This reduces the number of surface sites available for subsequent layers. Eventually, after three or four layers, the remaining binding sites are spaced far enough apart that back-biting is less likely, and a constant layer-by-layer deposition proceeds. Figure 3 also shows a plot of the fwhm of the $v_a(CH_2)$ band for the last layer after each deposition cycle. The fwhm starts out very broad, and decreases for the first few layers until a constant value is reached. The back-biting in the first layers causes the alkyl chains to be disordered. The alkyl chains assume a more uniform packing in the later layers, but even at this point the molecules are not consistently close-packed.

If the α,ω-diphosphonic acid is rigid, the back-biting observed with the 1,10-decanediyldiphosphonic acid cannot occur. Quaterthiophenediphosphonic acid (QDP) is a rigid molecule developed by Howard E. Katz and co-workers (19), and has been used in the one-layer-at-a-time build-up of zirconium phosphonate multilayers. We have used the zirconated LB template to investigate deposition of this molecule (25). QDP is deposited from an acidic 50:50 DMSO:H_2O solution at 25° C. A strong absorption at 390 nm allows the layer-by-layer deposition of QDP to be followed by UV-vis spectroscopy. Figure 4 contains a plot of the absorbance at 390 nm versus the number of layers of QDP. The increase in absorbance is linear, starting with the first layer, indicating that the same amount of material is deposited during each deposition cycle. XPS of one QDP layer indicates that 60-80 % of the template sites are bound by QDP, depending upon the electron attenuation lengths used for the analysis (25). This is reasonable because of the mismatch in size between the QDP molecule and the octadecylphosphonic acid molecules that make-up the template layer. The cross-sectional area per molecule in the LB template is 24 Å²/molecule (13), whereas the calculated cross sectional area for QDP is 28 Å²/molecule (29).

The layered nature of the QDP assembly is confirmed by X-ray diffraction. Figure 5 shows low angle diffraction from 10 layers of QDP assembled at the LB template. First and second order reflections corresponding to a (001) d-spacing of 20.19 Å are observed, which agrees well with the length of the QDP molecule, and shows that the molecules are oriented essentially perpendicular to the surface.

Conclusion.

In order to prepare an extended-lattice monolayer, the organization of the surface template layer is the crucial step. For zirconium phosphonate layers, it is found that the LB method produces an organized phosphonic acid surface layer that in turn can be used to produce monolayer or multilayer zirconium phosphonate films. Once the template layer is in place, self-assembling the capping layer generates a zirconium phosphonate bilayer that more closely resembles the solid-state structure than does a bilayer formed by capping the template with another LB film. If the LB template is used to prepare multilayer films, the well-characterized and ordered template layer

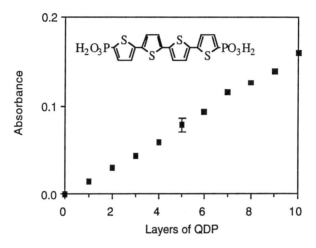

Figure 4. Absorbance at 390 nm monitoring the layer-by-layer deposition of QDP. (Adapted from ref. 25.)

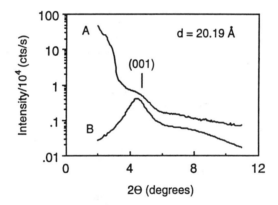

Figure 5. Low-angle X-ray diffraction from 10 layers of QDP self-assembled at the LB template. A) Experimental data. B) Fit of the experimental data with two Lorentzian curves after subtraction of the background. (Adapted from ref. 25.)

allows subsequent SA steps to be quantified. The results show that the organization of the binding sites in the template layer plays an important role in determining the organization of multilayered assemblies.

Acknowledgments.

We thank Dr. Howard E. Katz for providing us with a sample of QDP. Financial support has been provided by the National Science Foundation, and the donors of the Petroleum Research Fund, administered by the A.C.S.

Literature Cited.

(1)   Gaines, G. J. *Insoluble Monolayers at Liquid-Gas Interfaces*; Wiley-I nterscience: New York, 1966.
(2)   *Langmuir-Blodgett Films*; Roberts, G. G., Ed.; Plenum Press: New York, 1990.
(3)   Ulman, A. *An Introduction to Ultrahigh Organic Films: From Langmuir-Blodgett to Self-Assembly*; Academic Press: Boston, 1991.
(4)   Bigelow, W. C.; Pickett, D. L.; Zisman, W. A. *J. Colloid Sci* **1946**, *1*, 513-538.
(5)   Netzer, L.; Sagiv, J. *J. Am. Chem. Soc.* **1983**, *105*, 674-676.
(6)   Tillman, N.; Ulman, A.; Penner, T. L. *Langmuir* **1989**, *5*, 101-111.
(7)   Li, D.; Ratner, M. A.; Marks, T. J.; Zhang, C.-H.; Yang, J.; Wong, G. K. *J. Am. Chem. Soc.* **1990**, *112*, 7389-7390.
(8)   Lee, H.; Kepley, L. J.; Hong, H.-G.; Mallouk, T. E. *J. Am. Chem. Soc.* **1988**, *110*, 618-620.
(9)   Lee, H.; Kepley, L. J.; Hong, H.-G.; Akhter, S.; Mallouk, T. E. *J. Phys. Chem.* **1988**, *92*, 2597-2601.
(10) Akhter, S.; Lee, H.; Hong, H.-G.; Mallouk, T. E.; White, J. M. *J. Vac. Sci. Technol.* **1989**, *7*, 1608-1613.
(11) Hong, H.-G.; Sackett, D. D.; Mallouk, T. E. *Chem. Mater.* **1991**, *3*, 521-527.
(12) Cao, G.; Hong, H.-G.; Mallouk, T. E. *Acc. Chem. Res.* **1992**, *25*, 420-427.
(13) Byrd, H.; Pike, J. K.; Talham, D. R. *Chem. Mater.* **1993**, *5*, 709-715.
(14) Pike, J. K.; Byrd, H.; Morrone, A. A.; Talham, D. R. *J. Am. Chem. Soc.* **1993**, *115*, 8497-8498.
(15) Clearfield, A. *Comm. Inorg. Chem.* **1990**, *10*, 89-128.
(16) Bain, C. D.; Troughton, E. B.; Tao, Y.-T.; Evall, J.; Whitesides, G. M.; Nuzzo, R. G. *J. Am. Chem. Soc.* **1989**, *111*, 321-335.
(17) Porter, M. D.; Bright, T. B.; Allara, D. L.; Chidsey, C. E. D. *J. Am. Chem. Soc.* **1987**, *109*, 3559-3568.
(18) Nuzzo, R. G.; Zegarski, B. R.; Dubois, L. H. *J. Am. Chem. Soc.* **1987**, *109*, 733-740.
(19) Katz, H. E.; Schilling, M. L.; Chidsey, C. E. D.; Putvinski, T. M.; Hutton, R. S. *Chem. Mater.* **1991**, *3*, 699-703.
(20) Katz, H. E.; Scheller, G.; Putvinski, T. M.; Schilling, M. L.; Wilson, W. L.; Chidsey, C. E. D. *Science* **1991**, *254*, 1485-1487.
(21) Balachander, N.; Sukenik, C. N. *Langmuir* **1990**, *6*, 1621-1627.
(22) Crews, M. L. Masters Thesis, Florida, 1992.
(23) Byrd, H.; Pike, J. K.; Talham, D. R. *Thin Solid Films* **1993**, in press.
(24) Maoz, R.; Sagiv, J. *J. Colloid Interface Sci.* **1984**, *100*, 465-496.
(25) Byrd, H.; Whipps, S.; Pike, J. K.; Ma, J.; Nagler, S. E.; Talham, D. R. *J. Am. Chem. Soc.* **1994**, in press.
(26) Haller, G. L.; Rice, R. W. *J. Phys. Chem.* **1970**, *74*, 4386-4393.
(27) Tillman, N.; Ulman, A.; Schildkraut, J. S.; Penner, T. L. *J. Am. Chem. Soc.* **1988**, *110*, 6136-6144.
(28) Clearfield, A.; Smith, G. D. *Inorg. Chem.* **1969**, *8*, 431-436.
(29) Tasaka, S.; Katz, H. E.; Hutton, R. S.; Orenstein, J.; Fredrickson, G. H.; Wang, T. T. *Synthetic Metals* **1986**, *16*, 17-30.

RECEIVED March 25, 1994

Chapter 6

Shape-Selective Intercalation and Chemical Sensing in Metal Phosphonate Thin Films

Louis C. Brousseau, Katsunori Aoki, Huey C. Yang, and
Thomas E. Mallouk

Department of Chemistry, Pennsylvania State University, University
Park, PA 16802

Phosphonate salts of tetravalent, trivalent, and divalent metal ions
contain strong ionic-covalent bonds within the metal-oxygen
sheets that determine the details of their lamellar structures. The
divalent metal (Zn^{2+} and Cu^{2+}) compounds can be made
nanoporous by various techniques, and subsequent intercalation
by small molecules such as ammonia, amines, and other small
molecules forms the basis for size- and shape-selective piezoelectric
sensors. Several techniques have been developed for depositing
these materials as thin films on quartz crystal microbalance (QCM)
devices. The most successful of these, in terms of eliminating
interferences and speed of device response, involves layer-by-layer
growth of films through adsorption of their components from non-
aqueous solutions.

Metal phosphonate salts, a class of lamellar metal-organic solids, offer
considerable promise as structurally tunable media for shape-selective
binding and sensing of small molecules. The details of these layered
structures are determined by the identity of both the metal and the pendant
organic group(s). By choosing both components properly, solids with
molecule-size void spaces between layers can be created. Because these
materials are microcrystalline, their internal pore spaces have well-defined
size and shape. Moreover, vacant metal coordination sites can be created
in the micropores, so that the intercalation of small molecules exhibits both
steric (size and shape) and electronic (ligand binding) selectivity. These
properties can be exploited to create selective chemical sensors, when the
solid is interfaced to a measurement device such as a quartz crystal
microbalance (QCM). The challenges to be faced in designing practical
sensors of this type are engineering of the binding site - to impart
selectivity and to allow the device to function with the desired sensitivity -
and the elimination of non-specific ligation sites on the external surfaces of
crystallites.

0097–6156/94/0561–0060$08.00/0

Synthesis and structure of layered phosphates and phosphonates. The phosphate and phosphonate salts of most multivalent metal ions can be prepared by combining solutions of a soluble metal salt (chloride, nitrate, etc.) with H_3PO_4 or with a phosphonic acid (1-3). This simple procedure very often results in an insoluble lamellar solid, which can be either amorphous to x-rays or polycrystalline, depending on the rate of precipitation. In some cases the direct precipitation route does not give the desired material, and other reaction pathways, most commonly redox reactions, are used. For example, the vanadyl phosphonates are prepared by alcohol reduction of V_2O_5 in the presence of the appropriate phosphonic acid (4). Similarly, lamellar ferrous phosphonates do not form directly from ferrous sulfate solutions, but rather are prepared by reduction of FeOCl (5). Mo(V) polyoxo cluster phosphonates, which also adopt layered structures, can be assembled by a hydrothermal redox reaction involving Na_2MoO_4, Mo metal, and a phosphonic acid (6).

The crystal structures of metal phosphonates usually resemble those of the purely inorganic phosphate salts of the same metal or of another metal. The zirconium phosphonates $Zr(O_3PR)_2$, the most thoroughly studied of these materials, have essentially the same two-dimensional Zr-O-P network as the inorganic phosphate α-$Zr(O_3POH)_2 \cdot H_2O$ (7). This point is illustrated in Figure 1. Similar connectivity is found in compounds of formula $HFe(O_3PR)_2 \cdot H_2O$ (8), and a somewhat distorted variant occurs in $Ca(HO_3PR)_2$ (9). The vanadyl phenylphosphonate $VO(O_3PC_6H_5) \cdot 2H_2O$, on the other hand, adopts a structure (10) resembling $Mg(O_3POH) \cdot 3H_2O$ (11) rather than an inorganic vanadium phosphate. An important feature of all these compounds is that the bonding in the ionic-covalent metal-oxygen-phosphorus network is much stronger than the van der Waals interactions between pendant organic groups. The identity of the metal reliably determines the structure of the solid, and the organic groups, which are "along for the ride", may therefore be juxtaposed in a predictable manner.

Shape-selective intercalation reactions. In certain special cases, metal phosphonate compounds can be made with molecule-size void spaces between layers. This is accomplished either by using a "templating" approach - where a small molecule (alcohol or water) bound to the metal can be removed to leave an open coordination site - or by using a metal ion such as Cu^{2+} that prefers non-octahedral coordination. Johnson and coworkers (12) first demonstrated template effects in these materials with the compounds $VO(O_3PR) \cdot H_2O \cdot R'OH$. A benzyl alcohol (R'OH) molecule coordinated to the vanadium atom can be removed topochemically, to leave an open coordination site that binds primary alcohols selectively. The divalent metal (Zn, Co, Mn, Mg, Fe) phosphonates $M(O_3PR) \cdot H_2O$, shown in Figure 2, exhibit similar reactivity. In these compounds, a water molecule occupies one of six metal coordination sites, and may be removed topochemically by heating (13). This leaves an open coordination site surrounded by a rectangular "picket fence" of R groups, which impose steric restriction. The coordination chemistry of the metal ions in these

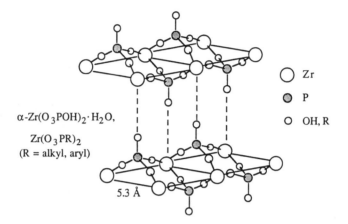

α-Zr(O$_3$POH)$_2 \cdot$ H$_2$O,

Zr(O$_3$PR)$_2$
(R = alkyl, aryl)

5.3 Å

○ Zr

◉ P

○ OH, R

Figure 1. Schematic drawing of the structure of α-zirconium phosphate and the zirconium phosphonate salts Zr(O$_3$PR)$_2$. The water of hydration is eliminated for clarity, and dotted lines indicate the registry of layers in α-Zr(O$_3$POH)$_2 \cdot$H$_2$O.

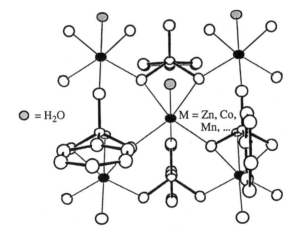

◉ = H$_2$O

M = Zn, Co, Mn, ...

Figure 2. View down the stacking axis of the M(O$_3$PC$_6$H$_5$)·H$_2$O (M = Zn, Co, Mn, Mg, Fe) structure. Water molecules (shaded) are removed topochemically upon heating, leaving the metal ions (black) coordinated by five phosphonate oxygen atoms. The phenyl rings restrict access of ligands to the open coordination site.

structures parallels that found in solution; that is, Mg and Mn bind oxygen-containing ligands, whereas Zn and Co prefer coordination by ammonia and amines.

Zinc methylphosphonate intercalates aliphatic amines with no branching at the α-position, for example *n*- and *iso*-butylamines; the *sec*- and *tert*-butylamine isomers are excluded because of steric restriction by the methylphosphonate group (13a). In the case of zinc phenylphosphonate (Figure 2), access to the site vacated by the water molecule is sufficiently restricted that only ammonia is intercalated, and aliphatic amines are excluded (13b). The excellent shape-selectivity of these reactions suggests the possibility of making sensors for amines and ammonia from zinc phosphonates, provided the intercalation event can be transduced to an observable (electronic, optical, mechanical, etc.) signal.

Monitoring intercalation of gas-phase molecules with quartz crystal microbalance (QCM) devices. Changes in several physical observables typically attend intercalation reactions. Both the mass and interlayer spacing (and hence the volume) of the sample increase; significant changes in electrical conductivity, dielectric constant, and magnetic susceptibility also occur for many host materials. In the present case, mass changes are conveniently transduced to an electrical signal by means of piezoelectric QCM devices (14). Mass (Δm) and frequency (Δf) changes for the QCM are related by the Sauerbrey equation (1), in which ρ_q and μ_q are the density and shear modulus of quartz (2.65 g/cm^3 and 2.95 x 10^{11} dyn/cm^3, respectively), A is the area of

$$\Delta f = \frac{-2\Delta m f_0^2}{A \sqrt{\rho_q \mu_q}} \qquad (1)$$

the gold electrodes contacting the quartz crystal, and f_0 is its resonant frequency. In our experiments, AT-cut quartz crystals (f_0 = 9 MHz) were used. Since frequency measurements made with these devices are reliable to $\pm$ 2 s^{-1}, changes in mass as small as a few nanograms can be reproducibly detected. The flow cell used to control exposure of the QCM devices to ammonia is shown schematically in Figure 3. In the case of other analytes, an excess (ca. 2 mL) of the liquid was injected into an argon-filled 100 mL round bottom flask, which was held at 0 °C. Argon was bubbled through the liquid, and the partial pressure of the analyte in the carrier gas at the QCM, which was held at room temperature, was then assumed to be equal to the vapor pressure at 0 °C.

Colloidal zinc phenylphosphonate can be prepared by stirring the microcrystalline solid in a solution containing a substoichiometric amount of $ZrOCl_2$ (15, 16). A mole ratio of 1 Zr/25 Zn is sufficient to prepare colloidal suspensions of the zinc salts. These zinc phenylphosphonate particles, capped with Zr^{4+}, adhere readily to gold surfaces primed with a monolayer of 4-mercaptobutylphosphonic acid. Gold QCM electrodes derivatized in

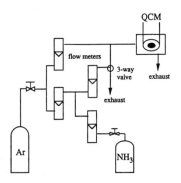

Figure 3. Schematic diagram of the apparatus used to mix ammonia and carrier gases, and to control gas flow to the QCM device.

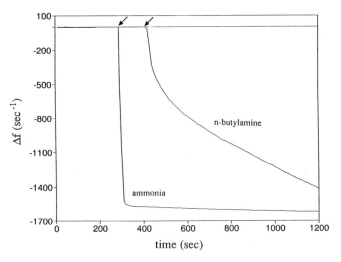

Figure 4. Plots of frequency change vs. time for exposure of anhydrous colloidal zinc phenylphosphonate/QCM devices to gas-phase ammonia and *n*-butylamine. Time of analyte injection is indicated by arrows.

this way, when heated above 100°C to convert the zinc phenylphosphonate to its anhydrous form, are responsive to ammonia gas, as shown in Figure 4. Exposure of the device to NH_3 gives a negative frequency change, which corresponds to the uptake of several micrograms of ammonia ca. 30 seconds. Purging the sample with argon at room temperature results in loss of approximately 50% of the weight gained; this is consistent with the irreversibility of binding at interlayer Zn^{2+} sites, and with ca. 50% non-specific adsorption on the external surface of the colloidal particles. Heating the device to 165° drives off intercalated NH_3, and restores the frequency to its initial value.

While the response of this device is rapid, the irreversibility of ammonia binding at room temperature places serious limitations on the use of zinc phosphonates in practical sensors. Another problem with devices derived from colloidal particles is the large external surface area, on which molecules other than the desired analyte can adsorb. This point is illustrated in Figure 4. Despite the fact that butylamine isomers are excluded from intercalation into bulk $Zn(O_3PC_6H_5)$, the mass change upon exposure of the device to either *n*- or *tert*-butylamine vapor is comparable to that found for ammonia. Adsorption of these interfering analytes is also quite slow, occurring over a period of hours.

Stepwise layer growth of divalent metal phosphonate films on the QCM.
The shape-selective intercalation of analytes is a bulk (interlayer) process. In order to minimize interferences arising from the adsorption of molecules on particle external surfaces, the ratio of bulk to external binding sites must be increased. Normally this might be accomplished by simply increasing the particle size; however, thicker layered particles would entail longer solid-state diffusion paths for analyte molecules, and could significantly slow the device response.

A strategy for minimizing the external surface area while keeping diffusion lengths very low is to grow the solid as a smooth film, one monolayer at a time, on the surface of the QCM electrodes. Layer-by-layer growth of metal phosphonates from aqueous solutions has been successfully demonstrated for both tetravalent (Zr^{4+}, Hf^{4+}) and trivalent (rare earth) metals (17). In these cases, binding of the metal to the phosphonate group is sufficiently strong that a monolayer (i.e., 10^{-10} moles on a QCM device) of metal ions adsorbed at a phosphonate surface is not removed by washing with water. In the case of divalent metal ions, unfortunately, the solublility of the phosphonate salt is sufficiently high in water that washing removes the growing film. However, films may be grown by sequential adsorption reactions in less polar solvents, as shown in Figure 5. Using ethanol-soluble metal salts such as perchlorates and acetates, multilayer films of excellent quality can be grown by sequentially dipping the QCM device in 5 mM solutions of M^{2+} (M = Zn, Cu) and alkyl- or arylbisphosphonic acids. The time required for each adsorption and washing step is ca. 10 min., which is significantly more rapid than that needed to grow films from Zr^{4+}.

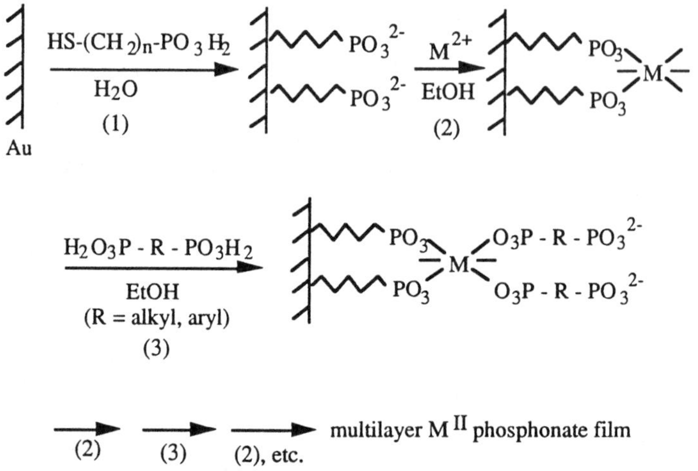

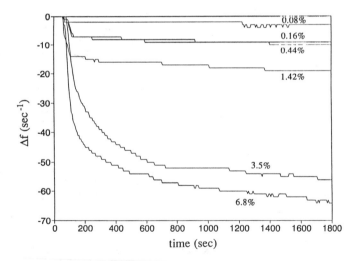

Figure 5. Procedure for layer-by-layer growth of divalent metal phosphonate films on gold surfaces. The sample is washed with ethanol between adsorption steps. Repetition of steps (2) and (3) gives multilayer thin films.

Figure 6. Frequency change vs. time for exposure of a 5-layer copper biphenylbisphosphonate QCM device to varying concentrations of ammonia in Ar. Concentrations indicated are volume percent.

The second problem presented by the zinc phenylphosphonate-based QCM devices is the irreversibility of analyte binding. The Zn^{2+} ion in anhydrous $Zn(O_3PR)$ is five-coordinate ; binding of an amine at the vacant sixth site is very favorable, since Zn^{2+} prefers octahedral coordination. Interestingly, the structures of $Cu(O_3PR) \cdot H_2O$ (R = CH_3, C_6H_5) compounds contain Cu^{2+} in square pyramidal coordination (18). Four of the five ligands surrounding the Jahn-Teller Cu^{2+} ion are phosphonate oxygen atoms, and the fifth is a water molecule. Amines bind weakly and reversibly to the open sixth coordination site. Whereas water also intercalates anhydrous $Zn(O_3PR)$ and represents a potential interference in materials containing open zinc coordination sites, the analogous copper compounds crystallize from aqueous solutions with the sixth coordination site free.

The problems of reversibility and selectivity were addressed simultaneously by growing copper phosphonate thin films by the sequential adsorption method. This method requires a bisphosphonic acid (adsorption step 3 in Figure 5), which bridges between metal-oxygen layers. Sequential adsorption of Cu^{2+} and **I** gave a layer repeat distance of 13.4 Å by ellipso-

$$H_2O_3P \diagdown \diagdown \diagup \diagup -PO_3H_2$$
I

metry, in close agreement with the layer spacing of 14.0 Å, determined by x-ray powder diffraction of bulk $Cu_2(O_3PC_{12}H_8PO_3) \cdot 2H_2O$.

Figure 6 shows traces of frequency vs. time for exposure of a 5-layer film (thickness = 74 Å) to varying concentrations of ammonia. The total frequency (mass) change about 25 times smaller than that found with colloidal zinc phenylphosphonate (Figure 4), consistent with the large difference in thickness of films prepared by the two methods. Note that equilibrium is reached within ca. 10 min., consistent with reversible binding to Cu^{2+}, and with a very short diffusion length through the film. The device response is best between about 0.5% and 5% ammonia by volume; above this value, all coordination sites are saturated, and the weight gained (60 ng) is in rough agreement with that calculated for a five-layer film of $Cu_2(O_3PC_{12}H_8PO_3) \cdot 2H_2O \cdot 2NH_3$ (35 ng). The discrepancy may be understood in terms of the surface roughness of the gold QCM electrodes, which was found from electrochemical experiments to be 2.0 ± 0.1.

Figure 7 compares the response of the copper biphenylbisphosphonate device to ammonia and butylamine isomers. The response to saturated vapors of both *n*- and *tert*-butylamine is quite interesting. The latter is excluded from intercalation, and the weight gained is significantly smaller than that seen for the same volume fraction of ammonia in the gas phase. Most of the adsorbed *tert*-butylamine is also removed by purging the cell with Ar. However, *n*-butylamine intercalates slowly, and the mass change is quite large, even at significantly lower partial pressure of amine. The

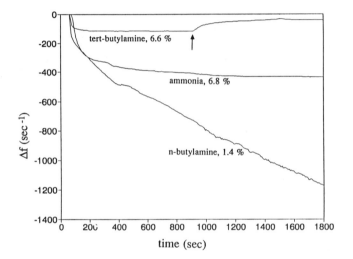

Figure 7. Frequency change vs. time for exposure of a 5-layer copper biphenylbisphosphonate QCM device to ammonia (6.8 % by volume in Ar), and gas phase *n*- and *tert*-butylamine (1.4 % and 6.6 % in Ar, respectively). Arrow indicates beginning of Ar purge for the *tert*-butylamine experiment.

weight gained upon 30-minute exposure to *n*-butylamine, 1380 ng, is in the range expected for stoichiometric binding at all the copper sites in the film (1620 ng, assuming a surface roughness factor of 2.0). This behavior represents a departure from the selectivity shown by zinc phenylphosphonate, which excludes even short-chain *n*-alkylamines from intercalation (13b). Apparently the slightly more open copper biphenylbisphosphonate structure, like $Cu(O_3PCH_3) \cdot H_2O$ (18), will accomodate *n*-alkylamines, provided the alkyl chain is small enough to be accomodated by the pillaring group **I**, while excluding *tert*-butylamine. Design of a truly selective ammonia sensor based on copper phosphonates will entail more steric restriction of this binding site.

Conclusions.   QCM devices containing copper and zinc phosphonates as chemically active materials respond rapidly (within 30 seconds in favorable cases) to gas phase ammonia and amines. In the case of the anhydrous zinc compounds, coordination of the analyte is irreversible at room temperature. The same analytes bind reversibly at five-coordinate Cu^{2+} sites, and the useful range for sensing ammonia in this case is 0.5 - 5 % by volume. Interferences associated with non-specific adsorption of amines on particle external surfaces can be minimized by growing the divalent metal phosphonates in layer-by-layer fashion on the gold electrodes of the QCM. In this way the external surface area is minimized without greatly increasing the thickness of the active particles.

At present the dynamic range of these sensors is narrow and is determined by the equilibrium constant for binding gas-phase ammonia to Cu^{2+}. Other metal ions, for example Co^{2+} and Mn^{2+}, are likely to have binding constants intermediate between those of Cu^{2+} and Zn^{2+}, and may be useful as components of more sensitive devices. The dynamic range of the devices might be extended by growing films from a mixture of divalent metal ions. Finally, we note that further steric restriction of the binding site is required in the case of the copper compounds, in order to eliminate interferences from *n*-alkylamines. Two possibilities exist for tuning the shape selectivity of the binding pocket: growing films from more bulky bisphosphonic acids, and substituting larger molecules (e.g., thiols, amines) for the coordinated water molecule in the structure. These possibilities will be tested in future experiments.

Acknowledgments.   We thank Prof. Eugene Smotkin for supplying the design of the oscillator circuit, and for helpful advice with the QCM experiments.   This work was supported by the National Science Foundation (CHE-9396243) and by the ACS Petroleum Research Fund. T.E.M. also thanks the Camille and Henry Dreyfus Foundation for support in the form of a Teacher-Scholar Award.

Literature Cited

(1) Clearfield, A. *Comments Inorg. Chem.* **1990**, *10,* 89.
(2) Alberti, G. *Acc. Chem. Res.* **1978**, *11*, 163.
(3) Cao, G.; Hong, H.-G.; Mallouk, T. E. *Acc. Chem. Res.* **1992**, *25*, 420.

(4) (a) Johnson, J. W.; Jacobson, A. J.; Brody, J. F.; Lewandowski, J. T. *Inorg. Chem.* **1984**, *23*, 3842; (b) Johnson, J. W.; Jacobson, A. J.; Butler, W. M.; Rosenthal, S. E.; Brody, J. F.; Lewandowski, J. T. *J. Am. Chem. Soc.* **1989**, *111*, 381; (c) Johnson, J. W.; Brody, J. F.; Alexander, R. M. ; Pilarski, B.; Katritsky, A. R. *Chem. Mater.* **1990**, *2*, 198.

(5) Bujoli, B.; Pena, O.; Palvadeau, P.; Le Bideau, J.; Payen, C.; Rouxel, J. *Chem. Mater.* **1993**, *5*, 583.

(6) Cao, G.; Haushalter, R. C.; Strohmaier, K. G. *Inorg. Chem.* **1993**, *32*, 127.

(7) (a) Clearfield, A.; Smith, G. D. *Inorg. Chem.* **1969**, *8*, 431; (b) Dines, M. B.; DiGiacomo, P. *Inorg. Chem.* **1981**, 20, 92; (c) Dines, M. B.; DiGiacomo, P.; Callahan, K. P.; Griffith, P. C.; Lane, R.; Cooksey, R. E. in *Chemically Modified Surfaces in Catalysis and Electrocatalysis*, J. Miller, Ed. (ACS Symp. Ser 192, **1982**), p. 223; (d) Alberti, G.; Costantino, U.; Környei, J.; Giovagnotti, M. L. L. *Reactive Polym.* **1985**, *4*, 1.

(8) (a) Bujoli, B.; Palvadeau, P.; Rouxel, J. *Chem. Mater.* **1990**, 2, 592; (b) Bujoli, B.; Palvadeau, P.; Rouxel, J. *C. R. Acad. Sci. Paris* **1990**, *310(II)*, 1213; (c) Palvadeau, P.; Queignec, M.; Vénien, J. P.; Bujoli, B.; Villieras, J. *Mat. Res. Bull.* **1988**, *23*, 1561.

(9) Cao, G.; Lynch, V. M.; Swinnea, J. S.; Mallouk, T. E. *Inorg. Chem.* **1990**, *29*, 2112.

(10) Johnson, J. W.; Jacobson, A. J.; Brody, J. F.; Lewandowski, J. T. *Inorg. Chem.***1984**, *23*, 3842.

(11) (a) Sutor, D. J. *Acta Crystallogr.* **1967**, *23*, 418; (b) Abbona, F.; Boistelle, R.; Haser, R. *Acta Crystallogr.* **1979**, *B35*, 2514.

(12) Johnson, J. W.; Jacobson, A. J.; Butler, W. M.; Rosenthal, S. E.; Brody, J. F.; Lewandowski, J. T. *J. Am. Chem. Soc.* **1989**, *111*, 381.

(13) Cao, G.; Mallouk, T. E. *Inorg. Chem.* **1991**, *30*, 1434; (b) Frink, K. J.; Wang, R.-C.; Colon, J. L.; Clearfield, A. *Inorg. Chem.* **1991**, *30*, 1439.

(14) Buttry, D. A.; Ward, M. D. *Chem. Rev.* **1992**, *92*, 1355.

(15) Brousseau, L.; Aoki, K.; Garcia, M. E.; Cao, G.; Mallouk, T. E. *NATO ASI Ser. C*, **400**, *Multifunctional Mesoporous Inorganic Solids*, Sequeira, C. A. C.; Hudson, M. J., Eds., 1993, pp. 225-236.

(16) Aoki, K.; Brousseau, L. C.; Mallouk, T. E. *Sensors and Actuators B*, **1993**, *14*, 703.

(17) (a) Lee, H.; Kepley, L. J.; Hong, H.-G.; Mallouk, T. E. *J. Am. Chem. Soc.* **1988**, *110*, 618; (b) Lee, H.; Kepley, L. J.; Hong, H.-G.; Akhter, S.; Mallouk, T. E. *J. Phys. Chem.* **1988**, *92*, 2597; (c) Akhter, S.; Lee, H.; Hong, H.-G.; Mallouk, T. E.; White, J. M. *J. Vac. Sci. Technol. A* **1989**, *7*, 1608; (d) Putvinski, T. M.; Schilling, M. L.; Katz, H. E.; Chidsey, C. E. D.; Mujsce, A. M.; Emerson, A. B. *Langmuir* **1990**, *6*, 1567; (e) Hong, H.-G.; Mallouk, T. E. *Langmuir* **1991**, 7, 2362; (f) Hong, H.-G.; Sackett, D. D.; Mallouk, T. E. *Chem. Mater.* **1991**, *3*, 521; (g) Katz, H. E.; Schilling, M. L.; Chidsey, C. E. D.; Putvinski, T. M.; Hutton, R. S. *Chem. Mater.* **1991**, *3*, 699; (h) Katz, H. E.; Scheller, G.; Putvinski, T. M.; Schilling, M. L.; Wilson, W. L.; Chidsey, C. E. D. *Science* **1991**, *254*, 1485; (i) Kepley, L. J.; Sackett, D. D.; Bell, C. M., Mallouk, T. E. *Thin Solid Films* **1992**, *208*, 132; (j) Umemura, Y.; Tanaka, K.-I.; Yamagishi, A. *J. Chem. Soc. Chem. Commun.* **1992**, 67.

(18) Zhang, Y.; Clearfield, A. *Inorg. Chem.* **1992**, *31*, 2821.

RECEIVED April 5, 1994

Chapter 7

High Surface Area Silica Particles as a New Vehicle for Ligand Immobilization on the Quartz Crystal Microbalance

Rick Cox[1], Dario Gomez[1], Daniel A. Buttry[1,3], Peter Bonnesen[2], and Kenneth N. Raymond[2,3]

[1]Department of Chemistry, University of Wyoming
Laramie, WY 82071–3838
[2]Department of Chemistry, University of California,
Berkeley, CA 94720

This paper describes a new strategy for the immobilization of ligands on the surface of a quartz crystal microbalance for the purpose of producing piezoelectric sensors for selective detection of heavy metals in solution. Immobilization is achieved by reacting the ligand of choice with the interfacial Si-OH groups of micron-sized, porous, high surface area silica particles and then anchoring a monolayer of these particles onto the surface of the QCM. The particles thus act as the vehicle by which ligand immobilization is achieved. Detection of the metal occurs when it binds to the ligand, producing a mass change of the QCM/particle layer composite resonator. The method is demonstrated by detection of UO_2^{2+} with a phthalamic acid derivative immobilized on 3 μm diameter Hypersil silica particles. The mass gain of the particle layer as a function of the concentration of UO_2^{2+} (in the range 1-10 μM) constitutes an adsorption isotherm, from which a binding constant for the reaction of the immobilized ligand with UO_2^{2+} of 2×10^5 M^{-1} is calculated.

The development of selective and sensitive chemical sensors for real-time measurements of chemical species in solution requires that two fundamentally different goals be achieved. First, the target analyte must be recognized (usually by binding it to some type of receptor that is coupled to the sensor) with some reasonable degree of chemical selectivity. Second, this recognition event must be transduced into some type of signal (often a voltage change) with sufficient sensitivity for the generation of a meaningfully large signal. This transduction event frequently involves causing a change in some property of the receptor, such as optical absorption, fluorescence, mass, or redox chemistry, with each approach having its own set of advantages and disadvantages for a given sensing application.

The use of mass sensitive, piezoelectric devices in chemical sensing applications requires that during the sensing event, the target analyte become bound at the surface of the device. This binding event leads to a change in the resonant frequency of the oscillating device (often, but not always, by virtue of a change in mass), that is used to infer the presence of the analyte. A variety of methods for capturing the target analyte and generating detectably large mass

[3]Corresponding authors

0097–6156/94/0561–0071$08.00/0
© 1994 American Chemical Society

changes has been explored, as discussed in a relatively recent review (*1*). For those cases in which the signal is derived solely from a mass change, the magnitude of the signal that is generated by the binding event is proportional to the amount of analyte that becomes bound, which is, in turn, related to the total amount of receptor present at the surface of the device. Thus, in order to maximize the sensor signal, it is frequently desirable to immobilize relatively large quantities of the receptor at the surface of the piezoelectric transducer. The focus of this contribution is the description of a new method for immobilization of large quantities of ligands for metals via their attachment to porous, high surface area silica particles, followed by the attachment of these derivatized particles to the transducer surface. These porous silica particles serve to increase the effective surface area of the QCM surface, thus allowing immobilization of greater quantities of ligand within a porous, but rigid, structure.

The use of these silica particles as the vehicle for receptor immobilization is demonstrated via detection of micromolar concentrations of UO_2^{2+} *in situ*, in aqueous solution using an immobilized phthalamic acid derivative. The detection of this species has relevance in a variety of contexts, including oceanographic studies (*2*), other natural waters (*3*), and occupational health (*4*). Based on a typical target concentration in the micromolar range for this species, a phthalamic acid derivative was chosen as the receptor ligand, since compounds of this type are known to form uranyl complexes with binding constants on the order of 10^4 - 10^5 M^{-1}(*5,6*). The relationship between the desired binding constant and the target concentration range is further elaborated below.

Experimental

All chemicals were of reagent grade or better, and were used as received. Solutions were made using water from a Barnstead Nanopure or a Milli-Q purification system. When required, tetrahydrofuran (THF) was dried by distillation from sodium benzophenone ketyl, and methanol by distillation from CaH_2. These solvents and their solutions were kept from exposure to the atmosphere during the ligand synthesis and immobilization steps described below. Gamma-aminopropyl-triethoxysilane (APS) was obtained from Aldrich and was distilled prior to use. Phthalic anhydride was purified by extraction and recrystallization from $CHCl_3$. Reagent grade $UO_2(NO_3)_2 \cdot 6H_2O$ was obtained from J.T. Baker, and was used as received. Hypersil silica gel (Shandon Southern Products, Ltd., 3±1 μm diameter, 170 m^2/g surface area, 120 Å pore size) was obtained from Phenomenex (Torrance, CA).

Ligand immobilization on the Hypersil was done after first activating the silica with a 1.5 hour treatment in 75-90 °C, 0.8 M HNO_3, and was based on previously published procedures (*7-9*). To produce the phthalamic acid silane derivative (N-[3-(triethoxysilyl)propyl]phthalamic acid) for reaction with the silica, phthalic anhydride (0.8 g) was dissolved in 25 mL dry THF, and recently distilled APS (0.8 g) was added neat, dropwise to the stirred solution over a 5 minute period. The reaction mixture was stirred for 5 hours, then transferred to a 25 mL flask along with 1.0 g of activated Hypersil. The solution was stirred for 72 hours, the particles isolated, and washed with THF. Yield is 1.3 g of particles (after careful drying). Elemental analysis of the particles gave 11.0% C, 1.31% H, and 1.25% N, with a C/N ratio = 10.3, to be compared with an expected value of 11.0. This gives a ligand loading of 0.892 millimoles of ligand per gram of particles. This is a very high loading, near the theoretical maximum for a silica surface. Based on this value, it is not clear whether this means that the immobilization procedure leads to formation of a complete, uniform coverage for the monolayer or that there is multilayer formation in some (more accessible) regions of the surface and little coverage in some (less sterically accessible) regions.

Particle immobilization on the Au electrode of the QCM (see ref. 10 for details of the QCM electrode preparation) was achieved by casting a 3000 Å adhesion layer of Glassclad-RC (Huls America), dusting the surface with particles, and gently pressing the particles into the adhesion layer. (Glassclad-RC is an oligomeric poly(dimethylsiloxane) material which can be cured at room or higher temperatures.) Thus, ca. 10% of the particle diameter is embedded within the adhesion layer. The QCM/particle composite resonators were then cured at 70 °C for 20-24 hours, followed by rinsing away any unattached particles.

Film thicknesses of the Glassclad-RC adhesion layers were measured with a Rudolph Research Auto-El ellipsometer. Conductance spectra (10,11) were obtained with a Hewlett-Packard 4192A impedance analyzer. Real-time measurements of the oscillation frequency of the QCM were made with a broadband oscillator circuit [10] built at UW and powered by a HP dual output power supply, a Philips PM6654 frequency counter, and a Kipp & Zonen XYY' recorder. Note that the broadband oscillator circuit is designed to track the series resonant frequency of the QCM resonator in real time as its mass changes due to metal binding, while the impedance analyzer is used to characterize the entire resonance spectrum of the resonator.

A standard flow cell (12) was used for the real-time measurements, with solution flow achieved with an Isco peristaltic pump. The flow rate of the various solutions used in all experiments reported here was 3.6 mL/min. The flow cell is designed to hold two crystals. One is referred to as the sensor crystal; it has a particle layer bearing the immobilized ligand. The other is referred to as the reference crystal; it has a particle layer without the immobilized ligand. The function of the reference crystal is to compensate for changes in the density and/or viscosity of the solution when its composition changes (12). This is particularly critical in the present case because the large void volumes of these particle layers lead to very large apparent mass changes when the density of the solution within the pores changes (11). Since the particle loading on each crystal is slightly different, their sensitivity to these density effects is also different. However, the relative responses of all sensor/reference crystal pairs to changes in solution density are determined by exposure to solutions of varying density (e.g. 0.1 mM and 1.0 mM $NaNO_3$, which do not interact specifically with the immobilized ligands) and observing the corresponding frequency changes. This allows a determination of the relative void volume of each particle layer. Then, in real-time experiments, metal uptake is always corrected for this "density offset" caused by exchange of the pore volume. The possibility of significant levels of non-specific adsorption of the uranyl species on the reference crystal was ruled out by virtue of control experiments in which the identity of the supporting electrolyte in the solution was varied. These experiments showed that the variation in the density offset was exactly as predicted for each electrolyte. The results make it unlikely that non-specific adsorption is important, because it is very unlikely that the extent of specific adsorption, if it were occuring, would be constant with variation in the solution composition.

The mass sensitivity of the 5 MHz AT-cut alpha-quartz crystals used here is 56.6 Hz cm^2 μg^{-1}. As will be seen below, we assume that these immobilized particle layers behave as rigid layers, so that this mass-to-frequency proportionality may be used to directly calculate mass changes from frequency changes (see ref. 11 for a detailed discussion of this point). Impedance analysis (to be reported elsewhere) as well as the extremely good agreement between predicted and observed mass changes for metal uptake based on these calculations (see below) both suggest this assumption to be a good one.

Results and Discussion

Figure 1 shows a schematic depiction of the 3 μm diameter silica particle layer, embedded within the polymeric adhesion layer on the Au QCM electrode. Simple geometric calculations based on the surface area of these particles and their apparent bulk density (including the void volume) show that the increase in effective surface area achieved by this method is ca. a factor of 75. Thus, 75 times more ligand may be immobilized per unit geometrical area of the QCM surface, leading to sensor signals ca. 75 times greater than those which would be observed for a single monolayer of ligand on a flat surface. Of course, this enhancement factor depends critically on the particle size and apparent surface area per gram.

Impedance analysis was used both to characterize the quality of the resonance of these composite resonators to determine suitability for use in the broadband oscillator system (11), and to determine the masses of the adhesion layer and the particle layer. The latter measurement is crucial for the quantitative comparison of predicted and observed mass gains due to metal binding, since, in combination with the elemental analysis, it gives the total loading of ligand per unit geometrical area of the QCM transducer surface. In the present case, the total ligand loading is 3.8×10^{-8} mol cm^{-2}, based on immobilization of 43 μg of particles per square centimeter. This leads to a predicted frequency decrease (mass gain) of 290 Hz for complete saturation of the ligand with uranyl cation, assuming a stoichiometry of two ligands per metal (5,6). (Note that this calculation does not take into consideration the ca. 10% of the particle diameter which is embedded within the adhesion layer, and is therefore not likely to be available to the metal, assuming that the material of the adhesion layer is present within the pore structure of the particle in this region. In addition, it does not allow for any waters of hydration, which may or may not be present after the drying process, a possibility about which we have no current information.) This value is in very good agreement with the 240 Hz observed for complete saturation of the ligand with metal, which was determined by exposing the sensor crystal to a 1 mM solution of uranyl nitrate, followed by brief rinsing with water, drying in a stream of dry nitrogen gas, and impedance analysis.

Figure 2 shows the real-time response of the sensor to changes in the composition of the solution within the flow cell. The data clearly show the frequency decrease (mass gain) of the reference QCM caused by simple incorporation of the uranyl solution (5×10^{-5} M $UO_2(NO_3)_2$, pH = 4.2) into the porous particle arrays. The QCM with the ligand immobilized onto the particles shows this same response, and also shows an additional frequency decrease (mass gain) from the ligation of UO_2^{2+} with the immobilized phthalamic acid derivative. Exchange of the solution back to pure water leads to a slow frequency increase (mass loss), presumably due to metal loss. Flushing the system with an acetic acid/sodium acetate solution at a pH of 2.75 (made by adjusting a 0.1 M NaAc solution with neat glacial acetic acid) produces an immediate, and much more rapid, frequency decrease (mass gain) of both QCM's from the solution density effect, with the magnitude of this response being comparable to that expected from the control experiments with sodium nitrate (vide supra). This addition of a low pH acetate solution also leads to a more rapid loss of mass (frequency increase) for the sensor QCM crystal. This is due to loss of UO_2^{2+} from the immobilized phthalamic acid ligand due to the very effective competition of the solution phase acetate ligand for the bound uranyl species and/or to a decrease in the stability constant for the complex between the immobilized ligand and the uranyl species due to the decrease in pH. Finally, flushing with pure water again leads to reestablishment of the original oscillation frequency, suggesting that all of the uranyl cation has been removed from the sensor crystal.

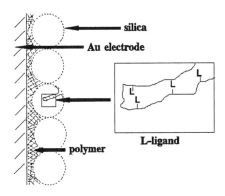

Figure 1.   Schematic depiction of a single layer of silica particles embedded in an adhesion layer on a Au electrode at the surface of a QCM.

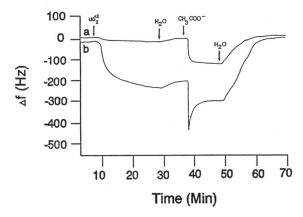

Figure 2.   Plots of frequency change versus time for a) the reference crystal on exposure to various solutions and b) the crystal with immobilized ligand. See text for details.

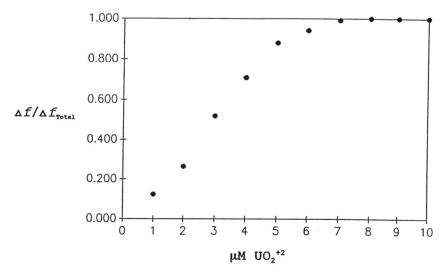

Figure 3.   Plot of the normalized frequency change (corrected for density effects) versus uranyl concentration for exposure of the crystal to various concentrations of uranyl nitrate.

The response time of the frequency decrease in Figure 2 is quite long. It is not clear whether this is due to slow diffusion of the uranyl species through the highly congested pore volume of the particles or to slow delivery of the uranyl species from solution at these low concentrations. Experiments in preparation with a new cell design that will result in an increase in the rate of delivery of solution phase species should help address this question. However, because of the long response times and the very slow rate of metal ion loss (unassisted by acetate) demonstrated in Figure 2, the uranyl solutions were exposed to the crystal for extended periods of time (typically 2-4 hours), to allow full equilibration between the immobilized ligand and the solution.

Figure 3 shows the results of a set of experiments in which the uranyl nitrate concentration was varied from 1 to 75 μM (only the data from 1-10 μM are plotted). Note that these responses were all corrected for the solution density effects discussed, and that these corrections are fairly large above ca. 25 μM. The data were actually obtained by sequentially exposing the sensor to increasing concentrations of uranyl nitrate. Variation in the order of exposure to the various concentrations was used to verify that the frequency change at a given concentration is independent of the order of exposure. These data were reproduced several times by regenerating the sensor crystal (i.e. removing the bound metal) using the acetate solution, as described above, and repeating the exposure to the various uranyl solutions.

The data in Figure 3 are plotted in the form of a normalized adsorption isotherm, and the value of Δf_{total} for this case was 240 Hz. The best-fit slope of the early, nearly linear section of the plot (between 0 and 5 μM UO_2^{2+}) gives a measure of the apparent binding constant of the immobilized ligand, in units of M^{-1}. In the present case, this value is 2×10^5 M^{-1}, which compares very favorably with the solution phase binding constants for phthalic acid and some of its derivatives which have been previously determined to fall in the range 10^4 - 10^5 M^{-1} (5,6). In spite of this qualitative agreement, it is clear from the significant curvature of the early portion of the plot in Figure 3 that the complexation sites on the surface are close enough to interact with one another, so that a simple Langmuir isotherm is not likely to be an adequate description of the adsorption behavior for this system.

Conclusions

This work has established that porous, micron-sized silica can be anchored on a QCM to give a sensor with a high apparent surface area. Further, it has been demonstrated that these high surface area silica particle arrays can be used as vehicles for the immobilization of ligands. Using this strategy for immobilization of a phthalamic acid derivative, it has been shown that detection of UO_2^{2+} can be achieved by virtue of the frequency changes which occur upon binding of the metal to the immobilized ligand. The metal uptake data as a function of metal concentration allow extraction of the formation constant for the metal complex. The value of 2×10^5 M^{-1} obtained in this way is in good agreement with expectations based on previously reported values for phthalic acid. These results suggest that this method of sensor fabrication should be of general value for cases where immobilization of large amounts of a given receptor is desired.

Acknowledgments

The work at the University of Wyoming was supported by the National Science Foundation through an EPSCoR-ADP grant. The work at the University of California at Berkeley was supported by the Office of Naval Research. Paul Walton is acknowledged for elucidating certain aspects of the solution phase behavior of phthalamic acid binding to UO_2^{2+}.

Literature Cited

1. Ward, M.D.; Buttry, D.A. *Science*, **1990**, *249*, 1000-1007.
2. Morse, J.W.; Shanbhag, P.M.; Saito, A.; Choppin, G.R. *Chemical Geology*, **1984**, *42*, 85-99.
3. Bermejo-Barrera, A.; Yebra-Biurrun, M.C.; Fraga-Trillo, L.M. *Anal. Chim. Acta*, **1990**, *239*, 321-23.
4. Decambox, P.; Mauchien, P.; Moulin, C. *Appl. Spec.*, **1991**, *45*, 116-18.
5. Palaskar, N.G.; Jahagirdar, D.V.; Khanolkar, D.D. *J. Inorg. Nucl. Chem.* **1976**, *38*, 1673-75.
6. Venkatnarayana, G.; Swamy, S.J.; Lingaiah, P. *Indian J. Chem.* **1985**, *24A*, 624-26.
7. Dudler, V.; Lindoy, L.F.; Sallin, D.; Schlaepfer, C.W. *Aust. J. Chem.* **1987**, *40*, 1557-63.
8. Gimpel, M.; Unger, K. *Chromatographia*, **1982**, *16*, 117-125.
9. Weethal, H.H. *Methods Enzymol.* **1976**, *44*, 134-48.
10. Buttry, D.A. In *"Electroanalytical Chemistry. A Series of Advances"*, Bard, A.J., Ed.; Marcel-Dekker:New York, 1991, Chapter 1.
11. Buttry, D.A.; Ward, M.D. *Chem. Rev.* **1992**, *92*, 1355-79.
12. Lasky, S.J.; Buttry, D.A., In *"Chemical Sensors and Microinstrumentation"*, Murray, R.W.; Dessy, R.E.; Heineman, W.R.; Janata, J.; Seitz, W.R., Eds.; ACS Symposium Series No. 403, 237-46, Chap. 16; American Chemical Society: New York, 1989.

RECEIVED March 25, 1994

Chapter 8

Selective Detection of $SO_2(g)$ at the P-Type Porous Silicon–Gas Interface

Photophysical Evaluation of Interfacial Chemistry

Andrew B. Bocarsly, Jonathan K. M. Chun, and Michael T. Kelly

Department of Chemistry, Frick Laboratory, Princeton University, Princeton, NJ 08544–1009

The electrooxidation of silicon in an aqueous HF electrolyte gives rise to a surface layer of "porous silicon" which luminesces in the visible upon excitation in the near UV or blue region of the optical spectrum. Our evaluation of this chemistry has uncovered a strong pH dependence on the luminescent process. High pH values give rise to substantial emission quenching; luminescence can be restored by exposing the quenched surface to acid in either the liquid or gas phase. We find that the Lewis base, SO_2, selectively and reversibly quenches emission of p-type porous silicon. This reaction is specific in that exposure of the quenched surface to a clean argon atmosphere restores the initial emission intensity. Likewise, other small gas phase molecules (O_2, CO, CO_2, NO, etc.) do not effect the observed photophysics. Quenching of the luminescence is observed with concentrations of less than 1% SO_2 in Ar. This sensitivity to SO_2 makes porous silicon a versatile and robust luminescent sensor material. The quenching mechanism is believed to involve the formation of a Lewis acid-base adduct between SO_2 and a protonated bridging silyloxo species.

SO_2 is released into the atmosphere by a variety of industrial and natural processes. Typical concentrations of this gas from stack emissions range from 100 to 5200 ppm requiring chemical scrubbing before release into the atmosphere.(1) SO_2 emissions pose a serious problem to the environment because they readily form acid rain in the higher atmosphere. Presently, few analytical procedures are available for the quantitation of this gas.(2) Rapid, selective, and economical detection of this gas would be useful for regulating its atmospheric discharge.

We have recently observed that photoluminescence (PL) from p-type porous silicon (p-PS) surfaces, generated on single crystal or polycrystalline silicon substrates via an oxidative HF etching process,(3,4) is selectively quenched by SO_2. An analytical relationship has been observed between SO_2 concentration and the luminescence intensity providing the possibility of a new sensing methodology.

In order to fully characterize the nature of the SO_2/p-PS interaction an understanding of the photophysics of p-PS is essential. Unfortunately, there is much current debate over the identity of the luminophore and the mechanism of emission associated with this system. Luminescence from porous silicon is surprising since monocrystalline Si does not exhibit room temperature photoluminescence. Two models

0097–6156/94/0561–0078$08.00/0

confinement model and the molecular luminophore model. Both models sucessfully account for certain aspects of the observed luminescence however, neither model as they currently stand can be considered completely satisfactory.

The quantum confinement model, originally put forth by Canham et. al.(5-7) argues that as a single crystal Si surfaces is etched to generate porous silicon, "macroscopic" and "microscopic"pores confine small domains (~3nm) of crystalline silicon between them. As a result of their small size the band gap of the particles is expected to increase. Besides the fact that small domains of material exist on the surface of porous silicon, other evidence for a quantum effect is the observed blue shift in PL emission spectra as the excitation wavelength is shifted to higher energy. SEM/TEM studies have shown that the porous layer is inhomogeneous, with smaller structures available near the surface and larger structures farther into the bulk Si substrate. Since higher energy wavelengths do not penetrate into the semiconductor as far as longer wavelengths do, high energy excitation only excites the top-most layers of the porous silicon. Correspondingly, "red" light penetrates farther into the bulk semiconductor. It can be argued that the blue shift in emission results from the excitation of different sized quantum particles within the porous silicon matrix. This explanation of the blue shifted emission also conveniently explains the width of the observed luminescence since it suggests that a distribution of particle sizes, quantum "boxes," are present in the p-PS sample.(4)

The surface molecular luminophore model postulates that the observed visible luminescence directly derives from a surface confined molecular species generated during the oxidative/HF etching of single crystal silicon. Polysilanes, silicon hydrides, and amorphous silicon also luminesce, and these investigators contend that the behavior of these compounds is consistent with the electrochemically produced porous silicon.(8-11) In addition, other molecular species like siloxanes and siloxenes have also been proposed since they have chemiluminescent properties or IR spectra somewhat similar to PS.(12,13) In both cases, visible luminescence emission, the blue-shift in emission spectra with high energy irradiation, and the diffuse band of emission, can be rationalized with an inhomogeneous molecular luminophore model.

We have previously reported that photoluminescence from p-PS is reveribly quenched by the presence of non-hydroxide containing Brönsted bases in either the gas or liquid phase.(14,15) The quenching process involves an acid-base reaction between the quencher and the ground state p-PS surface. Neither of the postulated models immediately accomodates this observation. However, the confinement model appears to be easily modified to take into account the observed chemistry.

Results and Discussion

Overview of Photoluminescence. In air or under argon, porous silicon emits a visible orange-red luminescence upon irradiation with wavelengths less than about 514nm as shown by the excitation spectra in Figure 1. The most prominent maximum is evident at ~400nm. The emission spectra, shown in Figure 1 exhibit a blue shift as the excitation energy shifts to higher energy as reported by other investigators. This peak shift is linear over the range of wavelengths surveyed (240-520nm). The observed emission is found to be quenched by a variety of materials, including Brönsted bases and $SO_{2(g)}$. The degree of reversibility of the quenching process along with the detailed kinetics of quenching are dependent on the quencher employed suggesting some mechanistic variation as a function of the quenching agent. In spite of these difference, our intial investigations assume that the observed quenching arises from a common set of surface "states", and thus, an understanding of the interfacial chemistry associated with Brönsted base gated emission is useful in determining the identity of the PS luminophore and in developing a general model of the quenching dynamics.

Acid/Base Reactivity of Porous Silicon. Immersing porous silicon in dilute acid solutions like 0.5M HCl or 0.5M H_2SO_4, results in an increase in the photoluminescent intensity. In contrast, treatment with strong aqueous bases like KOH or NaOH,

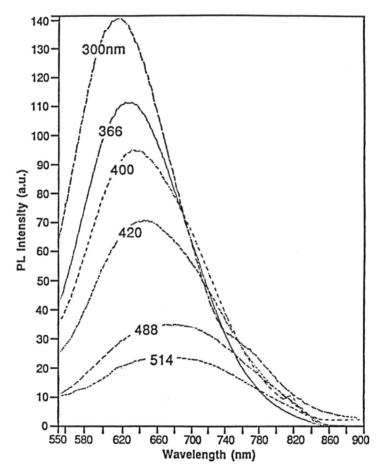

Figure 1. Luminescence emission with varying excitation wavelengths, under dry Ar, exhibiting a blue-shift of ~45nm with increasing excitation energy.

produces profuse H_2 evolution from the porous silicon surface and an irreversible and rapid loss of PL. Treatment with concentrated H_2SO_4 or HF (48%) do not restore the PL. The irreverisble quenching of PL by aqueous hydroxide is not unexpected since the chemistry of silicon etching by hydroxide is well known,(16) and proceeds with the liberation of H_2 according to the reaction:

$$Si\,(s) \; + \; 4OH^-\,(aq) \; \rightarrow \; SiO_4^{4-}\,(aq) \; + \; 2H_2\,(g) \tag{1}$$

Thus, hydroxide minimally destroys the substrate which supports the porous layer if not the porous layer itself.

Unlike addition of OH^-, however, exposure of the PS surface to a variety of other bases produced *reversible* quenching of the luminescence. For example, the vapors of diethylamine, isopropylamine, diisopropylamine,and ammonia totally quench the luminescence. PL is restored by exposure to water, mineral acids, or organic acids. Likewise, a variety of aqueous bases including amines, bicarbonate, carbonate, and cyanide, quench the luminescence reversibly. A correlation between pH and quenching effect is observed. The effect observed with the organoamines presented here has been confirmed by work done by Coffer et al.(17)

Although PL is easily quenched by bases and restored by acids, the photoluminescence is not uniformly quenched at all wavelengths. Samples exposed to a strong base, like diethylamine vapors, exhibit PL which is blue shifted from its unquenched condition. Within the framework of the molecular luminophore model, a single molecular luminophore is not expected to display this behavior, but if a large distribution of luminophores is available on the surface, then the blue shift could suggest pH-dependent and pH-independent species. A similar arguement can be made for quantum confined interfacial structures having different molecular surface components which regulated the emission process.

Surface Titration of p-Type Porous Silicon. Sodium bicarbonate was found to be a convenient reversible titrant, and thus, was used to quantify the observed alkaline quenching process. Figure 2 shows a typical collection of spectra recorded as a function of bicarbonate concentration. The PS was immersed in 0.01M HCl and back-titrated with 0.03M NaHCO₃ while monitoring the PL.Restoration of PL is easily accomplished by treatment with HCl. These emission spectra exhibited a blue shift from 680nm at pH = 3 to 610nm at pH = 6.7.

A titration curve is given by Figure 2(b) for the data shown in Figure 2(a). This curve was generated from the red emission data (800nm) to avoid the complication associated with the concurrent observation of pH dependent and independent luminophores. The observation of only one well defined wave in these curves indicate the presence of only one type of acidic proton associated with the luminescent process. Addition of aliquots of an HCl titrant to the bicarbonate quenched sample restored the luminescence producing a titration curve that within experimental error, reproduced the curve presented in Figure 2(b). Although within the observed experimental error no hysteresis was observed, the acid titration was found to be a rather sluggish process greatly increasing the observational error. Therefore, all quantitative titrations were carried out using a base titrant. Based on this data, a value of 3.0 ± 0.9 was obtained for the p-PS pK_a. The titration experiments also provided a means of calculating the number of active surfaced protons. A total of $\sim 10^{20}$ surface protons were liberated from the electrode at the equivalence point. Previous work on PS utilizing BET or SIMS measurements of surface active sites revealed that surface areas of 500-600 m^2/cm^3 typically exist on these materials.(18,19) Estimated surface areas for porous silicon pieces used in these experiments were calculated from the geometric area (~ 30 mm^2) and the porous silicon thickness ($\sim 10\mu m$) to yield a surface density on the order of 10^{16} protons/cm².

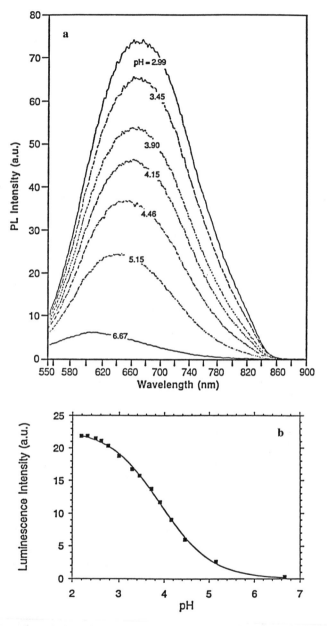

Figure 2. (a) Variation of the p-PS emission spectra with the addition of 0.03M bicarbonate. The silicon was initially immersed in a 0.1M HCl solution. The increase in pH produces PL quenching and a simultaneous blue shift in the emission. $\lambda_{ex} = 400$nm. (b) Titration curve generated from the data in part (a), monitored at 800nm. The solid curve is a calculated fit based on the acid-base equilibrium equation.

Time Resolved Luminescence Decay. The luminescent decay profiles of porous silicon are long lived and inhomogeneous across the emission band. This is evident in Figure 3a where the emission at 750nm clearly exhibits a longer lifetime than emission at 590nm. Variation in τ, the effective lifetime, with excitation wavelength is indicative of sample inhomogeneity and can occur if many energy levels are available in the solid. The long lived decay provides evidence for slow recombination kinetics required for an efficient luminescent process to exist.

The decay profiles cannot be fit to a single exponential, but previous work has demonstrated that a "stretched" exponential function describes the relaxation satisfactorily: (*9,20,21*)

$$I = I_0 exp[-(t/\tau)^b] \tag{2}$$

Where τ is the excited state effective lifetime and b is related to a fractal dimension.

For the surfaces under consideration τ ranges from 45 to 60 μs when measured at an emission wavelength of 750nm and an excitation wavelength of 400nm, and $b = 0.8 \pm 0.02$. Examining the PL decay for quenched porous silicon as a function of pH shows that the decay profiles are pH-independent. As shown in Figure 3b, little variation in τ is observed. This indicates that the observed quenching is static in nature, involving an interaction between the titrant base and the p-PS ground state. In addition this observation indicates that the surface proton is *not* the molecular luminophore. This latter conclusion is confirmed by the observation that deutration of the PS surface by base deprotonation followed by immersion in DCl or D₂O has no pronounced effected on the observed emission spectrum or the excited state lifetime.

Proton Gated Emission. Table I provides a partial listing of various PS surface species which have been suggested to be either directly or indirectly related to the luminophore. The pK_a provided in this Table is gleamed from the silica gel literature.(*22,23*)

Table I: The Acid-Base Properties of Silica Gel Surface Species

Surface Species	Expected pK_a
Silanol species − Si-OH	6-10
Silicone hydride species − Si-H, Si-H₂, etc.	>15
Silyl ether species[a] − Si-O-Si	<5[a]

[a] The pK_a of this species shifts negative as more oxygen moieties are added to the Si substituents.

Based on this data, the various silanols and surface hydrides which have previously been put forth either as the source of the molecular luminophore or as related to the luminophore must be discarded as the source of p-PS emission. (Samples used in the current study were stored in air for several weeks prior to collecting titration data. Thus, a relatively extensive library of oxidized surface functionalities is expected in the current study. This has been confirmed by IR spectroscopy of the samples employed(*24*)). This conclusion is further supported by the observation that silylation of the PS surface with chlorotrimethylsilane or methyltrichlorosilane does not effect the luminescence or associated quenching. This treatment is known to remove surface silanols, ruling out the possibility that this species is responsible for the observed quenching process. By elimination, the species shown in Scheme I below is proposed as the surface moiety which gates the emission process. Based on the static nature of the quenching process and the insensitivity of the emission to surface deuteration, this

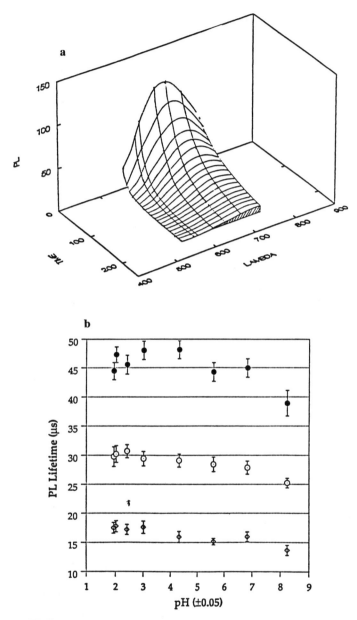

Figure 3. (a) Photoluminescent decay profiles monitored at different emission wavelengths for p-PS excited at 400nm. The time is given in μs. (b) Variation of PL decay lifetimes for p-PS with changes in solution pH. Filled circles are results for emission at 750nm, open circles at 700nm, and diamonds at 650nm.

species cannot be the actual luminophore. Rather, the presence of this species in its protonated form is apparently necessary to suppress available nonradiative pathways which compete with the emission process. A possible model which explains this behavior follows.

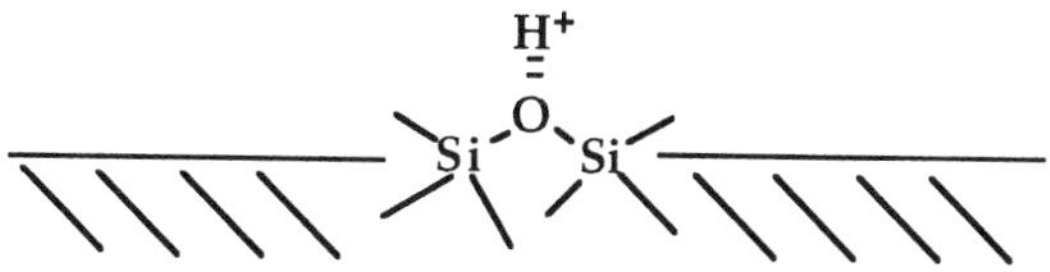

Scheme I

Sulfur Dioxide Based Quenching. Unlike the quenching observed in the presence of weak to moderate bases the quenching of p-PS by gas phase SO_2 is immediately restored by removal of the SO_2 source. Even under a dry Ar atmosphere, no acid is required to restore PL intensity after quenching. The PL response in SO_2 or Ar is illustrated in Figure 4. It is evident that the quenching response is rapid, <50 seconds, upon SO_2 addition. It is found that a "break in period of several quenching/restore cycles is necessary before the degree of quenching for a given exposure to SO_2 is reproducible. However, once the "break in" process has been executed a highly reproducible response is obtained over an extended time period. No PL quenching is observed for p-PS exposed to CO, CO_2, NO_x, H_2S, Me_2S or air. This is significant if porous silicon is to be used as a practical industrial sensor. No pre-separation is required to monitor exhaust gases.

Although qualitatively similar, differences do arise in comparing the quenching of PS by Brönsted bases verus SO_2. As shown in Figure 5a quenching with SO_2 causes a red shift in the emission profile in constrast to the blue shift noted with amine or bicarbonate quenching. Additionally, the time dependence of the quenching process is quite different for gas phase triethylamine compared to SO_2 as shown by the data in Figure 5b. These findings may indicate that SO_2 quenching occurs via a totally different mechanism than that observed for Brönsted bases.

The most likely alternate mechanism, initially suggested by Sailor,(25) involves energy transfer quenching via a dipole-dipole interaction. However, given that the dipole moment of SO_2 is identical to THF, a molecule which causes no significant quenching in our hands, this mechanism can be ruled out. Therefore, if a change in quenching mechanism is occuring the most likely target would be a different surface site for SO_2 binding than identified for the Brönsted base case. Having ruled out surface silanols the most likely target(s), in addition to the silyl ether sites are silicon sites which might provide for a direct Lewis base interaction as shown in Scheme II.

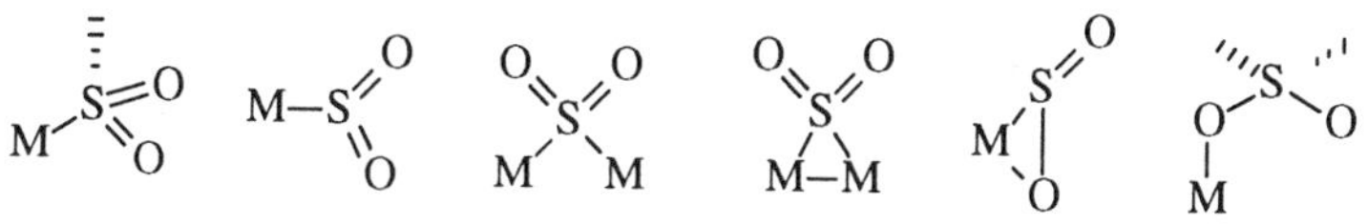

Scheme II

Porous Silicon for Environmental Monitoring. While the exact mechanism of the SO_2 interaction is still under investigation, the utility of p-PS as a sensitive chemical sensor of SO_2 is illustrated in Figure 6. Low concentrations of SO_2 (less than 5%) in Ar produce a large decrease in PL intensity of about 20-30%. For SO_2/Ar mixtures between 28 and 194 ppm, a change in PL intensity of ~2% was obtained. These changes were near the limit of detection for the fluorimeter employed due to a poor signal to noise response. In 100% SO_2 the PL of p-PS is reduced to ~70% of the initial

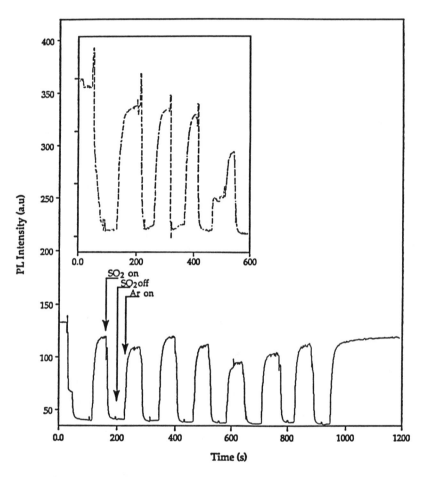

Figure 4. Response of porous silicon PL to SO_2 in Ar. As in the case of quenching by amines, SO_2 quenching was found to be unaffected by silylation of the PS surface. This again rules out a role for surface silanols. **Inset-** Initial response of the detector showing the "break-in" behavior.

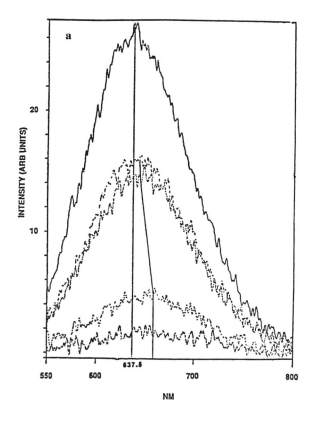

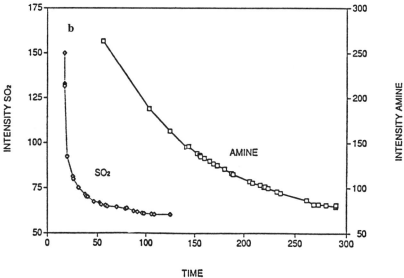

Figure 5 (a) The effect of SO_2 on the emission spectrum of p-PS as a function of increasing SO_2 concentration. (b) Quenching dynamics for p-PS exposed to SO_2 vs. triethylamine.

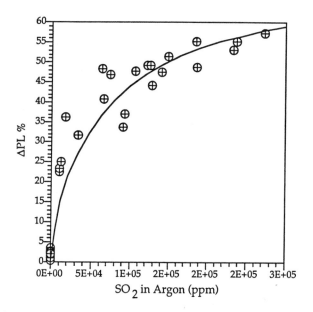

Figure 6. PL intensity vs SO_2 concentration for samples irradiated at 400nm and monitored at 650nm.

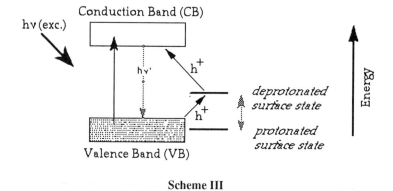

Scheme III

intensity. This data obviously implies a close correlation between the SO_2 concentration and the PL quenching. The sensitivity of PS to SO_2 is well within the concentration range for SO_2 in stack gases, which are typically an order of magnitude higher.

In addition to the reasonable sensitivity toward SO_2, which may, in fact, be extended to lower limits of detection with a modification of the experimental setup, p-PS is also relatively robust. PS survives many cycles of exposure to SO_2 and Ar without any obvious degradation of PL intensity. Experiments of modulated exposure of PS to SO_2 and Ar for over 6 hours at ~10min. cycles did not result in a significant decrease in the PL intensity. The longer-term stability has not been ascertained, but the PS luminescence is stable over more than a year and exhibits a base quenching phenomena over at least many months.

Conclusion

The experiments presented here suggest that sensitive surface interactions strongly affect the optical properties of porous silicon. Base quenching titrations clearly identify a surface confined acid which accounts for a majority of the PL emission. This pH-dependent luminescence exhibits a pK_a of 3-4. An isolated group of surface confined molecules such as surface hydrides or silanols, cannot be responsible for the PL observed in porous silicon based on the observed pK_a's. Therefore, a proton, bound to a silicon-bridging oxygen is implicated in the PL mechanism. However, available data does not necessarily indicate that this emission process is exclusive.

The static nature of the quenching process along with the lack of an H/D isotope effect strongly argues for a ground state quenching phenomena. Thus, the surface molecular model displayed in Scheme I is justifiably assigned to a ground state surface confined acid. Recent evidence for a pH-dependent PL quenching for quantum sized nanocrystalline silicon colloids is provided by Littau et al.(26) Therefore, based on the evidence presented here, a solid state luminophore must be invoked.

Scheme III illustrates a possible model for the luminescence behavior of p- PS consistent with the observed quenching phenomena. The luminescence emission is "gated" by a surface hole trap which is in communication with the valence band and has a pH dependent energy . This surface site is hypothesized to provide a site of non-radiative recombination. The deprotonated state must exist in the bandgap so that recombination of photoexcited electrons can bypass the radiative band-to-band transition, while the protonated state is expected to lie below the valence band edge.

This framework suggests that SO_2 may quench p-PS via a Lewis base interaction. However, subtle differences between Brönsted base quenching and SO_2 quenching open up the possiblity of a surface interaction between gas phase SO_2 and a silicon surface site. In addtion to variations in kinetics when comparing Brönsted base quenching to SO_2 quenching, one observes a much higher degree of reversibility in the case of SO_2 quenching making a sensing application possible. Initial results indicate that SO_2 can be detected with good stability and reproducibility down to 25ppm. Standard interferrents present in stack gas do not effect p-PS luminescence and therefore SO_2 quenching is quite selective. Current studies are aimed at ascertaining the limit of detection and sensor long term stability.

Acknowledgments

The U.S. Department of Energy, Office of Basic Energy Sciences is thanked for support of this work.

Literature Cited

(1)   Pasiuk-Bronikowska, W.; Ziajka, J.; Bronikowski, T. *Autoxidation of Sulphur Compounds*; Ellis Horwood: New York, 1992.

(2)   Wolfbeis, O. S.; Sharma, A. *Anal. Chim. Acta* **1988**, *208*, 53-58.
(3)   Shih, S.; Jung, K. H.; Hsieh, T. Y.; Sarathy, J.; Campbell, J. C.; Kwong, D. L. *Appl. Phys. Lett.* **1992**, *60*, 1863-1865.
(4)   Searson, P. C.; Macaulay, J. M.; Prokes, S. M. *J. Electrochem. Soc.* **1992**, *139*, 3373-3378.
(5)   Canham, L. T. *Appl. Phys. Lett.* **1990**, *57*, 1046-1048.
(6)   Lehmann, V.; Gösele, U. *Appl. Phys. Lett.* **1991**, *58*, 856-858.
(7)   Cullis, A. G.; Canham, L. T. *Nature* **1991**, *353*, 335-338.
(8)   Tsai, C.; Li, K.-H.; Kinosky, D. S.; Qian, R.-Z.; Hsu, T.-C.; Irby, J. T.; Banerjee, S. K.; Tasch, A. F.; Campbell, J. C.; Hance, B. K.; White, J. M. *Appl. Phys. Lett.* **1992**, *60*, 1700-1702.
(9)   Pearsall, T. P.; Adams, J. C.; Wu, J. E.; Nosho, B. Z.; Aw, C.; Patton, J. C. *J. Appl. Phys.* **1992**, *71*, 4470-4474.
(10)   Ito, T.; Kiyama, H.; Yasumatsu, T.; Watabe, H.; Hiraki, A. *Physica B: Cond. Mat.* **1991**, *170*, 535-539.
(11)   Xu, Z. Y.; Gal, M.; Gross, M. *Appl. Phys. Lett.* **1992**, *60*, 1375-1377.
(12)   McCord, P.; Yau, S.-L.; Bard, A. J. *Science* **1992**, *257*, 68-69.
(13)   Brandt, M. S.; Fuchs, H. D.; Stutzmann, M. *Solid State Comm.* **1992**, *81*, 307.
(14)   Chun, J. K. M.; Bocarsly, A. B.; Cottrell, T. R.; Benziger, J. B.; Yee, J. C. *J. Am. Chem. Soc.* **1993**, *115*, 3024-3025.
(15)   Chun, J. K. M.; Bocarsly, A. B.; Cottrell, T. R.; Benziger, J. B.; Yee, J. C. In *Microcrystalline Semiconductors - Materials Science and Devices*; Materials Research Society: Boston, MA, 1992; pp published.
(16)   Greenwood, N. N.; Earnshaw, A. *Chemistry of the Elements*; Pergamon Press: Oxford, 1990.
(17)   Coffer, J. L.; Lilley, S. C.; Martin, R. A.; Files-Sesler, L. A. In *Microcrystalline Semiconductors - Materials Science and Devices*; Materials Research Society: Boston, MA, 1992; pp published.
(18)   Canham, L. T.; Blackmore, G. W. In *Light Emission from Silicon*; Materials Research Society: Boston, MA, 1992; pp 63-67.
(19)   Herino, R.; Bomchil, G.; Barla, K.; Bertrand, C.; Ginoux, J. L. *J. Electrochem. Soc.* **1987**, *134*, 1994-2000.
(20)   Chen, X.; Henderson, B.; O'Donnell, K. P. *Appl. Phys. Lett.* **1992**, *60*, 2672-2674.
(21)   Xie, Y. H.; Wilson, W. L.; Ross, F. M.; Mucha, J. A.; Fitzgerald, E. A.; Macaulay, J. M.; Harris, T. D. *J. Appl. Phys.* **1992**, *71*, 2403-2407.
(22)   Gates, B. C.; Katzer, J. R.; Schuit, G. C. A. *Chemistry of Catalytic Processes*; McGraw Hill: New York, 1979, pp 45-49.
(23)   Gates, B. C. *Catalytic Chemistry*; John Wiley & Sons, Inc.: New York, 1992, pp 310-423.
(24)   Cottrell, T. R. Ph.D. Thesis, Princeton University, 1993.
(25)   Sailor, M. J.; Kavanagh, K. L. *Adv. Mater.* **1992**, *4*, 432-434.
(26)   Littau, K. A.; Szajowski, P. J.; Muller, A. J.; Kortan, A. R.; Brus, L. E. *J. Phys. Chem.* **1993**, *97*, 1224-1230.

RECEIVED April 19, 1994

Chapter 9

Chemical Sensors and Devices Based on Molecule–Superconductor Structures

J. T. McDevitt, D. C. Jurbergs, and S. G. Haupt

Department of Chemistry and Biochemistry, University of Texas, Austin, TX 78712–1167

The use of molecular materials for the development of novel electronic devices and sensors has attracted much attention recently. Virtually all previous molecular devices have been fabricated by organizing molecular species onto metal, semiconductor or insulator surfaces. In this chapter methods for preparing a series of composite molecule/superconductor devices are described. Thin films of the high-T_c superconductor, $YBa_2Cu_3O_{7-\delta}$, are coated with a selection of molecular dyes to form a series of light sensors. In this context, the dye layer serves both to enhance the responsivity to light of the composite device and to provide wavelength selectivity to the structure. In addition, the preparation and operation of a number of conductive polymer coated superconductor structures are described. Electrochemical techniques are exploited to alter the oxidation state of these polymers and, in doing so, it is found for the first time that superconductivity can be modulated in a controllable and reproducible fashion by a polymer layer. Thus, a new type of molecular switch for controlling superconductivity is demonstrated.

Chemical modification of solid-state surfaces with organic films is an attractive method for the development of new materials with enhanced properties. Organic coatings have been used to improve the environmental stability, the optical properties, the surface wettability, the friction coefficient and the bio-compatibility of a variety of technologically important substances (*1-3*). Moreover, organic films have been utilized in conjunction with conventional solid-state materials for the development of novel macromolecular electronic devices. Molecular transistors, Schottky diodes, metal-insulator-semiconductor diodes, MIS field effect transistors and light emitting diodes have all been prepared utilizing organic thin film structures (*4-9*).
Unlike conventional solid-state systems, macromolecular electronic devices display species specific responses to a variety of chemical stimuli. Consequently, much effort has been directed recently to the fabrication and study of chemical sensors based on molecule derivatized solid-state surfaces (*10-11*). Electrochemical

0097–6156/94/0561–0091$08.00/0
© 1994 American Chemical Society

sensors which measure changes in potential or current flow, acoustic sensors that produce changes in sound velocity or attenuation, organic polymers which exhibit large changes in conductivity upon exposure to various oxidants or reductants, and optical sensors that yield changes in light intensity or polarization have all been prepared utilizing such materials. With the discovery of superconductivity above 100K in the cuprate systems, new opportunities exist now for the development of hybrid molecule/superconductor structures. The focus of this chapter is placed on a discussion of the preparation and operation of the initial examples of molecule/high-T_c superconductor sensors and devices.

Relevant Superconductor Properties

Superconductors exhibit a number of properties such as zero resistance, the Meissner effect, Josephson tunneling, the proximity effect, and persistent currents (12) that make them well suited for use in electronic devices and sensors. Accordingly, superconducting electronic devices are particularly attractive due to the ultra-low power dissipation and ultra-fast response times that can be achieved from such substances (13).

Much of the interest in superconductor components revolves around the central issue that below a characteristic temperature, T_c, superconductors exhibit zero resistance to the flow of electricity (Figure 1A). Above this temperature, the material behaves as a normal metal wherein isolated electrons (or holes) carry the charge with finite resistance. Below T_c, however, the electrons form loosely associated pairs which are responsible for all the superconducting properties. At temperatures close to T_c, only a minute fraction of the conduction electrons form the Cooper pairs (Figure 1B). Under such circumstances, superconductivity is easily disrupted by heat, light, and magnetic fields. Creation of weakly coupled superconductor structures such as Josephson junctions, serves to further increase the sensitivity of the superconductor components. It is this sensitivity to external stimuli that provides the basis for the preparation of a variety of superconductor-based detectors and devices.

Fabrication of Molecule/Superconductor Structures

Recent work in the area of high-T_c research has yielded a number of high-T_c phases which possess transition temperatures above the boiling point of liquid nitrogen, 77K. In this regard, materials such as $YBa_2Cu_3O_{7-\delta}$ (T_c = 92K), $Bi_2Sr_2Ca_2Cu_3O_{10}$ (T_c = 110K), and $Tl_2Ba_2Ca_2Cu_3O_{10}$ (T_c = 125K) can be easily prepared as bulk ceramic pellet samples using conventional solid state synthetic approaches. Unfortunately, these ceramic compounds do not lend themselves to the preparation of useful electronic components. Rather, high-T_c thin films are the preferred geometry for such applications.

Fortunately, a number of processing methods have been developed or refined that have enabled researchers to prepare high quality films of $YBa_2Cu_3O_{7-\delta}$ which exhibit transition temperatures close to 90K and critical currents well in excess of 10^6 A/cm^2 at 77K (14). In this regard, laser ablation has become a popular research tool for the deposition of cuprate films because of the relative ease through which new materials can be deposited. The method consists of firing a pulsed excimer laser at a stoichiometric target of the material to be deposited and collecting the components which ablate onto a heated substrate. Under suitable conditions of laser power density, oxygen partial pressure, substrate temperature, deposition angle and substrate position, high quality superconducting films can be obtained in a single step.

Molecule/superconductor devices can be fabricated (15) according to the following steps (Figure 2). First, ~ 1000 Å of $YBa_2Cu_3O_{7-\delta}$ is deposited onto a clean

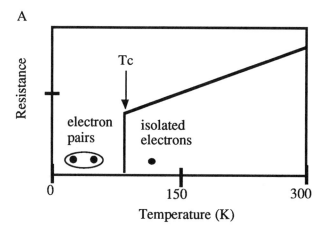

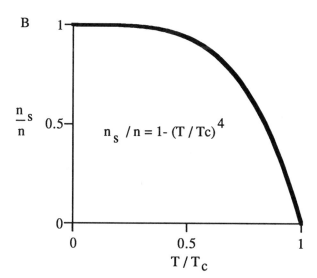

Figure 1. (A) Typical resistance vs. temperature curve for a high temperature superconductor sample. At temperatures above T_C, the material displays normal metallic properties whereby isolated electrons (or holes) carry the charge with finite resistance. Below T_C, however, the conductor is transformed into the superconducting state in which electron pairs carry the charge with zero resistance. (B) Curve showing the fraction of superconducting electron pairs to the total number of conduction electron plotted as a function of temperature. Here the temperature values are normalized to T_C.

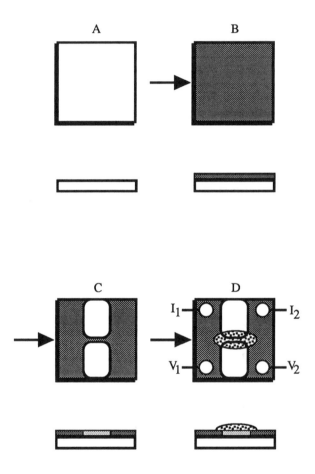

Figure 2. Schematic illustration showing the steps that are required to prepare molecule/superconductor structures: (A) In the initial step, a MgO (100) substrate is cleaned, (B) then ~1000 Å of $YBa_2Cu_3O_{7-\delta}$ is deposited by laser ablation. (C) A microbridge pattern is then created in the central portion of the film. (D) In the final step, the molecular layer is deposited over the microbridge junction.

MgO (100) surface using the laser ablation procedure. Second, a microbridge is patterned into the central portion of the film using either a laser etching method or conventional lithographic processing steps. In the final step, the molecular layer is deposited onto the microbridge by vacuum sublimation, spin coating from an anhydrous organic solvent, or by an electrochemical deposition procedure. Using these fabrication procedures, a large variety of molecule/superconductor structures can be prepared. Moreover, the methodology lends itself to independent tailoring of the molecular and superconductor components so that a plethora of different devices and sensors can be prepared for a variety of different applications.

Dye/Superconductor Optical Sensors

At the heart of all chemical sensors is a physical interaction in which molecular components interact with external stimuli to create a signal that can be monitored to gather information about the chemical nature of the environment in which the sensor is located. To foster more rapid developments in the field of chemical sensors, it is important now to explore new physiochemical processes (*16*). The ability of high temperature superconductors to be used as very sensitive detectors has been demonstrated previously (*17-20*). For example, superconductor structures have been utilized to prepare optical detectors which display rapid response times (~nanoseconds), high responsivity (10^{-3}-10^3 V/W), and a working wavelength range which spans from the ultraviolet to the far infrared (*17-20*). However, very little information exists related to the use of superconductor materials as components of chemically active devices. Only recently, have we reported details related to the fabrication and operation of the first molecule/superconductor devices (*21*). In this regard, it was demonstrated that molecular dyes can be utilized to alter the optical properties of superconductor junctions. In this section, the dye/superconductor structures will be discussed.

   In order to prepare well-behaved dye/superconductor light sensors, it is essential that dye deposition procedures be completed so as to avoid chemical damage of the fragile high-T_c thin film element. Consequently, it is important to electrically characterize the superconductor thin film before and after dye deposition. Thus, resistance versus temperature measurements of the superconductor microbridge are made using a 4-point contact method in which one current and one voltage lead are attached to either side of the bridge. Experiments of this type provide values for the superconducting critical temperature, T_c, as well as the room temperature resistivity, ρ_{RT}. These values can be utilized to assess the chemical purity and the quality of the superconductor element. Chemical damage of the high-T_c film can be noted from decreases in the measured value of T_c as well as from increases in the magnitude of ρ_{RT}. In addition to these measurements, determination of the values of the critical current, J_c, are made at various temperatures below T_c in order to assess the ability of the microbridge to carry current without resistance. This information is useful for the determination of the optimal operating current for the sensor.

   Typical materials used for the sensitizing layer include molecular dyes such as phthalocyanines, porphyrins, and rhodamine 6G. Coatings of this type can be prepared by a variety of techniques including vacuum sublimation, spin-coating, and spray deposition. Due to the smooth morphologies that can be achieved from the sublimation method, most of our research has been completed using sublimable dyes such as phthalocyanines and porphyrins which absorb light strongly in the visible region. A layer of H_2-phthalocyanine or H_2-octaethylporphyrin can be deposited by vacuum sublimation from a base vacuum of ~1×10^{-6} torr. Under such conditions, little change in the electrical properties of the superconductor are noted before and after the dye deposition.

Using the above mentioned information regarding the transition temperature and critical current values for the superconductor microbridge, the device is positioned in an optical cryostat on the surface of the cold stage (Figure 3). The structure is cooled to a temperature just below T_c and a direct bias current close to J_c is forced through the microbridge. Mechanically chopped monochromatic light is focused onto the junction and the in-phase voltage which develops across the microbridge is measured with a lock-in amplifier which is interfaced to a computer controlled data acquisition system. To maximize the signal to noise ratio and improve the performance of the sensor, the exact operating temperature and current are optimized using a monochromatic light source. The temperature and applied current are set at the determined levels and optical signals are acquired as a function of wavelength. The response is compared to an external sensor in order to eliminate variations in lamp intensity. Optical characterization of the devices is completed before and after dye deposition in order to assess the spectral changes which result from the presence of the dye sensitizing layer.

From comparisons of the response of the bare microbridge to that of a device coated with octaethylporphyrin (Figure 4A), it is apparent that the dye overlayer enhances the optical response of the superconducting microbridge in the visible region. Moreover, the featureless response of the bare superconductor microbridge changes dramatically upon deposition of the H_2-octaethylporphyrin dye layer as several sharp features appear in the optical data. These increases in responsivity at specific wavelengths can be seen more clearly after taking the ratio of the spectrum of dye-coated microbridge to that of the bare microbridge. This ratioed optical response is shown in Figure 4B along with the absorbance spectrum recorded on a UV-visible spectrophotometer. The peak enhancements in the optical response of the superconducting microbridge correspond closely to the absorption maxima of the H_2-octaethylporphyrin chromophore.

The dye/superconductor device is thought to function according to the following mechanism. Initially light is absorbed by the dye thereby resulting in an electronically excited state in the molecular phase. The dye rapidly relaxes back to the ground state thereby donating energy to the underlying superconductor. Once transferred to the superconductor, the excess energy serves to disrupt a portion of the superconducting electron pairs. Thus, the superconductivity in the microbridge is weakened, thereby leading to an increase in bridge resistance. The increase in resistance, however, is only a transient effect and disappears upon reformation of Cooper pairs as the excess energy within the superconductor is dissipated. The fact that the deactivation of the dye, the energy transfer, and the recombination of Cooper pairs are all relatively fast events leads to a situation where it is possible for high speed optical detection to be completed with the dye/superconductor structures.

Since a large number of dyes can be deposited onto the high-T_c structures, a variety of systems can be prepared with tailored spectral response characteristics. In this regard, we have prepared dye/superconductor systems that respond selectively to red, green, and blue light. Color specific light sensors of this type may find utility in the future for opto-electronic applications, color imaging, or robotic vision.

Conductive Polymer/Superconductor Structures as Chemical Sensors

The successful utilization of conducting polymers as elements in chemical sensors has been demonstrated recently by a number of researchers (22). One such example of the use of conductive polymers in this area is that of the Ion-Specific Field Effect Transistor (ISFET). These devices consist of a transistor element with a polymer coated gate. Large changes in the electrical and ionic conductivity which result when the polymer is oxidized by chemical or electrochemical means serve as the basis for chemical sensing method (23). The fact that conductive polymers can be cycled

Device located
in cryostat

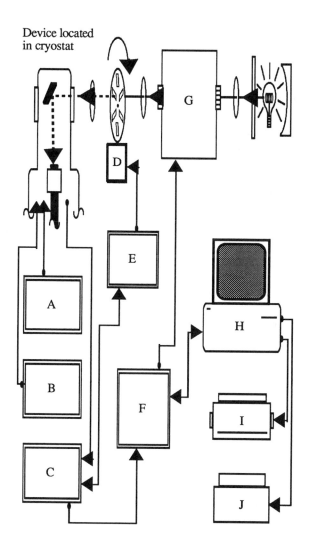

Figure 3. Schematic of instrumentation used for measuring the optical response of dye/superconductor structures: (A) temperature controller, (B) current source, (C) lock-in amplifier, (D) mechanical chopper, (E) chopper controller, (F) optical interface, (G) monochromator, (H) computer controller, (I) printer, (J) xy-plotter.

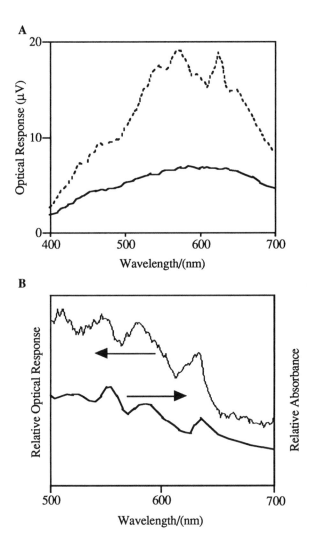

Figure 4. Optical response vs. wavelength for a $YBa_2Cu_3O_{7-\delta}$ microbridge device operating at a temperature just below T_c(mid): (A) unprocessed signals for bare junction (lower curve) and junction coated with 3300 Å of H_2-octaethylporphyrin (upper curve); (B) comparison between the absorbance spectrum of an H_2-octaethylporphyrin film (1.1 μm thick) as measured by UV-visible spectrophotometry (lower curve) and the normalized optical signal of the dye/superconductor device with the uncoated junction response subtracted (upper curve). Adapted from Reference 21 and reproduced with permission from the Journal of the American Chemical Society.

between their neutral (insulating) and oxidized (conductive) forms many times with little degradation of the host polymer makes these materials particularly attractive for chemical sensor applications. Changes in the current carrying capacity of the underlying channel which occur upon exposure of the polymer thin film to redox agents form the basis of the signal transduction event.

Recently, we have begun to explore the electronic interactions which occur between conductive polymers and superconductors at temperatures above and below T_c *(24-25)*. In this regard, we have found that the oxidation state of a conducting polymer can be used to control the flow of supercurrent in the underlying high-T_c microbridge and that this property can be exploited for the fabrication of a superconducting device that is analogous to the ISFET.

To prepare our polymer/superconductor sandwich devices, a high-T_c film of $YBa_2Cu_3O_{7-\delta}$ of thickness $\leq$ 1000 Å is patterned into a microbridge. The superconductor junction is then coated with a conductive polymer layer using either electrochemical techniques or solution processing steps. As with the dye/superconductor light sensor, care must be exercised when depositing the conductive polymer layer so that chemical damage of the high-T_c template is minimized. In this sense, the electrochemical window of the superconductor is limited to potentials less extreme then $\pm$ 1.0 V vs SCE *(15,26)*. In addition, the solvent system must be chosen so as to avoid damage of the superconductor surface. Aqueous or acidic solutions must be avoided because the cuprate superconductors corrode into nonconductive degradation products very rapidly when exposed to such fluids. On the other hand, we have found that the electrochemical preparation of polypyrrole from neat pyrrole and the spray coating of poly(3-hexylthiophene) (PHT) from tetrahydrofuran provide convenient methods to form well-behaved polymer/superconductor structures.

One of the important and novel features associated with the polymer/superconductor structures is that the oxidation state of the polymer dramatically influences the values for T_c and J_c of the underlying superconductor. Accordingly, when the polymer is oxidized to its conductive form, the transition temperature of the bridge can be suppressed by as much as 50K. The extent that the transition temperature is shifted is highly dependent on a number of factors. In this respect, the thickness of the superconductor and the oxidation mole fraction of the polymer appear to be the most important variables. Microbridges fabricated from very thin superconducting films display much larger shifts in T_c than similar structures prepared with thicker superconductor elements. Table I lists the maximum reversible shifts in transition temperature observed for a variety of $YBa_2Cu_3O_{7-\delta}$ thin film structures of variable superconductor thickness. In each case, the devices are coated with a constant thickness of ~2μm of polypyrrole.

The chemical nature of the oxidant is also an important factor which influences the behavior of the polymer/superconductor sensor. Utilization of corrosive oxidants tends to destroy the superconductor surface and irreversibly reduces the transition temperature of the microbridge through corrosion. We find that the nearly reversible modulation of superconductivity can be achieved through electrochemical treatments using quantenary ammonium salts such as tetraethylammonium tetrafluoroborate in acetonitrile. Further evidence for the chemical compatibility of these polymer/superconductor structures comes from their cyclic voltammetric responses which are nearly identical to those acquired at Pt electrodes *(27)*. Chemical dopants, such as $FeCl_3$, have been utilized successfully for the doping $Pb_{0.3}Bi_{1.7}Sr_{1.6}Ca_{2.4}Cu_3O_{10}$/PHT structures, but are too corrosive for $YBa_2Cu_3O_{7-\delta}$/PHT devices.

Having demonstrated that superconductivity can be modulated in a controllable and reproducible fashion based on changes of polymer oxidation level, it is now important to consider the underlying mechanism responsible for the effect. With such information in hand, it should be possible to develop more sophisticated devices and sensors for a variety of applications.

For the polymer/superconductor structures there are a number of important interactions that could be responsible for the observed behavior. For example, changes in the oxygen content have been shown previously to be promoted by electrochemical means and such changes can result in modest alterations of T_c. Consequently, we have conducted a number of electrochemical control studies in which devices were polarized without conductive polymer layers. No modulations of superconductivity were noted from these studies. Moreover, chemical degradation of the superconductor components have been shown previously to decrease both T_c and J_c. The return of superconductivity to higher temperatures which occurs upon reduction of the polymer layer, however, is not consistent with this explanation either.

Table I. Modulation of T_c for Polypyrrole / $YBa_2Cu_3O_{7-\delta}$ Thin Film Structures[a]

Superconductor Film Thickness/Å	T_c (Neutral Polymer)/K	Tc (Oxidized Polymer)/K	ΔT_c
200[b]	74	$\leq$24	$\geq$50
500	84	70	14
1500	80	79	1

[a] Each superconductor device was coated with 2μm polypyrrole film deposited from 0.25 M Et_4NBF_4/pyrrole by cycling 15 times between 0.21 and 1.0 V vs. SCE at 100mV/s. The polymer films were reduced by holding the superconductor at -0.5 V vs SCE for 3 min. in 0.1 M Et_4NBF_4/MeCN.

[b] For this entry, data was acquired for a 1000 Å film of $YBa_2Cu_3O_{7-\delta}$ that was deposited over a 1500 Å step edge. Reported thickness represents the estimated superconductor thickness over the step.

On the other hand, there are two possible explanations that are to be consistent with the acquired data. According to the first explanation, a "proximity effect" (28-29) occurs in the system in which quasi-particles (i.e., isolated electrons) from the metal migrate into the superconductor while Cooper pairs originating from the superconductor penetrate into the normal metal layer. The net result of such exchange of carriers is that superconductivity in the cuprate thin film is weakened by the presence of the thick normal layer. This effect is most pronounced directly at the interface between the two layers and the dilution of Cooper pairs is limited to ~1000 Å. Moreover, it has been shown that superconductors deposited on top of semiconducting samples, such as doped silicon, have significantly lower transition temperatures and that the magnitude of the effect is dependent upon the silicon doping concentration (30). Thus, it is possible that in our conductive polymer/superconductor system, the oxidized polymer is acting as a source of normal electrons and the observed suppression of T_c is due to this proximity effect. Consistent with this notion is the fact that upon reduction of the polymer to its nonconductive form, the transition temperature returns to its original value.

Alternatively, the superconducting field effect (31-33) may also play a role in the modulation of superconductivity. It is possible that the polymer layer is separated from the superconductor by an insulating barrier formed by corrosion products and the modulation is the result of an electric field effect. According to such an explanation, charging of the polymer would serve to create an electric field at the surface of the superconductor which could alter the superconductor carrier concentration in a way that is analogous to the field effect which operates in conventional metal-oxide-semiconducting devices. While shifts due to field effects are normally much less than those we have observed here, recent studies have shown that the modulation of superconductivity can be magnified by using films with weak link characteristics (*34*). We also find that the greatest observed shifts are obtained for superconductor films that are deposited onto step edge substrates. Defects in the substrate of this type have been shown to result in the creation of superconductor weak links.

Regardless of the exact mechanism, we have now demonstrated that it is possible to use a doped conductive polymer layer to control the transition temperature of a superconducting thin film. Because the polymer can be reversibly oxidized and reduced, it may be possible to incorporate these properties into a superconducting based sensor that is similar to the semiconducting ISFET .

Conclusions

In summary, methods have been devised to prepare a number of novel molecule/superconductor structures. The composite systems can be tailored for a variety of applications in which the superconductor and molecular components are chosen for a given purpose. In this chapter, the initial two examples of molecule/superconductor chemical sensor are reported.   Future work will undoubtedly lead to a better understanding of molecule/superconductor interactions which will foster further developments in the area of chemical sensors.

Ackowledgments

This research was supported by the National Science Foundation (with matching funds from the Electric Power Research Institute), the Texas Advanced Research Program, and the Welch Foundation.

Literature Cited

(1)   *Langmuir-Blodgett Films*; Roberts, G., Ed.; Plenum: New York, 1990.
(2)   Whitesides, G.M.: Laibinis, P.E. *Langmuir* , **1990**, *6* , 87.
(3)   *Chemically modified Surfaces in Catalysis and Electrocatalysis*; Miller, J.S., Ed.; ACS Symposium Series 192, American Chemical Society: Washington, D.C., 1982.
(4)   Chidsey, C. E.; Murray, R. W. *Science* **1986**, *231*, 25.
(5)   Wrighton, M. S., *Science* **1986**, *231*, 32.
(6)   Burroughs, J. H.; Bradley, D. D. C.; Brown, A. R.; Marks, R. N.; Mackay, K.; Friend, R. H.; Burns, P. L. ; Holmes, A. B. *Nature* **1990**, *347*, 539.
(7)   Sailor, M. J.; Klavetter, F. L.; Grubbs, R. H.; Lewis, N. S. *Nature* **1990**, *346*, 155.
(8)   Chao, S.; Wrighton, M. S. *J. Am. Chem. Soc.* **1987**, *109 (22)*, 6627.
(9)   Garnier, F.; Horowitz, G.; Peng, X.E.; Fickov, N. *Adv. Mater.* **1990**, *2*, 592.
(10) *Fundamentals and Applications of Chemical Sensors*; Schuetzle, D., Hammerle, R., Butler, J.W., Eds.; ACS Symposium Series 309, American Chemical Society: Washington, D.C., 1986.
(11) *Proceedings of the Symposium on Chemical Sensors II* ; Butler, M.; Ricco, A.; Yamazoe, N. Eds.; The Electrochemical Society: Pennington, N.J., 1993.
(12) *Introduction to superconductivity,* Tinkham, M., Ed., New York: McGraw-Hill, 1975.

(13)   *Principles of Superconductive Devices and Circuits,;*Van Duzer, T.; Turner, C.W.;Elsevier: New York, 1981.
(14)   Dijkkamp, D., Venkatesan, T., Wu, X.D., Shaheen, S.A., Jisrawi, N., Min-Lee,Y.H., McLean, W.L., Croft, M. *Appl. Phys. Lett.* **1987**, *51*, 619.
(15)   McDevitt, J.T., Riley, D.R., Haupt, S.G., *Anal. Chem.* **1993** *65*, 535A.
(16)   *Chemical Sensor Technology,* Seiyama, T., Ed.; Elsevier Science Publishing Company: New York, NY, 1988; Vol. 1.
(17)   Enomoto, Y.; Murakami, T. *J. Appl. Phys.* **1986**, *56*, 3807.
(18)   Kwok, H.S.; Zheng, J. P.; Ying, Q. Y. *Appl. Phys. Lett.* **1989,** *54*, 2473.
(19)   Forrester, M. G.; Gottlieb, M.; Gavaler, J. R.; Braginski, A. I. *Appl. Phys. Lett.* **1988**, *53*, 1332.
(20)   Enomoto, Y.; Murakami, T.; Suzuki, M. *Physica C* **1988**, *153-155*, 1592.
(21)   Zhao, J.; Jurbergs, D.; Yamazi, B.; McDevitt, J. T. *J. Am. Chem. Soc.* **1992**, *114*, 2737.
(22)   Wellinghoff, S.T. In *Polymers for Electronic Applications;* Lai, J.I., Ed.; CRC Press: Boca Raton, Florida, 1989, Chapter 4.
(23)   *Handbook of Conducting Polymers*; Skothhiem, T.A., Ed.; Marcel Dekker: New York, 1986.
(24)   Haupt, S.G.; Riley, D.R.; Jones, C.T.; Zhao, J.; McDevitt, J.T., *J. Am. Chem. Soc* **1993**, *115*, 1196.
(25)   Haupt, S.G.; Riley, D.R.; Zhao, J.; McDevitt, J.T. *J. Phys. Chem.* **1993**, *97*, 7796.
(26)   Peck, S.R.; Curtin, L.S.; McDevitt, J.T.; Murray, R.W.; Collman, J.P.; Little, W.A.; Zetterer, T.; Duan, H.m.; Dong, C.; Hermann, A.M. *J. Am. Chem. Soc.* **1992**, *114*, 6771.
(27)   Diaz, A. F.; Bargon, J., in *Handbook of Conducting Polymers;* Skotheim, T. A., .Ed.;Marcel Dekker: New York, 1986, Vol. 1; p81.
(28)   Meissner, H. *Phys. Rev.* **1960**, *117 (3)*, 672.
(29)   Hilsch, P. *Z. Phys.* **1962**, *167*, 511.
(30)   Hatano, M.; Nishino, T.; Kawabe, U. *Appl. Phys. Lett.* **1987**, *50*, 52.
(31)   Fiory, A.T.; Hebard, A.F; Eick, R.H.; Mankiewich, P.M.; Howard, R.E.; O'Malley, M.L. *Phys. Rev. Lett.* **1990**, *65*, 3441.
(32)   Kabasawa, U.; Asano, K.; Koayashi, T. *Jpn. J. Appl. Phys.* **1990,** *29*, L86.
(33)   Mannhart, J.; Bednorz, J.G.; Müller, K.A.; Schlom, D.G. *Z. Phys. B* **1991**, *83*, 307.
(34)   Mannhart, J.; Strobel, J.; Bednorz, J.G.; Gerber, Ch. *Appl. Phys. Lett.* **1993**, *62* , 630.

RECEIVED April 5, 1994

STRUCTURALLY TAILORED INTERFACES

Chapter 10

Synthesis and Characterization of Two-Dimensional Molecular Recognition Interfaces

Richard M. Crooks, Orawon Chailapakul, Claudia B. Ross, Li Sun, and Jonathan K. Schoer

Department of Chemistry, Texas A&M University, College Station, TX 77843–3255

We have used self-assembly chemistry to synthesize monolayer assemblies that function as molecular recognition interfaces. In the first part of this paper, we show that one-component self-assembled n-alkanethiol monolayers with carboxylic acid functionalized endgroups specifically adsorb vapor-phase acid-terminated molecules via hydrogen bonding or vapor-phase amine-terminate molecules via proton-transfer interactions. In the second part, we demonstrate that two-component monolayers, which consist of inert n-alkanethiol framework molecules and defect-inducing template molecules, can discriminate between solution-phase probe molecules based on their physical and chemical characteristics. By electrochemically etching the defects and then imaging the resulting surface by scanning tunneling microscopy the defect sites can be indirectly visualized.

Molecular recognition is the selective binding of a probe molecule to a molecular receptor. This binding interaction relies on both non-covalent intermolecular chemical interactions, such as hydrogen bonding or van der Waals forces, and steric compatibility, such as size or shape inclusion. At present, a detailed understanding of molecular recognition phenomena is hindered primarily by two experimental problems. First, in many natural systems the receptor is a large, flexible, and complex molecule with many potential binding sites, and as a result it is difficult to quantify the specific types and magnitudes of interactions that lead to probe binding. Second, there are only a few analytical methods that are sufficiently specific and sensitive that they can be used for studying individual molecular interactions in bound probe/receptor complexes. These and other difficulties associated with natural systems have resulted in the synthesis of simpler model receptors and characterization of their interactions with probe molecules (1-3).
    Two general strategies have been used for synthesizing and characterizing model receptors and their complexes with probe molecules. The first is based on interactions between small molecules: complexes formed between alkali-metal cations and cryptands or crown ethers are typical examples. This approach has the benefit of simplicity, and it is often possible to assign a recognition event to a particular type of intermolecular interaction. Polymeric receptors are better models

0097–6156/94/0561–0104$08.00/0

for natural systems than those based on small molecules, but they are considerably more difficult to synthesize and characterize since it is difficult to design in rigidity or binding sites that have time-independent conformations. Four different strategies have been used for synthesizing polymeric receptors (*1*). The first approach involves copolymerization of small-molecule receptors with a polymeric backbone, which results in receptors randomly dispersed along the polymer backbone. The second approach involves grafting of monomeric receptors onto preformed polymers, which results in structures that are very similar to those obtained by copolymerization. The third strategy results in a higher degree of organization between individual receptors and is thus a better mimic of natural systems. In this case, the receptors are synthesized together into a single polymer, and this material is then copolymerized with a spectator backbone. Finally, an even higher degree of organization and cooperation between receptors can be achieved by binding the probe molecule to the receptors, which also contain polymerizable groups. When this receptor complex is copolymerized with a chemically inert polymeric backbone, it induces the formation of cavities whose size and shape are determined by the preformed receptor complex. If the polymer is rigidified by heavy crosslinking, the receptor cavity is locked into place even after removal of the target molecule (*1*).

Our interest is focused on surface-confined molecular recognition interfaces, but our work is guided by the principles discussed above for small molecule-small molecule interactions and small molecule-polymer interactions in homogeneous solutions. There are three key advantages to studying molecular recognition on surfaces: (1) rigid receptor sites can be designed; (2) the synthetic chemistry is simplified; (3) the surfaces can be attached to transducers, which greatly simplifies analysis and may transform the molecular recognition interface into a chemical sensor.

Our studies of molecular recognition phenomena are based on the concept illustrated in Scheme I. The basic chemical building blocks are model organic surfaces consisting of self-assembling monolayers (SAMs) of organomercaptans (*4,5*). It has previously been shown that *n*-alkanethiols spontaneously adsorb to Au from dilute solutions of ethanol and other nonaqueous solvents, and that the resulting SAMs assume a close-packed ($\sqrt{3}$x$\sqrt{3}$)R30° overlayer structure on Au(111) and other textured Au surfaces. Spectroscopic studies indicate that monolayers formed from short organomercaptans are more disordered than those formed from longer-chain molecules, but all SAMs are quite robust in aqueous solutions and vapor-phase ambients. The best *n*-alkanethiol monolayers contain surprisingly few adventitious defect sites, even when prepared on ill-defined substrates (*6-8*).

As in all natural and synthetic approaches to molecular recognition, interactions in the surface-confined systems described here are promoted at the ambient/organic interface through both chemical and physical interactions. We have attempted to separate monolayer/molecule interaction phenomena into two distinct problems: one chemical and one physical; however, it is not possible at present to achieve this degree of segregation. Nevertheless, to the extent that it is possible, we view individual chemical and physical intermolecular interactions as tools that can be used in various combinations to synthesize more complex recognition apparatuses. The size and versatility of our "toolbox" is enhanced if we can quantitatively understand a range of monolayer/molecule interactions.

Results and Discussion

We have examined five types of chemical interactions that occur between monolayers and molecules: electrostatic binding, covalent linking, complexation interactions, proton transfer, and hydrogen bonding (*9-14*). These interactions are five of the six tools presently in our "toolbox"; the sixth is a physical recognition

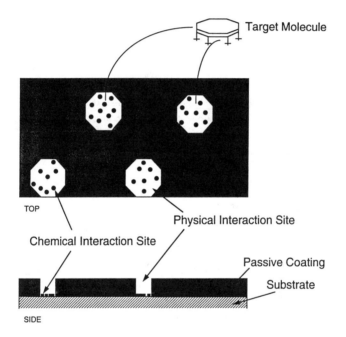

Scheme I.

strategy that is discussed later (*6*). We first consider two examples that illustrate chemical-interaction-based molecular recognition. The first example shows how carboxylic acid-terminated SAMs can be used to recognize vapor-phase acids via hydrogen bonding interactions. The second example shows how the same class of SAMs are used to recognize vapor-phase bases.

Chemical Interactions Between Probe Molecules and SAMs. In this section, we show that surface-confined monolayers of acid-functionalized organomercaptans interact with acidic or basic vapor-phase probe molecules by hydrogen-bonding (*12*) or proton-transfer (*13*) interactions, respectively. Hydrogen-bonding systems are typified by the interactions of n-alkanoic acids ($CH_3(CH_2)_nCOOH$, n = 0-14) with Au surfaces modified by 3-mercaptopropionic acid (Au/HS(CH_2)$_2$COOH), as shown in Scheme II.

   Hydrogen Bonding Interactions. Figure 1 shows FTIR external reflection spectroscopy (FTIR-ERS) data for Au/HS(CH_2)$_2$COOH and Au/HS(CH_2)$_2$CH$_3$ surfaces after and before exposure to a saturated vapor of myristic acid, $CH_3(CH_2)_{12}COOH$. Prior to $CH_3(CH_2)_{12}COOH$ modification, the Au/HS(CH_2)$_2$COOH spectrum, Figure 1b, indicates absorptions due to the acid C=O stretch and the enhanced α-CH$_2$ scissors mode at 1722 and 1410 cm^{-1}, respectively (*15-18*). After dosing, the presence of a second surface-confined $CH_3(CH_2)_{12}COOH$ layer is confirmed by the appearance of the methyl C-H stretching vibration at 2964 cm^{-1}, the increased intensity of the methylene C-H stretching vibrations at 2929 and 2858 cm^{-1}, and the doubling of the intensity of the C=O stretching vibration at 1717 cm^{-1} (Figure 1a).
   We performed control experiments by exposing a methylated SAM surface to vapor-phase n-alkanoic acids. The FTIR-ERS spectrum of a surface-confined monolayer of HS(CH_2)$_2$CH$_3$ is shown in Figure 1d. The peak at 2965 cm^{-1} is due to the asymmetric methyl C-H stretching vibration, and the peaks at 2935 and 2875 cm^{-1} are due to symmetric methyl C-H stretching vibrations. Other peaks attributable to hydrocarbon backbone modes are present at lower frequencies. The FTIR-ERS spectrum of the methyl surface after exposure to $CH_3(CH_2)_{12}COOH$, Figure 1c, is identical to the surface before acid dosing. This result clearly shows that only the acid-terminated SAM recognizes the vapor-phase acid.
   Closer examination of the FTIR-ERS data presented in Figure 1 provides additional evidence for hydrogen bonding between Au/HS(CH_2)$_2$COOH and $CH_3(CH_2)_{12}COOH$. The band at 1722 cm^{-1} in Figure 1b has been assigned to the C=O stretching vibration for a laterally hydrogen-bonded carboxylic acid terminal group, as shown in Scheme II (*15,16*). After $CH_3(CH_2)_{12}COOH$ exposure, the band shifts to 1717 cm^{-1} (Figure 1a). It has been shown previously that a 16 cm^{-1} shift in the C=O stretching frequency of n-alkanoic acids from about 1726 to 1710 cm^{-1} corresponds to a structural change from laterally hydrogen-bonded to a face-to-face dimer configuration (*15,17*). Based on the observed frequency shift of 5 cm^{-1}, we propose a model in which adsorbed $CH_3(CH_2)_{12}COOH$ is hydrogen-bonded to surface-confined HS(CH_2)$_2$COOH in both face-to-face and lateral configurations; that is, a dynamic superposition of the two configurations shown at the bottom of Scheme II.

   Proton Transfer Reactions. Figure 2 presents FTIR-ERS spectra for a Au/HS(CH_2)$_{10}$COOH monolayer after and before exposure to a saturated vapor of $CH_3(CH_2)_9NH_2$. Before amine exposure (Figure 2b), the asymmetric and symmetric C-H stretching vibrations, which arise from the methylene groups in the Au/HS(CH_2)$_{10}$COOH monolayer, are present at 2919 and 2849 cm^{-1}, respectively. As discussed earlier, the bands at 1739 and 1718 cm^{-1} are due to the C=O stretching vibrations of non-hydrogen-bonded and laterally hydrogen-bonded

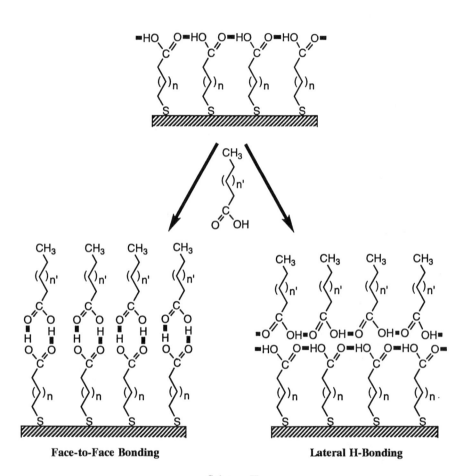

Face-to-Face Bonding   **Lateral H-Bonding**

Scheme II.

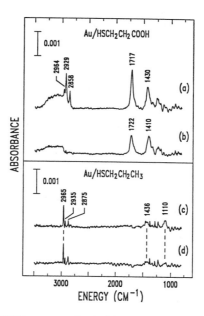

Figure 1. FTIR-ERS spectra of a Au/HS(CH$_2$)$_2$COOH surface before (b) and after (a) exposure to vapor-phase CH$_3$(CH$_2$)$_{12}$COOH and FTIR-ERS spectra of a Au/HS(CH$_2$)$_2$CH$_3$ surface before (d) and after (c) exposure to vapor-phase CH$_3$(CH$_2$)$_{12}$COOH.

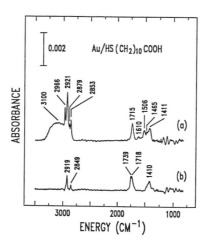

Figure 2. FTIR-ERS spectra of a Au/(CH$_2$)$_{10}$COOH monolayer before (b) and after (a) exposure to vapor-phase CH$_3$(CH$_2$)$_9$NH$_2$.

COOH terminal groups, respectively. The absence of a high-energy O-H stretching band is consistent with prior studies of acids in solid-state-like environments (15,19), and the effect may be compounded in the present case by an orientation effect.

The COOH-terminated monolayer surface recognizes vapor-phase $CH_3(CH_2)_9NH_2$, as indicated in Figure 2a by the increased intensity of the methylene stretching bands at 2921 and 2853 cm^{-1}, and the appearance of the asymmetric and symmetric methyl vibrations at 2966 and 2879 cm^{-1}, respectively. Importantly, the C=O band at 1739 cm^{-1} (Figure 2b), which corresponds to non-hydrogen-bonded COOH groups, disappears after amine exposure while most of the C=O band intensity at 1718 cm^{-1}, which corresponds to laterally hydrogen-bonded COOH groups, remains and shifts slightly to 1715 cm^{-1} (20). On the basis of these data, we conclude that non-hydrogen-bonded acid groups undergo a proton transfer reaction with NH_2 groups. This conclusion is supported by the disappearance of the C=O stretching band at 1739 cm^{-1} and bands traceable to COO^- (1610 cm^{-1}, ν_{COO^-}) and NH_3^+ (1506 cm^{-1}, δ_{NH3^+}) groups in Figure 2a. Since gravimetric measurements indicate that the surface coverage of $CH_3(CH_2)_9NH_2$ is about one monolayer, laterally hydrogen-bonded COOH groups, which are represented by the band at 1718 cm^{-1}, must undergo a hydrogen bonding interaction with $CH_3(CH_2)_9NH_2$. This conclusion is supported by the slight decrease in the C=O stretching frequency noted for hydrogen-bonded acid groups after amine exposure (20).

The FTIR-ERS spectrum shown in Figure 2a does not exhibit noticeable change for at least 10 h, indicating that the bilayer structure formed after amine exposure is quite stable. This result is in contrast to the hydrogen-bonded bilayers, which are stable for much shorter periods.

Physical Interactions Between Probe Molecules and SAMs. In addition to chemical interactions between monolayers and molecules, we have recently begun to evaluate approaches for studying physical recognition. Our principal strategy for implementing size and shape recognition, which was originally proposed by Sagiv more than a decade ago (21,22), is shown in Scheme III. Here, template molecules possessing the same geometrical properties as the molecules to be recognized are used to define interaction sites. The inert, self-assembling *n*-alkanethiol framework isolates template-induced physical recognition sites from one another. Template molecules may or may not be removed from within the framework depending upon the nature of the experiment. Finally, the template molecules can be dispersed on the substrate prior to the framework, or they may be simultaneously codeposited. We have used the latter approach.

There are a few indirect methods for characterizing molecule-size physical recognition sites, such as examining the extent of monolayer penetration by probe molecules as a function of their van der Waals radii and other chemical and physical properties (Scheme III, Frame 4). We have used an electrochemical version of this approach, which assumes that the defect sites define an array of ultramicro-electrodes, to analyze our composite SAMs (Scheme IV). In these experiments, the shape of the cyclic voltammetric wave is correlated to the size and number density of sites through which the probe molecules can penetrate, as shown on the right side of Scheme IV (6).

Chemical Characterization. The cyclic voltammetry results shown in Figure 3 were obtained from an electrode prepared by soaking Au foils in ethanol solutions containing various ratios of the defect-inducing organomercaptan template $HS(C_6H_4)OH$, 4-HTP, and the framework *n*-alkanethiol $HS(CH_2)_{15}CH_3$, $C_{16}SH$. Following deposition of the composite monolayer, we cycled the electrode potential

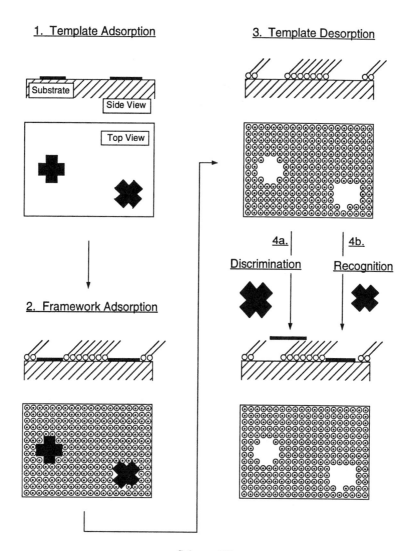

Scheme III.

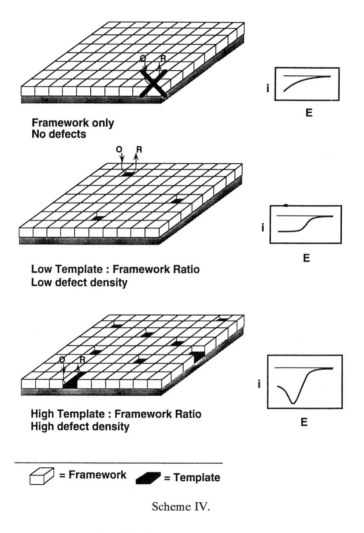

Scheme IV.

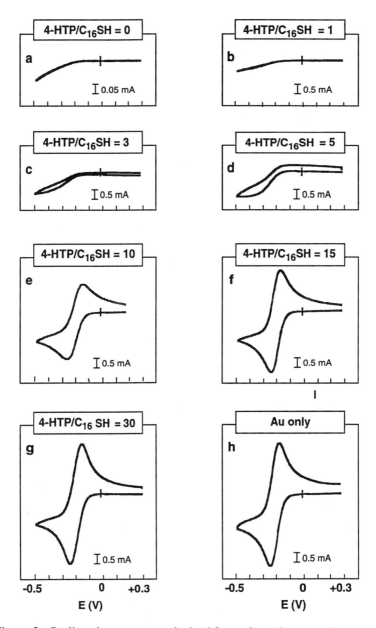

Figure 3. Cyclic voltammograms obtained for perforated, composite monolayers formed by immersing a Au surface into ethanol solutions containing various ratios of 4-HTP/C_{16}SH. The data were obtained in an aqueous electrolyte solution consisting of 5 mM $Ru(NH_3)_6^{3+}$ and 1.0 M KCl. The scan rate was 0.1 V/s.

between +0.3 and -0.5 V at 0.1 V/s in an aqueous electrolyte solution consisting of 5 mM $Ru(NH_3)_6^{3+}$ and 1.0 M KCl. In the presence of defect sites that have the correct combination of size and intermolecular interaction energies, the $Ru(NH_3)_6^{3+}$ probe molecules should penetrate the inert framework and undergo electron exchange with the underlying Au surface. If probe molecules cannot penetrate the monolayer framework, then they can only be reduced by electrons that tunnel through the $C_{16}SH$ layer. Since the distance of closest approach of the probe to the electrode surface is approximately the thickness of the monolayer, about 21 Å, the tunneling current should be small relative to that arising from direct electron transfer at template-induced defect sites.

Figure 3a shows the result obtained for a nominally defect-free $C_{16}SH$ monolayer surface. The roughly exponential shape of the cyclic voltammogram and the magnitude of the maximum cathodic current are consistent with electron tunneling through the film (23,24). Figure 3b shows the behavior of an electrode modified in a solution containing 4-HTP and $C_{16}SH$ present in a 1:1 ratio. Several points are noteworthy. First, the maximum cathodic current is about five times higher than that of the completely passivated electrode. We ascribe this current increase to surface defects induced by the template molecules; that is, the defects permit $Ru(NH_3)_6^{3+}$ to penetrate the monolayer framework. Second, the shape of the cyclic voltammogram is approximately sigmoidal, rather than exponential, and similar to that expected for an array of microelectrodes (25). This suggests that the template-induced defects are small and widely spaced relative to the diffusion layer thickness, since either large defects or closely-spaced small defects will result in peak-shaped cyclic voltammograms that are characteristic of linear diffusion (Scheme IV). Third, since the concentrations of template and framework molecules in the deposition solution are identical, and since it is clear that only a very small fraction of the molecules on the Au surface are template molecules, it follows that the much longer framework molecules compete more effectively for surface adsorption sites than the template molecules.

Figures 3c and 3d are consistent with the qualitative interpretation of Figure 3b. The shapes of these voltammograms arise from radial diffusion of $Ru(NH_3)_6^{3+}$ to small, widely dispersed defect sites on the electrode surface. This conclusion is confirmed by the scan rate dependence of the data shown in Figure 3d: scan rates between 10 and 1000 mV/s result in only a doubling of the maximum limiting current, i_{lim}, which is not consistent with the ten-fold increase anticipated for linear diffusion (26). As the concentration of the template molecules in the deposition solution is increased relative to the framework molecules, i_{lim} increases and there is a clear departure from pure radial diffusion into a mixed linear/radial regime. Mixed diffusion behavior is especially evident in Figure 3e, but when 4-HTP/$C_{16}SH$ = 15, nearly ideal linear diffusion obtains (Figure 3f). When 4-HTP/$C_{16}SH$ = 30, the cyclic voltammetry obtained using the modified surface (Figure 3g) is indistinguishable from that of a naked Au surface (Figure 3h).

Figure 4 presents data analogous to those shown in Figure 3, except that the solution-phase redox probe molecule is $Fe(CN)_6^{3-}$. This set of data follows the general trends discussed for Figure 3. For example, there is a clear progression from electron transfer via tunneling through the monolayer film to direct electron transfer at the electrode surface governed first by radial, and then by linear, diffusion as the 4-HTP/$C_{16}SH$ ratio increases. Qualitatively, the only difference between the cyclic voltammograms shown in Figures 3 and 4 is that the transition from radial to linear diffusion occurs at a higher 4-HTP/$C_{16}SH$ value for $Fe(CN)_6^{3-}$ than for $Ru(NH_3)_6^{3+}$. Since the perforated monolayers used to generate Figures 3 and 4 are the same there is only one possible explanation for this behavior: some of the defect sites that admit $Ru(NH_3)_6^{3+}$ do not admit $Fe(CN)_6^{3-}$. That is, although the total number and average size of the defects is fixed, there are differences in the intermolecular interactions between the probe molecules and at least some of the

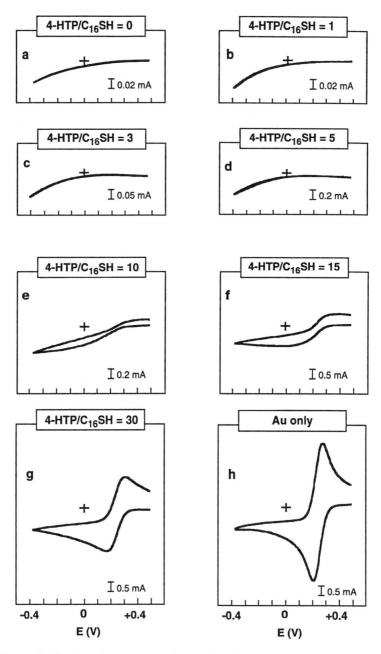

Figure 4. Cyclic voltammograms for perforated, composite monolayers formed by immersing a Au surface in ethanol solutions containing various ratios of 4-HTP/C_{16}SH. The data were obtained in an aqueous electrolyte solution consisting of 5 mM Fe(CN)$_6^{3-}$ and 1.0 M KCl. The scan rate was 0.1 V/s.

molecular recognition defect sites. This important observation indicates that the defect sites discriminate between probe molecules. These data also confirm that the template-induced defects are of molecular dimension; if they were not, then the shapes of the voltammograms obtained using both $Ru(NH_3)_6^{3+}$ and $Fe(CN)_6^{3-}$ probe molecules would be identical.

We have used several probe molecules other than $Ru(NH_3)_6^{3+}$ and $Fe(CN)_6^{3-}$ to evaluate intentionally perforated organic monolayer surfaces, and the results of these studies are summarized in Table I and Figure 5. On the basis of these data, we conclude that while the size and shape of defects is an important factor in determining the extent of probe penetration, chemical characteristics such as the permanent molecular charge, are of equal or greater importance.

Table I. Physical Properties of Probe Molecules Used in this Study.

Probe Molecule[f]	Hydrated Diameter (Å)	Diffusion Coefficient (10^{-6} cm^2/s)	Heterogeneous Rate Constant (cm/s)
$Mo(CN)_8^{4-/3-}$	9.0[a]	4.8	0.5[b]
$Fe(CN)_6^{4-/3-}$	5.2	8.3[c]	0.15[c]
$Fe(bpy)(CN)_4^{2-/1-}$	5.7	7.7[c]	0.43[c]
$Fe(bpy)_2(CN)_2^{0/1+}$	10.1	4.3[c]	0.63[c]
$Ru(NH_3)_6^{3+/2+}$	6.2	7.1[d]	>1[e]

a. Otashima, K.; Kotato, M.; Sugawara, M.; Umezawa, Y. *Anal. Chem.* **1993**, *65*, 927.
b. Saji, T.; Maruyama, Y.; Aoyagui, S. *J. Electroanal. Chem.* **1978**, *86*, 219.
c. Saji, T.; Yamada, T.; Aoyagui, S. *J. Electroanal. Chem.* **1975**, *61*, 147.
d. Licht, S.; Cammarata, V.; Wrighton, M. S. *J. Phys. Chem.* **1990**, *94*, 6133.
e. Endicott, J. F.; Schroeder, R. R.; Chidester, D. H.; Ferrier, D. R. *J. Phys. Chem.* **1973**, *77*, 2579.
f. (bpy) is the bipyridyl ligand.

Scanning Tunneling Microscope Characterization. The electrochemical approach just described is clearly an important means for evaluating template-induced defects, but the results are difficult to interpret in terms of the purely chemical and physical characteristics of the defect sites. Clearly, a more direct approach for visualizing defects is desirable. We have attempted to use scanning tunneling microscopy (STM) to directly image template-induced defects, but our results have been ambiguous for at least two reasons. First, there are many structures on the surface of SAM-covered Au substrates that appear by STM to be defect sites whether templates are present or not (*27-29*). We have recently shown that these features are due to monoatomic pits in the Au surface, and while not electroactive, they appear similar to intentionally formed defect sites (*28*). Second, we have found that the STM tip changes the structure of SAM surfaces as a function of the number of scans regardless of the imaging conditions employed (*29*). This effect tends to enlarge the defects rendering their initial size and shape impossible to determine.

As an alternative to direct visualization, we have considered the two indirect STM-based methods illustrated in Scheme V. The fundamental problem with imaging organic surfaces is that they are soft, and thus easily damaged by the STM tip. The two processes shown on the bottom of Scheme V avoid this problem. On the right side of Scheme V a metal is electrochemically deposited into the template-

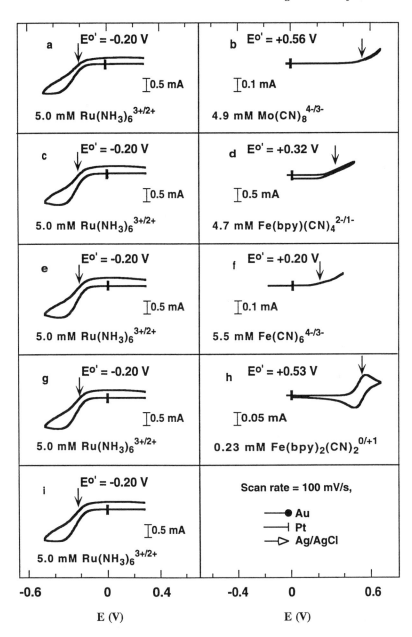

Figure 5. Cyclic voltammograms for a perforated, composite monolayer formed by immersing a Au surface in an ethanol solution containing a ratio of 4-HTP/C_{16}SH = 5. The data were obtained in aqueous electrolyte solutions containing 1.0 M KCl and the indicated probe molecules. Cyclic voltammograms were obtained in the sequence given in the figure; data for Ru(NH$_3$)$_6^{3+}$ were obtained after each of the other probe molecules to insure that the characteristics of the modified electrode surface did not change during the course of the experiment. The scan rate was 0.1 V/s. Some properties of the probe molecules are given in Table I.

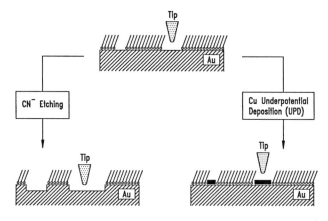

Scheme V.

defined structure, which creates resilient positive images of the original defects. This approach is promising, but there are a number of technical difficulties that make data interpretation difficult at the present time (*30*). A less ambiguous approach is shown on the left side of Scheme V. Here , we use electrochemical methods to etch the Au substrate only in those regions that contain a suitable template molecule. While this approach permits us to evaluate the number density of defects and their size after etching, it does not permit us to directly evaluate the defect size prior to etching. We now present our preliminary findings concerning the etching method.

In preparation for using this combined STM/CN⁻-etching method to study intentionally perforated monolayers, we correlated the electrochemical response obtained from SAMs containing etched adventitious defects to data calculated from STM images. We prepared $C_{16}SH$ SAMs by soaking flame-annealed Au(111) surfaces in dilute ethanol solutions of the mercaptan for about 1 day (*29,30*). We then etched the SAMs three times in an aqueous 0.1 M KCN/0.1 M Na_2HPO_4 solution (etching was carried out by stepping the substrate potential to positive potentials three times: 0.15 V for 30 s, 0.20 V for 20 s, and then 0.20 V for 10 s). This process results in etching of the Au substrate only in those regions not passivated by the SAM; that is, in regions of adventitious defects. After etching, we used a NanoScope III STM (Digital Instruments, Santa Barbara, CA) to image the surface. Image analysis reveals the average size and number density of the etched defect sites.

Figure 6a shows an inverted image of an atomically flat SAM-modified Au(111) surface after etching in CN⁻. We quantitatively analyzed the etched surface by counting the number density of pits and measuring their average radius in several different locations. We found that the average defect radius was 6.8±1.6 nm. The number density of defects was 0.9×10^9/cm², which is in good agreement with our prior results obtained using a less direct electrochemical method (*6*).

We performed cyclic voltammetric analyses for the etched surface shown in Figure 6a. These data, which are shown in Figure 6b, were obtained by immersing the electrode in an electrolyte solution consisting of 5 mM $Ru(NH_3)_6^{3+}$ and 1.0 M KCl and then measuring the resulting current over the potential range 0.3 to -0.5 V (vs. Ag/AgCl). The value of i_{lim} is nearly independent of scan rate, indicating good ultramicroelectrode behavior, and the sigmoidal shape indicates that the diffusion layers of the electrodes do not overlap significantly (*6*). We compared the experimentally determined value of i_{lim}, 0.5 µA, to the value calculated from the STM image, 4.8 µA, using the relationship given in eq 1 for the limiting current at a very small disk electrode. Here, n = 1 eq/mol, F is the Faraday, r is the average defect radius, ρ is the total number of defects, D is the diffusion coefficient of

$$i_{lim} = 4nFrDC\rho \qquad (1)$$

$Ru(NH_3)_6^{3+}$, and C is its concentration. It is interesting and somewhat disconcerting to note that the calculated and electrochemically measured values of i_{lim} differ by approximately one order of magnitude. At the present time we believe this is a consequence of four factors: (1) the etch pits are shaped more like cylinders than disks, so eq 1 is not the best expression to use for calculating a theoretical value of i_{lim}; (2) if the factors that govern diffusion to nanometer-scale electrodes are not the same as those that govern diffusion to electrodes of micron-scale dimensions, then eq 1 is inappropriate (*31*); (3) some of the etch pits (or a portion of all or some of the etch pits) observed by STM may not be electroactive; (4) if there is communication between etch pits, then the electrochemically measured value of i_{lim} will be surpressed. We are presently considering all of these possibilities in order to reconcile the STM and electrochemical data. Qualitatively,

a.

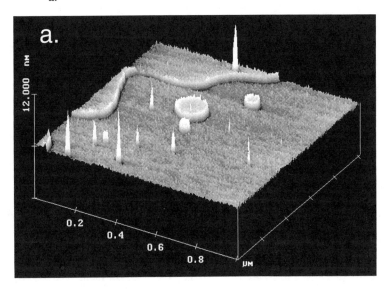

b.

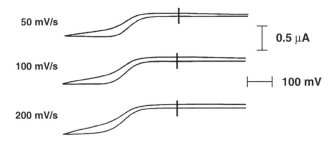

Figure 6. (a) STM image of a $C_{16}SH$ SAM containing adventitious defects obtained after CN^- etching. The image is inverted on the axis normal to the surface to emphasize the size, number density, and depth of the etch pits (note that the vertical axis is expanded relative to the two horizontal axes). The tip/substrate bias was +300 mV and the tunneling current was 150 pA. (b) Cyclic voltammograms obtained in a 5 mM $Ru(NH_3)_6^{3+}$/1.0 M KCl aqueous solution. The electrode area was 6 x 10^{-4} cm^2 and the scan was made between 0.3 and -0.5 V.

however, we believe this method is useful for correlating the number of molecule-sized, template-induced defects to electrochemical results. Indeed, our preliminary studies of 4-HTP-induced defects show that higher concentrations of 4-HTP in the deposition solution result in STM images containing higher surface concentrations of etch pits.

Conclusions

We have shown that SAMs can be used to construct organic surfaces that discriminate between probe molecules on the basis of their chemical and physical characteristics. While our current interest is in studying the fundamental interactions that exist between monolayers and molecules that lead to this level of discrimination, our ultimate goal is to combine two or more chemical or physical interaction phenomena into individual recognition sites. Analysis of these more complex structures will lead to a better understanding of how individual interactions combine in natural systems to yield highly specific binding sites.

Acknowledgments

We gratefully acknowledge the Office of Naval Research and the National Science Foundation (CHE-90146566) for supporting this work. We also thank Dr. Antonio J. Ricco of Sandia National Laboratories for helpful discussions, insightful comments, and data obtained using surface acoustic wave devices that support the findings presented herein. JKS acknowledges an IBM Manufacturing Research Fellowship.

Literature Cited

1. Wulff, G. In *Polymeric Reagents and Catalysts*; Ford, W. T., Ed.; ACS Symposium Series 308; American Chemical Society: Washington, D.C., 1986; pp 186-230, and references therein.
2. Lehn, J.-M. *Angew. Chem. Int. Ed. Engl.* **1990**, *29*, 1304.
3. Rebek, J. Jr. *Acc. Chem. Res.* **1990**, *23*, 399.
4. Nuzzo, R. G.; Allara, D. L. *J. Am. Chem. Soc.* **1983**, *105*, 4481.
5. Dubois, L. H.; Nuzzo, R. G. *Annu. Rev. Phys. Chem.* **1992**, *43*, 437, and references therein.
6. Chailapakul, O.; Crooks, R. M. *Langmuir* **1993**, *9*, 884.
7. Chidsey, C. E. D.; Loiacono, D. N. *Langmuir* **1990**, *6*, 682.
8. Creager, S. E.; Hockett, L. A.; Rowe, G. K. *Langmuir* **1992**, *8*, 854.
9. Sun, L.; Johnson, B.; Wade, T; Crooks, R. M. *J. Phys. Chem.* **1990**, *94*, 8869.
10. Sun, L.; Thomas, R. C; Crooks, R. M.; Ricco, A. J. *J. Am. Chem. Soc.* **1991**, *113*, 8550.
11. Kepley, L. J.; Crooks, R. M.; Ricco A. J. *Anal. Chem.* **1992**, *64*, 3191.
12. Sun, L. Kepley, L. J.; Crooks, R. M. *Langmuir* **1992**, *8*, 2101.
13. Sun, L.; Crooks, R. M.; Ricco, A. J., *Langmuir* **1993**, *9*, 1775.
14. Xu, C.; Sun, L.; Kepley, L. J.; Crooks, R. M. *Anal. Chem.* **1993**, *65*, 2102.
15. Nuzzo, R. G.; Dubois, L. H.; Allara, D. L. *J. Am. Chem. Soc.* **1991**, *112*, 558.
16. Ihs, A.; Liedberg, B. *J. Colloid Interface Sci.* **1991**, *144*, 282.
17. Yarwood, J. *Spectroscopy* **1990**, *5*, 34.
18. Chidsey, C. E. D.; Loiacono, D. N. *Langmuir* **1990**, *6*, 682.
19. Bellamy, L. J. *The Infra-red Spectra of Complex Molecules*, 3rd ed., Chapman and Hall: London, 1975.

20. Vinogradov, S. N.; Linnell, R. H. *Hydrogen Bonding*; Van Nostrand Reinhold Co.: New York, 1971; p 74.
21. Sagiv, J. *Isr. J. Chem.* **1979**, *18*, 339.
22. Sagiv, J. *Isr. J. Chem.* **1979**, *18*, 346.
23. Chidsey, C. E. D. *Science* **1991**, *251*, 919.
24. Finklea, H. O.; Hanshew, D. D. *J. Am. Chem. Soc.* **1992**, *114*, 3173.
25. Wightman, R. M.; Howell, J. O. *Anal. Chem.* **1984**, *56*, 524.
26. Bard, A. J.; Faulkner, L. R. *Electrochemical Methods*; Wiley: New York, 1980; p 222.
27. Sun, L. Crooks, R. M. *Langmuir* **1993**, *9*, 1775.
28. Chailapakul, O.; Sun, L.; Xu, C.; Crooks, R. M. *J. Am. Chem. Soc.*, in press.
29. Ross, C. B.; Sun, L.; Crooks, R. M. *Langmuir* **1993**, *9*, 632.
30. Sun, L.; Crooks, R. M. *J. Electrochem. Soc.* **1991**, *138*, L23.
31. Smith, C. P.; White, H. S. the University of Utah, personal communication.

RECEIVED March 25, 1994

Chapter 11

Channel Mimetic Sensing Membranes Based on Host–Guest Molecular Recognition by Synthetic Receptors

Kazunori Odashima[1], Masao Sugawara[2], and Yoshio Umezawa[2]

[1]Faculty of Pharmaceutical Sciences, University of Tokyo, Bunkyo-Ku, Tokyo 113, Japan
[2]Department of Chemistry, School of Science, University of Tokyo, Bunkyo-Ku, Tokyo 113, Japan

Ordered multi- or monolayers composed of synthetic receptor molecules (long alkyl chain derivatives of macrocyclic polyamines, cyclodextrins, and calixarenes) were prepared as "channel mimetic sensing membranes". Host-guest complexation by these molecules induced changes in the permeability through intermolecular voids in the membrane and further through intramolecular channels of the hosts. Membrane permeabilities were evaluated by cyclic voltammetry using electroactive markers such as $[Fe(CN)_6]^{4-}$, $[Mo(CN)_8]^{4-}$, $[Ru(bpy)_3]^{2+}$, $[Co(phen)_3]^{2+}$, and p-quinone. Particularly noteworthy are the results obtained by horizontal touch cyclic voltammetry with a condensed monolayer of a β-cyclodextrin derivative, confirming its intramolecular channel function. With this condensed monolayer, a channel mimetic selectivity was observed, which is based on differences in the ability of each guest to decrease the membrane permeability by blocking the intramolecular channels of the cyclodextrin molecules.

Recent advances in host-guest molecular recognition chemistry for the design and synthesis of host molecules to mimic bioreceptor functions have stimulated a number of research groups to develop novel types of sensing membranes where synthetic hosts are used as sensory elements. In particular, many successful applications of acyclic ligands as well as of crown ethers to liquid membrane type ion-selective electrodes (ISEs) for alkali and alkaline earth metal ions have been reported (1, 2). More recently, we have developed some new types of liquid membrane ISEs for organic guests and metal cyano complexes (3-9), using macrocyclic polyamines (3-5), macrocyclic dioxopolyamines (6), a cytosine-pendant triamine (7), a calix[6]arene hexaester (8), or a β-cyclodextrin derivative (9) as the sensory element. These ISEs exploit different modes of host-guest interaction for the potentiometric discrimination of various types of guests. However, in contrast to chemical sensing by ISEs based on membrane potential changes, there are still few systematic investigations of chemical sensing based on membrane permeability changes despite the possibility that it provides a promising principle of chemical sensing.

Chemical Sensing Based on Membrane Permeability Changes

Chemical sensing based on membrane permeability changes is expected to be a most promising approach toward highly sensitive and selective sensing membranes, because

0097–6156/94/0561–0123$08.00/0
© 1994 American Chemical Society

an increase or reduction of the membrane permeability as a result of the selective host-guest complexation can lead to a signal amplification. Such a mode of control of membrane permeability is effected in a most sophisticated and efficient manner in ligand-gated ion channels of biomembranes. We have been aiming at a novel type of sensing systems designated as "ion channel sensors", which has an inherent function of signal amplification based on the control of membrane permeability (10). For this purpose, a comprehensive investigation of both artificial (11-14) and biological (15-17) systems has been carried out. In this chapter, we describe a series of approaches to the development of "channel mimetic sensing membranes" based on totally artificial systems. Several types of channel mimetic sensing membranes containing synthetic hosts (1~7) were prepared and the permeability changes of these membranes induced by organic guests (8~20) were evaluated by cyclic voltammetry using appropriate electroactive compounds as permeability markers. The structures of the hosts and guests used in these studies are shown in Figure 1.

Channel Mimetic Sensing Membranes Based on the Control of Permeability through Intermolecular Voids

As a simplest starting point for channel mimetic sensing membranes, charged and ordered membranes containing different kinds of sensory elements were deposited directly onto glassy carbon electrodes by the Langmuir-Blodgett (LB) method. In these systems, charged lipids and valinomycin were used as sensory elements (11). Complexation between these membrane molecules and inorganic cations or anions added as guests to sample solutions induced changes in the cyclic voltammograms of electroactive marker ions such as $[Fe(CN)_6]^{4-}$ and $[Ru(bpy)_3]^{2+}$ (bpy = 2,2'-bipyridine). These voltammetric changes can be ascribed to a guest-induced

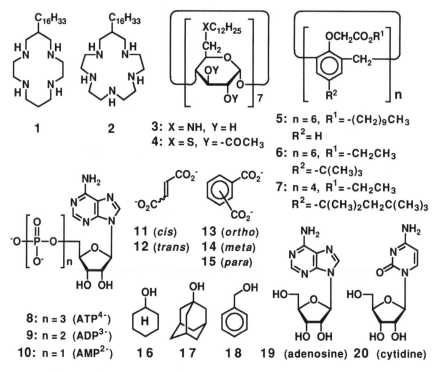

Figure 1. Structures of synthetic hosts (1~7) and guests (8~20).

modulation of the permeability for these markers as a result of a change in the net charge and/or packing structure of the ordered membranes. In all of these examples, the permeability through the intermolecular voids between the membrane molecules is controlled. Figure 2a shows a schematic representation of a cationic guest-induced increase in the permeability for an anionic marker ion in the case of a membrane composed of an anionic lipid plus valinomycin (K^+ ion receptor). Surface pressure-molecular area (π-A) isotherms at the air/water interface provided further evidence that a guest-induced change not only in the membrane charge but also in the membrane packing is important for the permeability control (*13*).

Organic Anion-Sensitive Membranes Composed of Polyamine Type Hosts. Chemical sensing of *organic* guests on the basis of the above principle was attained by the use of polyamine type hosts that, by multiple protonation, function as polycationic hosts capable of forming strong complexes with organic anions (ref 12 and Nagase, S.; Kataoka, M.; Odashima, K.; Umezawa, Y.; Kimura, E., unpublished results). As the polyamine hosts, long alkyl chain derivatives of macrocyclic tetra- and pentaamines (**1, 2**) and polyamino-β-cyclodextrin (**3**) were used.

Macrocyclic Polyamine Membranes. Ordered membranes composed of a macrocyclic polyamine (**1** or **2**) and L-dipalmitoyl-α-diphosphatidylcholine (L-DPPC) (1:1 molar ratio, five layers) were deposited directly onto glassy carbon electrodes by the horizontal lifting LB method (membranes 1 and 2, respectively). These lipophilic macrocyclic polyamines, developed by Kimura and coworkers (*3, 18*), can be multiply protonated when the membranes come into contact with aqueous solutions. The polycationic property thus acquired is a prerequisite for both the complexation with anionic guests and the control of ion permeation. This property is clearly reflected in the pH profile of these membranes, showing an increase (decrease) in the permeability for $[Fe(CN)_6]^{4-}$ ($[Ru(bpy)_3]^{2+}$) with decreasing pH (figure not shown). These permeability changes can be ascribed to an increasing electrostatic attraction (repulsion) between the anionic (cationic) marker ion and the protonated macrocyclic polyamine bearing a pH-dependent amount of positive charge.

The same principle of permeability control by the membrane charge applies for the responses to anionic guests. The permeability for $[Fe(CN)_6]^{4-}$ ($[Ru(bpy)_3]^{2+}$) decreased (increased) with increasing concentration of an anionic guest. These observations can be reasonably explained by a decrease in the electrostatic attraction (repulsion) between the protonated polyamine host and the anionic (cationic) marker ion due to partial neutralization of the positive membrane charge by complexation of the membrane host with the anionic guest. Figure 2b shows schematically the guest-induced decrease in the membrane permeability when an anionic marker ion is used. The magnitude of permeability change induced by adding a guest depended on the structure of the guest, resulting in a voltammetric discrimination of guests. The selectivity of response was quantitatively evaluated on the basis of a selectivity factor, which indicates the relative decrease of permeability induced by each guest. The selectivity factors determined for three different groups of anionic guests in the case of using $[Fe(CN)_6]^{4-}$ (1.5 x 10^{-3} M) as a marker ion are listed in Table I. The solution pH for the cyclic voltammetric measurements was set to 6.0 (1.0 x 10^{-2} M AcONa-AcOH buffer containing 5.0 x 10^{-2} M K_2SO_4) so that the polyamine hosts were sufficiently protonated while the guests could occur sufficiently in their anionic forms.

Membranes 1 and 2 showed similar selectivities for all of the three groups of organic anions tested. For both membranes, the responses were in the orders of ATP^{4-} (**8**) > ADP^{3-} (**9**) > AMP^{2-} (**10**); **11** (*cis*) > **12** (*trans*); and **13** (*ortho*) > **14** (*meta*) > **15** (*para*) (Table I). For the adenine nucleotides, both membranes showed a greater response to the guests with a greater negative charge. For the dicarboxylate isomers, a greater response was observed for the guests having a shorter distance between the two anionic groups. The selectivities displayed by membranes 1 and 2 seem to be controlled by electrostatic interactions between the protonated polyamine hosts and the anionic guests. These voltammetric selectivities are qualitatively consistent with the selectivities of complexation in aqueous solutions (*18, 19*) as well as with the selectivities of potentiometric response in liquid membranes displayed by protonated macrocyclic polyamine hosts (*3-5*). Selectivities were higher for membrane 2 containing the pentaamine host (**2**), which is more highly protonated than the tetraamine host (**1**) at a given pH.

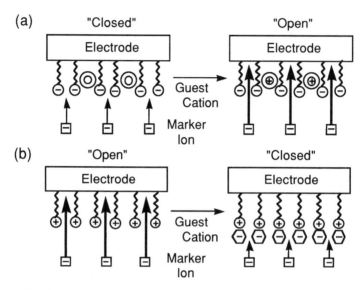

Figure 2. Schematic representations of the modes of guest-induced permeability change for an anionic marker ion. (a) Cation-responsive membrane containing an anionic lipid plus valinomycin. (b) Anion-responsive membrane composed of a protonated polyamine host. For simplicity, the LB membranes (actually 4~6 layers) are represented as monolayers. (Reproduced with permission from reference 10. Copyright 1991 Elsevier.)

Table I.   Selectivity Factors for Anion-Responsive Channel Mimetic Sensing Membranes Containing Polyamine Hosts[a]

guest anion	membrane 1 (host **1**)	membrane 2 (host **2**)	membrane 3 (host **3**)
8 (ATP^{4-})	1.00	1.00 (1.00)[b]	1.00 (1.00)[b]
9 (ADP^{3-})	0.87	0.78	0.44
10 (AMP^{2-})	0.38	0.19	0.15
11 (*cis*)	1.00	1.00 (0.69)[b]	1.00 (0.31)[b]
12 (*trans*)	0.06	0.03	0.41
13 (*ortho*)	1.00	1.00 (0.44)[b]	0.55 (0.77)[b]
14 (*meta*)	0.89	0.37	1.00
15 (*para*)	0.10	0.02	0.83

SOURCE: Adapted from ref 12 (membranes 2 and 3) or unpublished results by Nagase, S.; Kataoka, M.; Odashima, K.; Umezawa, Y., Kimura, E (membrane 1).
[a] pH = 6.0.   [b] The values in the parentheses are the cross selectivity factors for the primary guests of each group.

Polyaminocyclodextrin Membrane.   An ordered membrane composed of polyamino-β-cyclodextrin 3 (four to six layers) was deposited directly onto a glassy carbon electrode by the horizontal lifting LB method (membrane 3). Host 3 as well as host 4, developed by Tagaki and coworkers (20), is a new type of cyclodextrin derivative with an amphiphilic nature. The basic structure with long alkyl chains introduced on only one side (primary hydroxyl side) of the cavity affords these cyclodextrin derivatives an interfacial receptor function. Since host 3 is also a polyamine type host, a control of permeability for marker ions by the membrane charge is also effected in membrane 3 composed of this host (Figure 2b). However, since host 3 has a cavity of *ca.* 7.5 Å diameter that is capable of accommodating an organic guest, the membranes containing this host are expected to show a response selectivity which is different from those containing simple macrocyclic polyamines.

The selectivity factors for membrane 3 at pH 6.0 are shown in Table I. For the adenine nucleotides (8~10) and the *cis/trans* isomers of dicarboxylate (11, 12), membrane 3 showed similar selectivities as membranes 1 and 2 containing macrocyclic polyamines. However, for the positional isomers of phthalate (13~15), the response selectivity was different for these two types of membranes. Whereas membranes 1 and 2 showed responses in the order of 13 (*ortho*) > 14 (*meta*) > 15 (*para*), membrane 3 interestingly showed a different response order, *i.e.*, 14 (*meta*) > 15 (*para*) > 13 (*ortho*), a selectivity which is quite different from that expected on the basis of simple electrostatic effects (Table I). This different selectivity is possibly due to a host-guest complexation involving not only electrostatic interactions but also inclusion into the β-cyclodextrin cavity, which is capable of recognizing differences in the steric structures of guests.

Cation-Sensitive Membranes Composed of Calixarenes.   Calixarenes (ester, amide, or ketone derivatives) provide another class of host compounds, which show interesting complexation selectivities based on inclusion of inorganic or organic guests into their well-defined cavities. We have recently started a systematic investigation on the control of membrane permeabilities by use of ordered membranes composed of calixarene esters (either an LB membrane or a monolayer formed at the air/water interface) (Yagi, K.; Khoo, S. B.; Namba, M.; Sugawara, M.; Odashima, K.; Umezawa, Y.; Sakaki, T.; Shinkai, S., unpublished results). The calixarene hosts used include hexamers 5 (8) and 6 (21, 22) as well as tetramer 7 (23). All of these hosts were shown to be useful as sensory elements of liquid membrane electrodes for organic and inorganic cations (8, 23, 24).

Preliminary experiments showed that a metal cation guest (Cs^+ and Na^+ for the hexamers and tetramer, respectively) induced an increase in the permeability of the membranes composed of host 6 or 7 for an anionic marker ion such as $[Fe(CN)_6]^{4-}$ and a decrease in the permeability for cationic marker ions such as $[Ru(bpy)_3]^{2+}$ and $[Co(phen)_3]^{2+}$ (phen = 1,10-phenanthroline). These results can be explained, again, by an electrostatic effect involving a positive membrane charge acquired by the complexation between a calixarene host and a metal cation guest. A cationic guest-induced increase in the membrane permeability was also observed in the case of using a neutral marker (*p*-quinone). A plausible explanation for this result is that the packing density of the monolayer decreased due to electrostatic repulsion between the positively charged host-guest complexes. This could lead to an increase in the intermolecular voids and hence an increase in the membrane permeability.

Channel Mimetic Sensing Membranes Based on the Control of Permeability through Intramolecular Channels

Interesting results have been obtained on the control of permeability through intermolecular voids. However, considering the high efficiency and selectivity of permeation through biological ion channels based on a discrete molecular entity with a well-defined channel structure, it would be quite significant for the development of artificial "channel mimetic sensing membranes" to attain a control of permeability through *intramolecular channels.*

Long Alkyl Chain Derivative of Cyclodextrin as a Channel Mimic.
Cyclodextrin derivatives **3** and **4** are quite interesting since their basic structure should allow them to function not only as receptor sites but also as intramolecular channels. However, the sensing membrane composed of the cyclodextrin derivative (**3**) described in the previous section was based on the control of permeability through the intermolecular voids between the cyclodextrin molecules (Figure 3a); the marker ion used ($[Fe(CN)_6]^{4-}$) cannot pass through the β-cyclodextrin cavity due to steric reasons. Therefore, the potential intramolecular channel function of the cyclodextrin derivative has not been verified. The following experiments were conducted to examine the possibility of the intramolecular channel function, which would allow the control of membrane permeability by blocking the cyclodextrin cavity (channel entrance) with a guest molecule (Figure 3b).

Horizontal Touch Cyclic Voltammetry with a Condensed Monolayer of a Cyclodextrin Derivative. To obtain experimental evidence supporting that such a mode of permeability control is possible, an approach based on horizontal touch cyclic voltammetry was carried out for a condensed monolayer of **4**, which was formed at the air/water interface in a Langmuir trough (*14*). This technique, first used by Fujihira (*25*) and recently sophisticated by Bard (*26, 27*), enables the investigation of the electrochemical properties of oriented monolayers at varying packing densities under an appropriately controlled surface pressure.

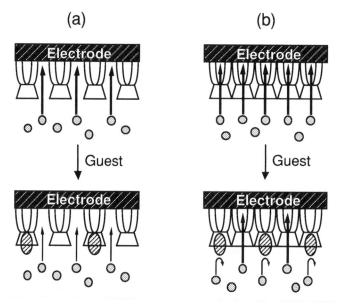

Figure 3. Schematic representations of the possible modes of controlling the permeability through an ordered layer of a long alkyl chain derivative of cyclodextrin by host-guest complexation. (a) Control of permeation through intermolecular voids. (b) Control of permeation through intramolecular channels. (Adapted from ref 14.)

On the basis of π-A isotherm measurements, a controlled surface pressure of 50 mN m^{-1} was applied for the formation of a condensed monolayer of cyclodextrin derivative **4**. At this surface pressure, the molecular area of **4** in the membrane is limited to *ca.* 210 Å^2 molecule^{-1} if a complete integrity of the monolayer is assumed. Since this molecular area corresponds to the basal area of β-cyclodextrin, the intermolecular voids that could cause membrane leakage are expected to be minimized. The permeability of this condensed membrane both in the presence and absence of a guest in the subphase solution was compared for the electroactive markers that may pass (*p*-quinone) or cannot pass ([Co(phen)$_3$]$^{2+}$, [Mo(CN)$_8$]$^{4-}$) through the β-cyclodextrin cavity (channel entrance). Highly oriented pyrolytic graphite (HOPG) was used as a working electrode because the surface of a freshly cleaved HOPG contains a fair extent of atomically smooth planes (*28*). As guests supposed to block the channel and control the membrane permeability, uncharged organic molecules have been chosen to avoid complications by electrostatic effects on the guest-induced permeability change, as described in the previous section (*11-13*). All voltammetric measurements were carried out at pH 6.0 (0.1 M AcONa-AcOH buffer) and 17.0 °C.

Permeability of Markers through the Cyclodextrin Monolayer. Permeability through the monolayer of cyclodextrin derivative **4** has been estimated from the accessibility of electroactive markers to the electrode surface, which in turn can be estimated from the peak potential and area of cyclic voltammograms (Figure 4). The calculation of the voltammogram area has been made by an integration of either the oxidation or reduction peak that corresponds to the initial process of the redox cycle, *i.e.*, the oxidation peak in the cases of [Co(phen)$_3$]$^{2+}$ and [Mo(CN)$_8$]$^{4-}$, and the reduction peak in the case of *p*-quinone.

(a) [Co(phen)$_3$]$^{2+}$ as a Permeability Marker. Figure 4a shows the cyclic voltammograms obtained with [Co(phen)$_3$]$^{2+}$ (1.0 × 10^{-4} M) as a permeability marker. The presence of the cyclodextrin monolayer caused a dramatic decrease in the area of the oxidation peak (curve A vs B). The decrease in the area of the oxidation peak was estimated to be 85%. Furthermore, changes in the cyclic voltammogram upon addition of cyclohexanol (**16**) as a guest (up to 2.0 × 10^{-2} M) were very small (curve B vs C~E) as compared to the corresponding changes when *p*-quinone was used as a marker (*vide infra*; Figure 4c). These observations can be explained by considering the large size of the marker [Co(phen)$_3$]$^{2+}$ (*ca.* 13 Å minimum diameter), making it incapable of passing through the β-cyclodextrin cavity (*ca.* 7.5 Å diameter). Consequently, the access of this marker to the electrode surface is efficiently inhibited by the cyclodextrin monolayer, and in addition, the complexation with guest **16**, which blocks the channel, does not affect the marker permeability (Figure 4a).

(b) [Mo(CN)$_8$]$^{4-}$ as a Permeability Marker. Figure 4b shows the cyclic voltammograms obtained with [Mo(CN)$_8$]$^{4-}$ (1.0 × 10^{-3} M) as a permeability marker. Similarly as in the case of [Co(phen)$_3$]$^{2+}$, a marked inhibition of electron transfer was observed in the presence of the cyclodextrin monolayer (curve A vs B; 70% decrease in the area of the oxidation peak). The change in the cyclic voltammogram upon addition of guest **16** was very small (curve B vs C). These observations, again, can be explained by considering the steric incapability of the marker [Mo(CN)$_8$]$^{4-}$ (*ca.* 9 Å minimum diameter) to pass through the β-cyclodextrin cavity (Figure 4b).

Despite the fact that [Co(phen)$_3$]$^{2+}$ as well as [Mo(CN)$_8$]$^{4-}$ is much too large to pass through the β-cyclodextrin cavity (channel entrance), and that the intermolecular voids between the membrane forming cyclodextrin molecules should be minimized under a controlled surface pressure of 50 mN m^{-1}, a complete inhibition of electron transfer was not attained for either of these markers. A possible explanation for this discrepancy may be given by considering defects in the monolayer that could be generated by the contact with the HOPG electrode. This would be favored by the edged periphery of the HOPG block and/or by the unbalanced cross sectional areas within the molecule of cyclodextrin derivative **4**. In the long alkyl chain derivative of cyclodextrin (**4**), the cross sectional areas of the cyclodextrin moiety and the long alkyl chains (in total) are *ca.* 200 and 140 Å^2, respectively.

(c) *p*-**Quinone as a Permeability Marker.** Figure 4c shows the cyclic voltammograms obtained with *p*-quinone (1.0×10^{-3} M) as a permeability marker. The presence of the cyclodextrin monolayer caused some decrease in the area of the cyclic voltammogram (curve A vs B). But in this case, the magnitude of change was much smaller (31% decrease in the reduction peak) than in the cases of $[Co(phen)_3]^{2+}$ and $[Mo(CN)_8]^{4-}$. In addition, the reduction peak remained within the investigated potential window though the peak shifted to a more negative potential.

The electron transfer to *p*-quinone is known to be very slow; the heterogeneous electron transfer rate constant (k^0) of the *p*-quinone/hydroquinone redox couple in an aqueous solution is two to three orders of magnitude smaller than those of the $[Co(phen)_3]^{3+/2+}$ and $[Mo(CN)_8]^{3-/4-}$ couples. For such a marker with a slower electron transfer kinetics, a greater shift as well as broadening of the voltammogram peak is expected to occur in the presence of the cyclodextrin monolayer (see ref 29 for relevant discussions). As a result, a part of the voltammogram peak might be pushed out of the potential window that is set for the voltammetric measurement in the absence of the cyclodextrin monolayer (bare HOPG electrode). This aspect leads to an underestimation of the permeability for *p*-quinone as long as the permeability is estimated from the voltammogram area in the presence of the cyclodextrin monolayer relative to that for the bare electrode. Despite such an unfavorable factor for *p*-quinone, a well-defined voltammetric peak was still observed within the potential window and

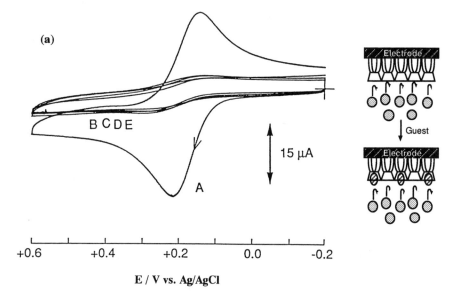

(a)

B C D E

15 µA

A

+0.6   +0.4   +0.2   0.0   -0.2

E / V vs. Ag/AgCl

Figure 4. Cyclic voltammograms of (a) $[Co(phen)_3]^{2+}$, (b) $[Mo(CN)_8]^{4-}$, and (c) *p*-quinone as permeability markers (pH 6.0). Curve A: in the absence of a cyclodextrin monolayer on a buffer solution containing no guest. Curve B: in the presence of a condensed monolayer of cyclodextrin derivative **4** on a buffer solution containing no guest. Curve C~E: in the presence of a condensed cyclodextrin monolayer on a buffer solution containing cyclohexanol (**16**) as a guest in concentrations of 5.0×10^{-3}, 1.0×10^{-2}, and 2.0×10^{-2} M, respectively. For curve C in Figure 4b, the concentration of **16** was 1.0×10^{-2} M. Schematic representations of the behavior of each marker in the presence and absence of the guest are also shown for each figure. (Adapted from ref 14.)

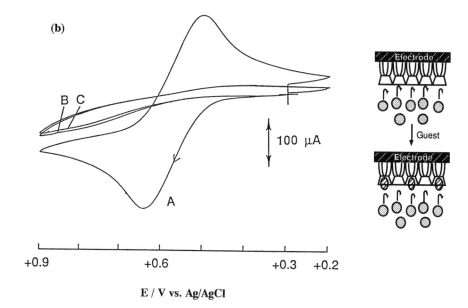

E / V vs. Ag/AgCl

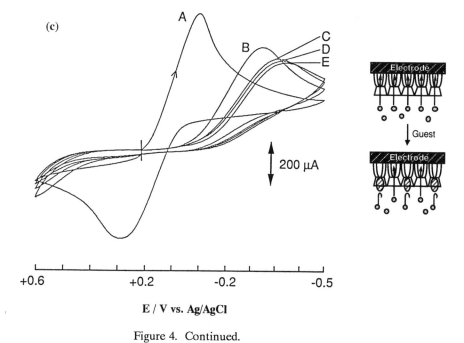

E / V vs. Ag/AgCl

Figure 4. Continued.

the decrease in voltammogram area was much smaller than in the cases of $[Co(phen)_3]^{2+}$ and $[Mo(CN)_8]^{4-}$.

These results clearly indicate a much more feasible access of p-quinone to the electrode surface. To interpret these results, it must be taken into account that p-quinone ($ca.$ 6.3 Å shorter width) can sterically pass through the β-cyclodextrin cavity (channel entrance) and that the cyclodextrin monolayer is sufficiently condensed so that the intermolecular voids are minimized. These points, taken together, suggest that p-quinone permeates through the intramolecular channel formed by the cyclodextrin derivative **4** (Figure 4c). Since the difference between the voltammograms for p-quinone and the other two bulky markers is quite pronounced, an incomplete integrity of the cyclodextrin monolayer in contact with the HOPG electrode ($vide\ supra$) does not invalidate the above discussion.

Addition of cyclohexanol (**16**) to the subphase solution caused a further decrease in the voltammogram area (curve B vs C in Figure 4c). Furthermore, the extent of decrease became greater by increasing the concentration of this guest in the subphase (curve C → D → E). These observations can be most reasonably explained by a channel blocking effect of the guest, leading to inhibition of the permeation of p-quinone through the β-cyclodextrin cavity (Figure 4c).

Dependence of Marker Permeability on Surface Pressure. The dependence of the cyclic voltammogram area on the surface pressure applied to the cyclodextrin monolayer was examined. By increasing the surface pressure from 10 to 50 mN m^{-1} or, in other words, by decreasing the molecular area from 260 to 210 Å^2 molecule^{-1}, the voltammogram areas for all three markers decreased almost linearly. However, an important point to be emphasized here is that the magnitudes of the decrease were different; the decrease for p-quinone was much smaller than that for $[Co(phen)_3]^{2+}$ and $[Mo(CN)_8]^{4-}$. This is also consistent with the capability of p-quinone and incapability of the other two bulky markers to sterically pass through the β-cyclodextrin cavity.

Another point to be noted is the fact that a dramatic increase in the voltammogram area upon a decrease in the surface pressure was observed for both of the two bulky markers, regardless of whether the marker is hydrophilic ($[Mo(CN)_8]^{4-}$) or hydrophobic ($[Co(phen)_3]^{2+}$). These results indicate that the main factor controlling the permeability through the intramolecular channel is the steric bulkiness rather than the hydrophobicity of the marker. Such an aspect seems to be characteristic of the cyclodextrin monolayer and contrasts to the properties of monolayers of simple alkane derivatives, in which the permeability (through intermolecular voids) is controlled mainly by the hydrophobicity and not the steric bulkiness of the marker (*30*).

Selectivity of Permeability Control. In the present study, the voltammetric responses in the presence of several uncharged guests (**17~20**) other than cyclohexanol (**16**) were also examined, using p-quinone (1.0 × 10^{-3} M) as a marker for the permeability change. For the guests examined, some differences were observed in the magnitude of decrease in the voltammogram area as shown in Table II. Since the guest-induced decrease in the voltammogram area was negligible for 3.0 × 10^{-4} M of guest **16**, the magnitude of the voltammetric response can be considered as **17** >> **16** > **19** > **20** > **18**. This can be regarded as a channel mimetic selectivity, which is based on differences in the ability of these guests to decrease the membrane permeability by blocking the intramolecular channel of the cyclodextrin molecule. The greatest decrease in the voltammogram area was observed for 1-adamantanol (**17**), reflecting the greatest inhibition of marker permeability by this guest. Such a selectivity is consistent with the much stronger complexation between the native β-cyclodextrin and guest **17** as compared with the other guests in aqueous solutions (*31, 32*).

Table II.   Selectivity of Guest-Induced Permeability Decrease for a Channel Mimetic Sensing Membrane Composed of a Condensed Monolayer of β-Cyclodextrin Derivative (4)[a]

guest	concentration of guest [M]	relative area of cyclic voltammogram [%][b]	guest-induced decrease [%]
without guest	–	69.2	–
16	1.0 x 10^{-2}	62.5 ± 1.5	6.7
19	1.0 x 10^{-2}	63.7 ± 1.2	5.5
20	1.0 x 10^{-2}	64.6 ± 0.3	4.6
18	1.0 x 10^{-2}	65.3 ± 0.7	3.9
17	3.0 x 10^{-4}	64.5 ± 1.3	4.7

SOURCE: Adapted from ref 14.   [a] pH = 6.0; surface pressure = 50 mN m^{-1}. [b] Calculated as a percentage of the voltammogram peak area relative to that measured in the absence of the cyclodextrin monolayer (bare electrode). The average values of two or three runs are shown.

Concluding Remarks

In contrast to chemical sensing based on membrane potential changes, which has been extensively studied and found wide analytical applications as ion-selective electrodes, chemical sensing based on membrane *permeability* changes is still in its initial stage despite a promising possibility of mimicking the remarkable transmembrane signaling function of ion channels in biomembranes.   This chapter has focused on our approaches toward "channel mimetic sensing membranes" based on ordered multi- or monolayers composed of totally artificial hosts, which are very promising not only because of their chemical stability but also of the versatility in molecular design.
Starting from simplest prototypes, a number of channel mimetic sensing membranes have been developed, which display interesting selectivities for organic guests based on a control of permeability through intermolecular voids. Furthermore, the intramolecular channel function of a long alkyl chain derivative of cyclodextrin was confirmed by horizontal touch cyclic voltammetry with a condensed monolayer formed at the air/water interface. By exploiting this system, a new type of channel mimetic sensing membrane has been developed, which displays a selectivity based on differences in the ability of each guest to decrease the membrane permeability by blocking the intramolecular channel of the cyclodextrin. A similar mode of permeability control is found in the efficient action mechanisms of neurotoxins such as tetrodotoxin and saxitoxin in the inhibition of biological ion channels. Further design of host molecules to improve the complexation ability and selectivity as well as to introduce more suitable channel components may lead to a more efficient signal amplification to yield higher sensitivities.

Literature Cited

(1)   Ammann, D.; Morf, W. E.; Anker, P.; Meier, P. C.; Pretsch, E.; Simon, W. *Ion-Selective Electrode Rev.* **1983**, *5*, 3-92.
(2)   Kimura, K.; Shono, T. In *Crown Ethers and Analogous Compounds*; Hiraoka, M., Ed.; Studies in Organic Chemistry 45; Elsevier: Amsterdam, 1992; Chapter 4 (pp 198-264).
(3)   Umezawa, Y.; Kataoka, M.; Takami, W.; Kimura, E.; Koike, T.; Nada, H. *Anal. Chem.* **1988**, *60*, 2392-2396.

(4)   Kataoka, M.; Naganawa, R.; Odashima, K.; Umezawa, Y.; Kimura, E.; Koike, T. *Anal. Lett.* **1989**, *22*, 1089-1105.
(5)   Naganawa, R.; Kataoka, M.; Odashima, K.; Umezawa, Y.; Kimura, E.; Koike, T. *Bunseki Kagaku* **1990**, *39*, 671-676.
(6)   Naganawa, R.; Radecka, H.; Kataoka, M.; Tohda, K.; Odashima, K.; Umezawa, Y.; Kimura, E.; Koike, T. *Electroanalysis* **1993**, *5*, 731-738.
(7)   Tohda, K.; Tange, M.; Odashima, K.; Umezawa, Y.; Furuta, H.; Sessler, J. L. *Anal. Chem.* **1992**, *64*, 960-964.
(8)   Odashima, K.; Yagi, K.; Tohda, K.; Umezawa, Y. *Anal. Chem.* **1993**, *65*, 1074-1083.
(9)   Odashima, K.; Hashimoto, H.; Umezawa, Y. Submitted for publication.
(10) Odashima, K.; Sugawara, M.; Umezawa, Y. *Trends Anal. Chem.* **1991**, *10*, 207-215.
(11) Sugawara, M.; Kojima, K.; Sazawa, H.; Umezawa, Y. *Anal. Chem.* **1987**, *59*, 2842-2846.
(12) Nagase, S.; Kataoka, M.; Naganawa, R.; Komatsu, R.; Odashima, K.; Umezawa, Y. *Anal. Chem.* **1990**, *62*, 1252-1259.
(13) Sugawara, M.; Sazawa, H.; Umezawa, Y. *Langmuir* **1992**, *8*, 609-612.
(14) Odashima, K.; Kotato, M.; Sugawara, M.; Umezawa, Y. *Anal. Chem.* **1993**, *65*, 927-936.
(15) Uto, M.; Michaelis, E. K.; Hu, I. F.; Umezawa, Y.; Kuwana, T. *Anal. Sci.* **1990**, *6*, 221-225.
(16) Minami, H.; Sugawara, M.; Odashima, K.; Umezawa, Y.; Uto, M.; Michaelis, E. K.; Kuwana, T. *Anal. Chem.* **1991**, *63*, 2787-2795.
(17) Minami, H.; Uto, M.; Sugawara, M.; Odashima, K.; Umezawa, Y.; Michaelis, E. K.; Kuwana, T. *Anal. Sci.* **1991**, *7 (Supplement)*, 1675-1676.
(18) Kimura, E. In *Crown Ethers and Analogous Compounds*; Hiraoka, M., Ed.; Studies in Organic Chemistry 45; Elsevier: Amsterdam, 1992; Chapter 8 (pp 381-478) and the references cited therein.
(19) Potvin, P. G.; Lehn, J. M. In *Synthesis of Macrocycles. The Design of Selective Complexing Agents*; Izatt, R. M., Christensen, J. J., Eds.; Progress in Macrocyclic Chemistry 3; John Wiley & Sons: New York, 1987; Chapter 4 (pp 167-239).
(20) Tagaki, W. *Yukagaku* **1988**, *37*, 394-401 and the references cited therein.
(21) Chang, S.-K.; Cho, I. *J. Chem. Soc., Perkin Trans. 1* **1986**, 211-214.
(22) Arnaud-Neu, F.; Collins, E. M.; Deasy, M.; Ferguson, G.; Harris, S. J.; Kaitner, B.; Lough, A. J.; McKervey, M. A.; Marques, E.; Ruhl, B. L.; Schwing-Weill, M. J.; Seward, E. M. *J. Am. Chem. Soc.* **1989**, *111*, 8681-8691.
(23) Sakaki, T.; Harada, T.; Deng, G.; Kawabata, H.; Kawahara, Y.; Shinkai, S. *J. Inclusion Phenom. Mol. Recognit. Chem.* **1992**, *14*, 285-302.
(24) Cadogan, A.; Diamond, D.; Smyth, M. R.; Svehla, G.; McKervey, M. A.; Seward, E. M.; Harris, S. J. *Analyst* **1990**, *115*, 1207-1210.
(25) Fujihira, M.; Araki, T. *Chem. Lett.* **1986**, 921-922.
(26) Zhang, X.; Bard, A. J. *J. Am. Chem. Soc.* **1989**, *111*, 8098-8105.
(27) Miller, C. J.; Bard, A. J. *Anal. Chem.* **1991**, *63*, 1707-1714.
(28) Chang, H.; Bard, A. J. *Langmuir* **1991**, *7*, 1143-1153.
(29) Amatore, C.; Savéant, J. M.; Tessier, D. *J. Electroanal. Chem. Interfacial Electrochem.* **1983**, *147*, 39-51.
(30) Bilewicz, R.; Majda, M. *Langmuir* **1991**, *7*, 2794-2802.
(31) Du, Y.-Q.; Nakamura, A.; Toda, F. *J. Inclusion Phenom. Mol. Recognit. Chem.* **1991**, *10*, 443-451.
(32) Matsui, Y.; Mochida, K. *Bull. Chem. Soc. Jpn.* **1979**, *52*, 2808-2814.

RECEIVED April 5, 1994

Chapter 12

Passivation and Gating at the Electrode–Solution Interface via Monomolecular Langmuir–Blodgett Films

Mechanism of Alkanethiol Binding to Gold

Marcin Majda

Department of Chemistry, University of California,
Berkeley, CA 94720

To define molecularly active sites at the electrode surface, monomolecular Langmuir-Blodgett films are being designed to carry out two functions: passivation and gating. This new strategy requires us to produce monomolecular LB films of exquisitely low defect level. To accomplish this, we have investigated the mechanism of octadecanethiol ($C_{18}SH$) binding to gold under potentiostatic conditions, in order to better understand $C_{18}SH$ passivating properties. Our approach involves constant potential current measurements taken during the LB transfer of monolayers that contain $C_{18}SH$ from the air/water interface onto gold substrates. Thiol-gold coupling involves a potential-dependent partial electron transfer from sulfur to gold of 0.26 electron per thiol at -0.3 V, and 0.4 electron at 0.7 V vs. SCE. The requirements of electro-neutrality and zero capacitance of the emersed alkanethiol-coated gold surface lead to a postulate that an equivalent number of protons are released during the $C_{18}SH$ LB transfer.

Structural control of the electrode/solution interface is a complex problem of fundamental importance in electrochemical sciences (*1*). To achieve some elements of such control, it would be desirable to impart molecular character onto the otherwise "naked" electrode surface so that, as a result, it might acquire desired catalytic properties, gain some elements of molecular selectivity, or exhibit other desirable molecular characteristics. To accomplish this, electrochemists have explored numerous possibilities of coating the electrode surface with thin films (from a single monolayer to micrometers in thickness) of a wide variety of materials (*2*). This area of electrochemistry, often referred to as the chemical modification of electrodes, is the subject of a number of recent reviews (*1-5*).

In our recent research, we hope to achieve a molecular level definition of the sites of electroactivity. To this end, we relied on monomolecular Langmuir-Blodgett (LB) films designed to carry out two functions: passivation and gating (*6*). The basic elements of the overall scheme involved in this project are presented schematically below.

0097–6156/94/0561–0135$08.00/0
© 1994 American Chemical Society

Scheme I   Scheme II

The electrode surface is coated with a monolayer LB film composed of two types of molecules. The long alkyl chain molecules, a 70:30 mol% mixture of octadecanethiol ($C_{18}SH$) and octadecyl alcohol ($C_{18}OH$), are used to block access to the electrode surface. These molecules serve also as a matrix for incorporation of a small number of "gate molecules", the second monolayer component, designed to open access to the electrode surface and thus to define sites of electroactivity in the otherwise passivating film. The key idea of gated access involved in this scheme is this: since species in solution can approach the electrode surface only through the gate sites, we can rely on the chemical structure of the gate molecules in order to induce some elements of selective behavior by the electrode. The passivating properties of mixed $C_{18}SH/C_{18}OH$ LB monolayers and an experimental illustration of the gated access to the electrode surface idea were reported in the literature (6,7).

We envision two general classes of selective behavior. **Scheme I:** electrochemically active analyte species (A) in solution are "recognized" by the gate molecules on the basis of size, shape, charge or chirality. Recognition would involve weak interactions or binding between G and A that would lead to changes in the analyte's electroactivity that would allow its determination in the presence of other species in solution of similar reactivity. This might involve a shift of the analyte's formal redox potential or a change of its rate of electron transfer so that its electrochemical signal could be distinguished from that of other species in solution. **Scheme II:** analyte species in solution engage in selective binding with the gate molecules (again due to selective interactions stemming from compatibility of sizes, shapes, charges, or chirality) that results in blocking of the gate sites. This is then detected by an electroactive spectator species (P) by measuring a decrease of its reduction (or oxidation) current. Unlike Scheme I, analyte species need not, in this case, be electroactive. This substantially expands the generality of this scheme.

Sensitivity of electrochemical measurements in these two schemes would be highest if the surface density of the gate molecules were sufficiently low so that they could act as independent molecular size micro-electrodes. This requires the surface concentration of the gate molecules to be below ca. 10^{-15} mol/cm^2 (below ca. 10^{-3} mol%). Under these conditions the individual gate sites are separated by distances comparable or larger than the radii of the hemispherical diffusion zones formed around individual gate sites (8). The observed current is then equal to the sum of currents collected at the individual gate sites, and is thus very sensitive to their "state" (open/closed).

In order to work in this region of gate site concentrations, the level of intrinsic pin-hole defects in the passivating monolayer films has to be below that level. This defines, for us, standards of passivation. To meet these standards, we investigated the effect of a number of parameters on the level of pin-hole defects in the passivating LB films. The effect of monolayer composition and the LB transfer pressure were examined and described previously (7). In this report, we focus on the effect of the electrode potential on the binding of $C_{18}SH$ to gold. We describe a method which combines Langmuir-Blodgett transfer with electrochemical measurements and which led us to a quantitative understanding of the mechanism of alkanethiol binding to gold. In spite of a wide-spread interest and extensive research in numerous laboratories devoted to the area of self-assembled thiol monolayers (*9-15*), mechanistic aspects of alkanethiol binding to gold are not well understood. Prevailing reaction mechanisms proposed in the literature involve oxidative coupling of alkanethiolate, RS^-, to gold, $Au(I)$, on the gold surface and a concurrent release of H_2 or H_2O according to the following (*9,16,17*):

$$RSH + Au(0)_s \longrightarrow RS\text{-}Au(I)_s + 1/2H_2 \tag{1}$$

$$2RSH + 2Au(0)_s + 1/2O_2 \longrightarrow 2RS\text{-}Au(I)_s + H_2O \tag{2}$$

Our measurements suggest that alkanethiol forms a coordination-type bond with the gold surface transferring a fraction, δ, of an electron:

$$R-\underset{H}{S} + Au_s \longrightarrow R-\underset{H}{S^{+\delta}}\cdots Au_s + \delta e^- \tag{3}$$

That fraction, the partial charge number of $C_{18}SH$ was determined to be dependent on the electrode potential ranging from 0.26 at -0.3 V vs. SCE to 0.40 at 0.70 V vs. SCE. The partial electron transfer from thiol to gold is likely to weaken thiol's S-H bond. This, in turn, is likely to result in at least partial dissociation of thiol's proton.

Experimental Section

Chemicals. n-Octadecanol (Aldrich, 99%) and n-octadecanethiol (Aldrich, 98%) House distilled water was passed through a four cartridge Barnstead Nanopure II purification train. The final resistivity of water used in the experiments was in the range 18.1-18.3 Mohm cm.

Electrode fabrication procedures. Electrodes were produced by vapor-deposition of about 100 nm thick gold films (99.95 % Au, Lawrence Berkeley Laboratory) on microscope glass slides pre-coated with a 6 to 10 nm thick film of Cr. The geometrical pattern of the vapor deposited electrodes is shown and discussed in Figure 1. The surface area of its central rectangular fragment is 0.20 cm². Immediately before the LB experiments, gold electrodes were cleaned with several organic solvents and then were immersed into a fresh hot (ca. 80 °C) chromic acid solution for about 25 s. The chromic acid treatment step was followed by an extensive rinsing with Nanopure water. Upon completion of this step, the electrodes were dried in a stream of argon and transferred immediately (in order to minimize their contamination) into the plexiglass enclosure of the LB instrument. Before each LB transfer experiment, the surface of a gold substrate was electrochemically reduced and conditioned by voltammetric scanning in the potential range from -0.300 V to 0.600 V at 50 mV/s.

Langmuir-Blodgett procedures. The Langmuir-Blodgett experiments were carried out with a KSV Minitrough instrument operated under computer control. Routine Langmuir-Blodgett procedures and the LB transfer protocol were described

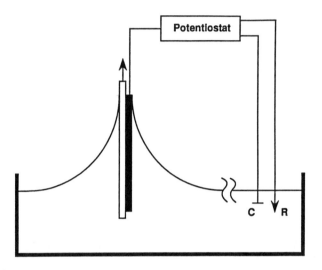

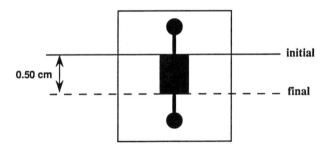

Figure 1. Schematic diagram illustrating Langmuir-Blodgett transfer under potentiostatic conditions. A gold-coated glass slide is acting as a working electrode in a three-electrode potentiostatic circuit. Below, an inset shows the pattern of the vapor deposited gold film. The central rectangular area (A = 0.20 cm^2) is coated with an LB monolayer as the substrate is withdrawn from the subphase. The two lines mark the initial and the final position of the water meniscus in the LB experiments.

previously (*18*). The LB transfers of $C_{18}OH$ and $C_{18}SH/C_{18}OH$ monolayers were carried out at 20 mN/m at a rate ranging from 0.5 mm/min to 50 mm/min. Standard electrochemical instruments and set-up were described previously (*18*).

Results and Discussion

Our experimental approach is based on the classical Langmuir-Blodgett transfer (*19*) of monolayers from the air/water interface onto the surface of vapor-deposited gold films on glass slides. The monolayers consisted of either pure octadecane alcohol ($C_{18}OH$) or of octadecanethiol/octadecanol ($C_{18}SH/C_{18}OH$) mixtures. (These are typically 70:30 mol%. Pure $C_{18}SH$ monolayers are not stable on the water surface above ca. 10 mN/m (*7*).) Both types of monolayer assemblies were spread on 0.05 M $HClO_4$ subphase and transferred at 20 mN/m at a rate of 10 mm/min. The LB transfers were carried out under potentiostatic conditions with the gold substrate acting as a working electrode. A schematic diagram of the experimental set-up is shown in Figure 1. The geometric pattern of the vapor-deposited gold substrates designed for these experiments is also shown in Figure 1. The lines coinciding with the top and the bottom edge of the central rectangular part of the electrode mark the initial and final positions of the water meniscus in the LB experiments. Thus, the top circular area was always kept above the water surface and used to make electrical contact while the bottom circular area was always submerged in the aqueous subphase.

Figure 2 shows two typical current-time transients recorded during LB transfer experiments. Let us consider first the case of a pure $C_{18}OH$ monolayer transferred at 0.20 V vs SCE. At this potential, the gold surface is free of gold oxide and carries a small negative excess charge since, as we show below, the value of the potential of zero charge (E_{zc}) was found to be 0.36 V vs. SCE. During an LB transfer, the dynamic contact angle was rather high (ca. 60°) indicating that $C_{18}OH$ is transferred directly onto the gold surface (rather than on a thin film of water) (*20*). Thus, the rectangular fragment of the gold substrate is emersed dry. The observed positive current (flowing from the electrode to the outside circuit) is a result of the need to remove excess double-layer charge, initially stored on the gold surface when the substrate is submerged, when its surface area is decreased under constant potential conditions. Clearly, the dry gold surface coated with the $C_{18}OH$ monolayer has negligible capacitance compared with the clean gold surface contacting an aqueous electrolyte. The magnitude of this discharging current is proportional to the initial charge density, σ^M, and the rate of change of the surface area, dA/dt:

$$i = \sigma^M \, dA/dt \qquad (4)$$

Indeed, we found the current to be linearly proportional to the rate of substrate withdrawal, as long as the latter is less than 50 mm/min. This is shown in Figure 3 (closed circles). The negative deviation at the highest transfer rate indicates a breakdown of the experiment as some electrolyte is trapped at the emersed gold surface (in other words, LB transfer of $C_{18}OH$ no longer results in a complete removal of the double-layer). Integration of the current associated with $C_{18}OH$ LB transfer gives the initial excess charge density on the gold surface at a particular potential. These data are shown in Figure 4 (triangles). The readily available value of $E_{zc} = 0.36$ V vs. SCE is consistent with the literature data for gold (111) surfaces in

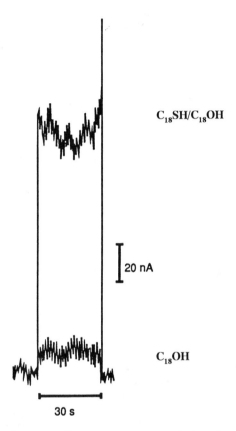

Figure 2: Typical current–time transients recorded during LB transfer of $C_{18}SH/C_{18}OH$ (70:30 mol%) and $C_{18}OH$ monolayers at 20 mN/m, from 0.05 M $HClO_4$ subphase onto gold-coated glass slides held at 0.20 V vs. SCE. LB transfer rate: 10 mm/min (corresponding to $dA/dt = 0.4$ cm²/min). Since 0.20 V is a potential negative of E_{zc}, and because $C_{18}SH$ chemisorption involves a partial electron transfer from thiol to gold, current in both transients is anodic.

non-specifically adsorbable electrolytes (*21*). Vapor-deposited gold films were shown, indeed, to exhibit predominantly the (111) crystallographic texture (*22*).

A much higher current is observed when a $C_{18}SH/C_{18}OH$ (70:30 mol%) monolayer is transferred onto gold substrates as shown in Figure 3. It is clear that an additional process contributes to the overall current besides the double-layer discharge described above. Unlike aliphatic alcohols, for which physisorption on gold does not involve partial electron transfer (*23*), chemisorption of alkylthiols is naturally expected to result in much stronger interactions. Hence, in this case, we postulate that the additional charge flow is due to the electron transfer from $C_{18}SH$ molecules to gold, as the former bind to the substrate surface (see equation 3 above). Integration of this current gives the total charge passed (Figure 4 closed squares), while subtraction of the double-layer charge from these data gives a net charge due to $C_{18}SH$ binding to gold. The latter is plotted vs. substrate potential in Figure 4 (open squares). These results indicate that $C_{18}SH$ binding to gold involves a partial electron transfer that varies roughly linearly with potential from 0.26 at -0.3 V to 0.40 at 0.7 V vs. SCE. Measurements at more positive potentials are affected by gold oxide formation and will be reported elsewhere (*24*)

Can we be sure that our technique allows us to integrate all the charge involved in $C_{18}SH$ binding to gold on the time scale of the LB transfer experiments which take only 30 s? In other words, do all $C_{18}SH$ molecules transferred to gold bind to its surface and is our technique a good approximation of equilibrium conditions of a prolonged self-assembly process? To address these concerns, we first recall recent results of Bain et al. who showed that the self-assembly of octadecanethiol on gold from 1 mM ethanol solutions results in essentially a limiting value of contact angle in less than 1 min (*14*). Nevertheless, in order to expand the time scale of the LB experiments, we performed additional potentiostatic LB experiments at slower transfer rates. These data are shown in Figure 3 (open circles). The linearity of the observed current with the LB transfer rate demonstrates that the charge transferred as a results of $C_{18}SH$ binding to gold is independent of the rate of substrate withdrawal. A negative deviation observed at 50 mm/min is again an artifact resulting from entrapment of a thin film of water on the gold surface during LB transfer. We have also measured the dependence of the charge due to $C_{18}SH$ coupling to gold on the mole fraction of $C_{18}SH$ in mixed $C_{18}SH/C_{18}OH$ LB monolayers. The mole fraction of the thiol was varied from 70 mol% to 10 mol% (six data points). A linear plot of the charge vs. $C_{18}SH$ mole fraction with zero intercept was obtained with a correlation coefficient 0.985. These results substantiate our notion that the potentiostatic LB transfer technique gives us the total charge resulting from $C_{18}SH$ chemisorption on gold.

Finally, we return again to the mechanistic aspects of the chemisorption process as presented in equation 3. The partial electron transfer from thiol to gold would leave a net positive charge on the sulfur as the alkanethiol LB monolayer is deposited on the gold substrate. This is not allowed since, following emersion, the capacitance of the gold/octadecanethiol interface is essentially zero and no other ions can be transferred from the aqueous subphase to match that charge. This argument leads to the postulate that alkanethiol chemisorption must be accompanied by a release of an equivalent quantity of protons to maintain the electro-neutrality of the LB film on gold as it is emersed from the subphase. As mentioned above, partial electron transfer from thiol to gold is likely to weaken the S-H bond and enhance proton dissociation. However, in view of the electro-neutrality argument, the extent of thiol

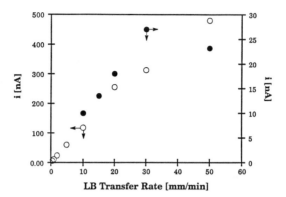

Figure 3: Plots of current vs. transfer rate recorded during LB transfer of: $C_{18}OH$ monolayer from 0.05 M $HClO_4$ at 0.20 V vs. SCE (closed circles).; and $C_{18}SH/C_{18}OH$ monolayer from 0.05 M $HClO_4$ at 0.00 V vs. SCE (open circles).

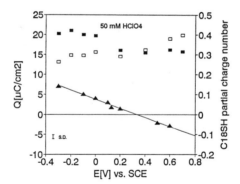

Figure 4: Plots of charge densities vs. substrate potential recorded in the course of the LB transfer experiments. Closed squares: total charge recorded with $C_{18}SH/C_{18}OH$ (70:30 mol%) monolayers. Triangles: charge recorded with $C_{18}OH$ monolayers corresponding to the double-layer charge at the gold/solution (0.05 M $HClO_4$) interface. Open squares: charge resulting from $C_{18}SH$ chemisorption (and $C_{18}SH$ partial charge number)–the difference between the total charge and the double-layer charge.

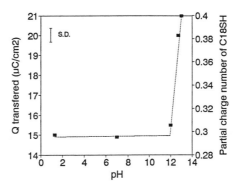

Figure 5: Plot of the charge transferred due to $C_{18}SH$ chemisorption on gold and the $C_{18}SH$ partial charge number as a function of pH of the subphase in the LB transfer experiments. Data obtained at -0.20 V vs. SCE in $HClO_4$, $NaClO_4$ and KOH electrolytes.

dissociation is governed only by the partial electron transfer. To examine this argument, we carried out the LB transfer experiments at -0.2 V from solutions of variable pH. As expected, we found that the partial charge number is independent of pH except when the latter exceeds 12 (see data in Figure 5). Since that pH coincides with an estimate of $C_{18}SH$ pK_a value (25), the observed increase of the partial charge number could be explained by an induction effect.

The coupling of the partial electron transfer and proton dissociation via the electro-neutrality requirement holds only for the case of the LB experiments described here. Thus, the extent of proton dissociation following alkanethiol self-assembly is not subject to this limitation and may proceed to a larger extent. Indeed, vibrational spectroscopic data reporting absence of the thiol proton would suggest its complete dissociation.

In summary, we believe that these experiments and data provide a better understanding of alkanethiol interactions with gold. One question which we have not been able to address so far is the variation of the $C_{18}SH$ partial charge number with potential in a more negative range than that covered in Figure 4. Will the linearity of that dependence lead to conditions where alkanethiol binding to gold is not possible? Another question, closer to the primary goal of our research outlined in the Introduction, is whether controlled potential LB transfer experiments can result in structurally improved monolayers on gold and bring us closer to accomplishing our goal of passivation and gating at the electrode/solution interface.

Acknowledgment. Research described in this article was supported financially by the National Science Foundation under Grant CHE-9108378.

Literature Cited

1. Bard, A.J., Abruña, H.D., Chidsey, C.E., Faulkner, L.R., Feldberg, S.W,. Itaya, K., Majda, M., Melroy, O., Murray, R.W., Porter, M.D., Soriaga, M.P., White, H.S. *J. Phys. Chem.* **1993**, *97*, 7147.

2. Murray, R.W. in *Molecular Design of Electrode Surfaces*, Murray, R.W., Ed.; Techniques of Chemistry Series, John Wiley & Sons: New York, 1992, Vol. 22, Chapter 1; p. 1.
3. Murray, R.W. in *Electroanalytical Chemistry*, Bard, A.J., Ed.; Marcel Dekker, New York, 1984, Vol. 13; p. 191.
4. Abruña, H.D., in *Electroresponsive Molecular and Polymeric Systems*; Skotheim, T., Ed.; Marcel Dekker: New York, 1988; p.97.
5. *Molecular Design of Electrode Surfaces*, Murray, R.W., Ed.; Techniques of Chemistry Series; John Wiley and Sons: New York, 1992; Vol. 22.
6. Bilewicz, R., Majda, M. *J. Am. Chem. Soc.* **1991**, *113*, 5464.
7. Bilewicz, R., Majda, M. *Langmuir* **1991**, *7*, 2794.
8. Amatore, C., Savéant, J.-M., Tessier, D. *J. Electroanal Chem.* **1983**, *147*, 39.
9. Ulman, A. in *An Introduction to Ultrathin Organic Films: From Langmuir Blodgett to Self-Assembly*; Academic Press: San Diego, 1991.
10. Bain, C.D., Whitesides, G.M. *Angew. Chem.* **1989**, *101*, 506.
11. Dubois, L., Nuzzo, R.G. *Ann. Rev. Phys. Chem.* **1992**, *43*, 437.
12. Nuzzo, R.G., Fusco, F., Allara, D.L. *J. Am. Chem.Soc.* **1987**, *109*, 2358.
13. Porter, M.D., Bright, T.B., Allara, D.L., Chidsey, C.E.D. *J. Am. Chem. Soc.* **1987**, *109*, 3559.
14. Bain, C.D., Troughton, E.B., Tao, Y.-T., Evail, J., Whitesides, G.M., Nuzzo, R.G. *J. Am. Chem. Soc.* **1989**, *111*, 321.
15. Troughton, E.B., Bain, C.D., Whitesides, G.M., Nuzzo, R.G., Allara, D.L. Porter, M.D. *Langmuir* **1988**, *4*, 1147.
16. Bain, C.D., Biebuyck, H.A., Whitesides, G.M. *Langmuir* **1989**, *5*, 723.
17. Whitesides, G.M., Laibinis, P.E. *Langmuir* **1990**, *6*, 87.
18. Charych, D.H., Landau, E.M., Majda, M. *J. Am. Chem. Soc.* **1991**, *113*, 3340.
19. *Langmuir Blodgett Films*; Roberts, G. Ed.; Plenum Press: New York, 1990.
20. Gaines, G.L. *Insoluble Monolayers at Liquid Gas Interfaces;* Interscience Publishers, J. Wiley & Sons: New York, 1991; Chapter 8, pp. 326-333.
21. Hamelin, A. in *Modern Aspects of Electrochemistry*, Conway, B.E., White, R.E., Bockris, J.O'M., Eds.; No. 16; Plenum Press: New York, 1985, Chapter 1, p. 1.
22. Zei, M.S., Nakai, Y., Lehmpfuhl, G., Kolb, D. *J. Electroanal. Chem.* **1983**, *150*, 201.
23. Richter, J., Lipkowski, J. *J. Electranal. Chem* **1988**, *251*, 217.
24. Krysinski, P., Majda, M. *J. Am. Chem. Soc.* submitted.
25. Irving, R.J., Nelander, L., Wadso, I. *Acta Chem. Scand.* **1964**, *18*, 769.

RECEIVED March 25, 1994

Chapter 13

Design and Electrochemical Characterization of Lipid Membrane–Electrode Interfaces

Naotoshi Nakashima and Toshiyuki Taguchi

Department of Applied Chemistry, Faculty of Engineering, Nagasaki University, 1–14 Bunkyo-cho, Nagasaki 852, Japan

An acidic phosphate lipid which contains thiol moieties at a hydrophobic terminal position was synthesized, and monolayer electrodes of this lipid were fabricated on gold electrodes *via* chemisorption from an ethanolic solution by altering lipid concentration and immersion time. The electrochemistry of the lipid monolayers on gold was examined by using ferricyanide and 1-trimethylammoniomethylferrocene as redox-active markers. The ion gating property of the monolayer responsive to pH and ion stimulus is described.

Assembled organic monolayers on gold electrodes are of considerable current interest (1). Our interest in this field is to design and develop functional metal electrode surfaces on which have been deposited organized synthetic lipid membranes (2-4). Characteristics of liposomes and synthetic bilayer membranes of acidic phospholipids in aqueous solutions have been investigated extensively in relation to membrane functions such as ion recognition, ion transport, phase separation and cell fusion (5-7). From aspects of application and utilization of biomembrane functions, developments of monolayer and bilayer electrodes having biomembrane-mimetic structures and properties should be attractive and important. A preliminary account of the formation of monolayer membranes of a mercaptan-containing phosphate lipid, **1**, by the spontaneous assembly onto gold and their ion gate properties have been given in the literature (3). In this paper, we describe the results that further characterize this system by using ferricyanide and 1-trimethylammoniomethylferrocene as electroactive markers.

$$HS-(CH_2)_{11}-O \diagdown \atop HS-(CH_2)_{11}-O \diagup P \diagup{\diagup O \atop \diagdown OH}$$

1

$$\left[\text{Fe} \right] -CH_2N^+(CH_3)_3 \quad Br^-$$

2

0097–6156/94/0561–0145$08.00/0

Experimental

Synthesis.   Synthesis of bis (11-mercaptoundecyl) phosphoric acid, **1**, has been described very briefly in the literature (3); here the detailed synthetic procedures including a reaction scheme are described.

Synthetic Scheme:

Bis (11-bromoundecyl) phosphoric acid.   11-Bromoundecanol (22 g, Jansen Chem. Co.) and phosphorus oxychloride (4.5 g) were dissolved in benzene (70 ml) and the mixture was refluxed for 24 hrs. After evaporating the solvent, 20 ml of water was added to the residue followed by stirring overnight at room temperature. Chloroform (25 ml) extraction and solvent evaporation gave a viscous oil, which was recrystallized twice from ethyl acetate. Yield 4.6 g, mp 69-71°C. Anal. Calcd for $C_{22}H_{45}O_2PBr_2$. %C, 46.82; %H, 8.04. Found. %C, 46.74; %H, 7.88.

Bis(11-thiobenzoylundecyl) phosphoric acid.   Bis(11-bromoundecyl)phosphoric acid (2.0 g), thiobenzoic acid (2.0 g) and triethylamine (1.5 g) were dissolved in ethanol (50 ml) and the mixture was refluxed for 48 hrs under a nitrogen atmosphere. Addition of ethyl acetate (40 ml) to the reaction mixture at ambient temperature gave a precipitate, which was filtered off. Evaporating the solvent gave a oily product which was dissolved in ethanol (2.0 ml) followed by the addition of 1N HCl (15 ml) to get a precipitate, which was recrystallized from ethyl acetate (20 ml) twice. Yield 2.0 g, mp 71-72°C. Anal. Calcd for $C_{36}H_{55}O_6PS_2$. %C, 63.69; %H, 8.16. Found. %C63.17; %H8.06.

Bis (11-mercaptoundecyl) phosphoric acid.   Bis (11-thiobenzoylundecyl) phosphoric acid (0.5 g) and hydrazine hydrate (1.0 g) were dissolved in methanol (30 ml) and then the mixture was refluxed for 4 hrs under a nitrogen atmosphere. The reaction mixture was allowed to cool to room temperature, and 1N HCl was added to produce a precipitate, which was separated and was recrystallized twice from ethyl acetate (20 ml). Yield 0.2 g, mp 67-48°C. The product was identified with IR, ^{1}H-NMR, TLC and elemental analyses. Anal. Calcd for $C_{22}H_{47}O_4PS_2$. %C, 56.14; %H, 10.06. Found. %C, 55.85; %H, 9.87; N, not detected.

Chemicals.   Potassium ferricyanide was used as received. 1-Trimethyl-ammoniomethyl ferrocene bromide, **2**, (formula given above) (Tokyo Kasei Co.) was recrystallized from ethanol before use.

Preparation of modified electrodes.   The experimental procedure for making monolayer electrodes using **1** is as follows. A gold disk electrode (Bioanalytical Systems, diameter 1.6 mm) was polished well with alumina (Beuhler, particle sizes: 1-, 0.3- and then 0.025 μ m) and was rinsed with acetone, Millipore water and then ethanol, followed by immersion in ethanolic solution of **1** for a given time at 25℃. The electrode was rinsed twice with ethanol and then was allowed to air-dry.

Electrochemical Measurements.   Cyclic voltammetry was performed with a Toho-Giken PS-06 in 0.1M KCl aqueous solution at 25℃ in the presence of an electroactive marker. A saturated calomel electrode (SCE) and a platinum wire were used as the reference electrode and the counter electrode, respectively.

Results and Discussion

Response to pH.   Figure 1 shows cyclic voltammograms both for a **1** monolayer electrode prepared by 24 hrs immersion of a polished gold electrode in a 1 mM ethanolic solution of **1** and for a bare gold electrode, both in the presence of **2** (2 mM). In the measured pH range (pH 2.1-10.2), no electrochemical communication of **2** was observed at the modified electrode, indicating that the monolayer blocked the electrochemistry of **2**. On the other hand, as shown in Figure 2, a monolayer electrode prepared by dipping a gold electrode in a 1 mM solution of **1** in ethanol for 20 min gave strong pH-dependent cyclic voltammograms; i.e., evident oxidation / reduction currents observed at alkaline and neutral pHs almost disappeared at acidic pHs. It is interesting to note that the small pH change from 6.0 to 7.2 caused large change in the voltammograms. Observed pH dependence was reversible with some hysteresis (Figure 3). Apparent pKa of a monolayer of **1** determined from the pH profile of the cyclic voltammograms is 6.5-7, about 5 pKa units higher than that of a dialkylphosphoric acid in aqueous solution (pKa, ca. 1.7) and 2 to 3 pKa units higher than that of a monolayer of **1** with the lower surface coverage (3). A similar pKa shift of liposomal membranes of dialkylphosphatidic acid in aqueous solution is reported (7). Ward et al. (8) found pH-dependent quartz crystal microbalance (QCM) responses for self-assembled monolayers of 15-mercaptohexadecanoic acid. The pKa of a monolayer of the carboxylic acid determined by the QCM method is about 8 which is greater than values observed for the carboxylic acid in aqueous solution (pKa $\approx$ 4.5).

As we previously reported (3), when ferricyanide was used as an electroactive marker, a rather loosely packed monolayer of **1** on gold gave voltammograms whose intensities showed pH dependence opposite to that described in the preceding paragraph; i.e., faradaic currents of ferricyanide were inhibited in the higher pH region but was evident at lower pHs. Together with the results described above, electrostatic repulsion or attraction between the electroactive markers and the phosphate of the monolayer is believed to be a main factor for the pH-dependent voltammograms, although the surface coverages of the gold electrodes are not the same. The monolayer

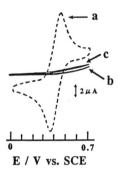

Figure 1. Cyclic voltammograms for a bare electrode (trace a) and for a **1**
monolayer electrode (trace b: pH 2.1; trace c: pH 10.2). The solutions
are 2 mM of **2** in 0.1 M KCl. The monolayer electrode was prepared by
24 hrs dipping in 1mM ethanolic solution of **1**. Scan rate, 200 mV / s.

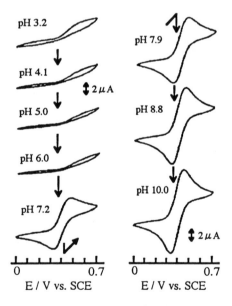

Figure 2. Cyclic voltammograms for a **1** modified electrode at the given pHs.
The monolayer electrode was prepared by dipping a gold electrode for
20 min in a 1 mM ethanolic solution of **1**. Other experimental
conditions are the same as in Figure 1. Scan rates, 200 mV / s.

electrode formed by dipping for 20 min blocked the electrochemistry of ferricyanide at all measured pHs from 2.0 to 11.0, so that it was impossible to evaluate the blocking ability toward the cationic and the anionic markers with the same monolayer electrode. A schematic illustration for the pH-responsive electrochemistry is shown in Figure 4.

Response to Ions. Umezawa and coworkers (9) described the voltammetric response of the Langmuir-Blodgett (LB) films (three layers or five layers) of didodecylphosphoric acid and dioctadecyldimethylammonium bromide toward alkali earth metal ions and perchlolate anion, respectively. It is important to pay attention to the stability of LB films when measurements are conducted in aqueous solutions, because lipids, especially synthetic lipids, are water-soluble though solubility is rather low, and because hydrophobic / hydrophilic interactions between monolayers and the first monolayer with an electrode surface are believed to be small. On the other hand, mercaptan compounds form "coordination" linkage between gold and thiolate moieties; therefore, assembled monolayers on gold are stable for long periods of time (i.e., months) even in aqueous solutions. The stability is one of the most important factors required for ion sensors. Rubinstein et al. (10, 11) reported ion-selective voltammetric responses of assembled tetradentate ligand monolayers on gold electrodes. In this study, again two electroactive markers, ferricyanide and ferroceneammonium, 2, were used to examine the ionic responses of monolayers of 1 with different surface coverages on gold.

The cyclic voltammogram of ferricyanide (2 mM) at a monolayer electrode of 1 prepared by long time dipping (i.e., 10 hrs) showed almost no response to Ca^{2+} (3mM). On the other hand, a loosely packed monolayer of 1 on Au was found to show a large response. As shown in Figure 5, with the addition of Ca^{2+} to monolayer of 1 prepared by short time dipping (3 min) of a gold electrode in a 10^{-5} M solution of 1 in ethanol, well-defined cyclic voltammograms of ferricyanide with large peak separations appeared. Response time was very fast (< 5 sec). The added Ca^{2+} is believed to chelate with the phosphate group of the monolayer at the monolayer surface; however, the binding energy of the linkage would be rather small in comparison to three dimention systems, because simple rinsing in pure water was enough to remove the binding ion; therefore, responses to other ions can be estimated with the same monolayer electrode. Voltammetric responses of the electrode toward Mg^{2+}, Ba^{2+}, Sr^{2+}, La^{3+}, Na^+, K^+ and Cs^+ were measured, and the observed maximum currents (capacitive currents are included) near zero to -0.2V were plotted as a function of added ions (Figure 6). Trivalent ion La^{3+} was found to be most sensitive; the sensitivity of the alkali earth metal ions is: $Ba^{2+}>Ca^{2+}>Sr^{2+}>Mg^{2+}$. Addition of Na^+, K^+ and Cs^+ did not influence the voltammograms.

When redox compound, 2, was used as a marker, the addition of Ca^{2+} caused a decrease in the faradaic currents of the voltammograms. Figure 7 shows results for two different monolayer electrodes of 1: one is prepared by 40 min dipping and the other by 15 hrs dipping in a 1 mM solution of 1 in ethanol. Figure 8 shows the

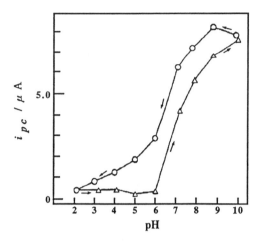

Figure 3. Cathodic peak currents of cyclic voltammograms of a **1** monolayer electrode as a function of pH. The pH profile for increasing pH ($\triangle$) is from the voltammograms given in Figure 2. Experimental conditions are the same as in Figure 2.

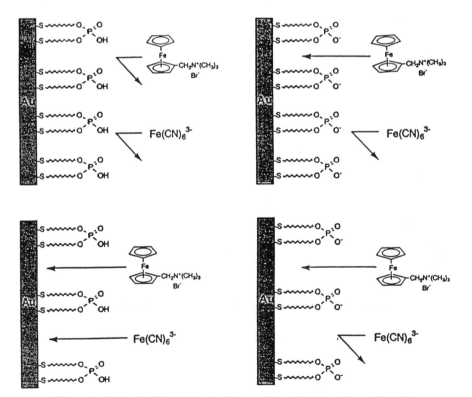

Figure 4. Schmatic illustration for the reaction of ferricyanide and **2** at the monolayer membrane / electrode interfaces.

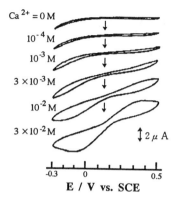

Figure 5. Influence of Ca^{2+} addition on cyclic voltammograms of a **1** monolayer electrode. The solutions are 2 mM $K_3Fe(CN)_6$ in 0.1M KCl (pH 9.0). Scan rate, 200 mV / s. The monolayer electrode was fabricated by dipping a gold electrode for 3 min in 10^{-5}M solution of **1** in ethanol.

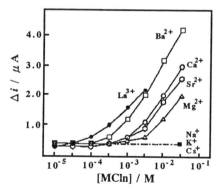

Figure 6. Currents of voltammograms of a **1** monolayer electrode as a function of added ion concentration. Experimental conditions are the same as in Figure 5. Details, see the text.

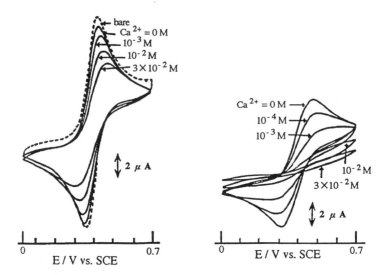

Figure 7. Influence of Ca^{2+} addition on cyclic voltammograms of **1** monolayer electrodes prepared by 40 min (left) and 15 hrs (right) dipping of a gold electrode in a 1mM solution of **1** in ethanol. The solutions are 2 mM of **2** in 0.1M KCl. pH 9.0. Scan rates, 200mV / s.

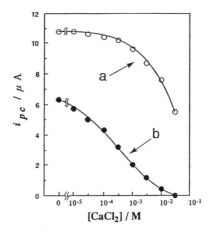

Figure 8. i_{pc} vs. Ca^{2+} concentration for the data in Figure 7. a: Electrode dipping for 40 min, b: electrode dipping for 15 hrs.

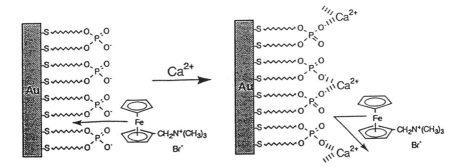

Figure 9. Possible molecular mechanism for the ion-induced voltammetric response at the lipid monolayer / electrode interface.

cathodic currents as a function of the concentration of Ca^{2+}. It is evident that the responses to Ca^{2+} of the electrode formed by dipping for a longer time in a lower concentration (10^{-5} M) in comparison with the short time dipping electrode.

A possible molecular mechanism for the ion responsive change in voltammogram is shown in Figure 9. Chelation of the phosphate of the monolayer with Ca^{2+} decreases the negative charge density of the monolayer surface, which leads to the decrease in electrostatic attraction between the cationic marker and the phosphate monolayer on gold. On the other hand, Ca^{2+} binding causes a decrease in the electrostatic repulsion between the phosphate and ferricyanide; the packing density of the monolayer, in this case, is very loose and the electroactive ion can permeate into the monolayer. The effect of the conformational change of the monolayer induced by the addition of the ions would be needed to consider, but at the present stage, the contribution of this factor to the change in the voltammograms is unknown.

Concluding Remarks

The ion gated lipid monolayers having different surface coverages which were responsive to pH and ion stimulus were formed on gold electrodes *via* a simple chemisorption method. Stability and delicate modulation of the lipid monolayer assembly on the gold electrodes would also be an interesting aspect of application.

References

1. For reviews see: J. D. Swalen, D. L. Allara, J. D. Andrade, E. A. Chandross, G. Garoff, J. Israelachvili, T. J. McCarthy, R. Murray, R. F. Pease, J. F. Rabolt, K. J. Wynne and H. Yu, *Langmuir*, **1987**, *3*, 932 ; A. Ulman, *"An Introduction to Ultrathin Organic Films: From Langmuir-Blodgett to Self-Assembly"*, Academic Press, Boston, MA, pp 237-304, 1991.

2. N. Nakashima, Y. Takada, M. Kunitake and O. Manabe, *J. Chem. Soc., Chem. Commun.*, **1990**, 845.
3. N. Nakashima, T. Taguchi, Y. Takada, K. Fujio, M. Kunitake and O. Manabe, *J. Chem. Soc., Chem. Commun.*, **1991**, 232.
4. N. Nakashima, K. Abe, T.Hirohashi, M. Kunitake and O. Manabe, *Chem. Lett.*, **1993**, 1021.
5. S. Ohnishi, *Adv. Biophys.*, **1975**, *8*, 35; R. Nayar, L. D. Mayer, M. J. Hope and P. R. Cullis, *Biochem. Biophys. Acta*, **1984**, *777*, 343.
6. N. Nakashima, R. Ando, H. Fukushima and T. Kunitake, *Chem. Lett.*, **1985**, 1503 and references cited therein.
7. H. Trauble, M. Teubner, P. Wooley and H. Eibl, *Biophys. Chem.*, **1976**, *4*, 319.
8. J. Wang, L. M. Frostman and M. D. Ward, *J. Phys. Chem.*, **1992**, *96*, 5224.
9. M. Sugawara, K. Kojima, H. Sazawa and Y. Umezawa, *Anal. Chem.*, **1987**, *59*, 2842.
10. S. Steinberg, Y. Tor, E. Sabatani and I. Rubinstein, *J. Am. Chem. Soc.*, **1991**, *113*, 5176.
11. S. Steinberg and I. Rubinstein, *Langmuir*, **1992**, *8*, 1183.

RECEIVED March 25, 1994

Chapter 14

Molecular Modeling and Chemical Sensor Response

Michael Thompson, M. Donata Frank, and David C. Stone

Department of Chemistry, University of Toronto, 80 St. George Street, Toronto, Ontario M5S 1A1, Canada

The rational design of selective receptor sites on transducer surfaces is an important goal in chemical sensor technology. A potentially valuable tool is the semiempirical molecular orbital calculation of interaction energies, particularly involving hydrogen bonds, for surface-bound molecular entities of a highly simplified format with gas phase probes. The strengths and limitations of this approach with respect to compound basis sets, parameterization, choice of method etc. are discussed. The ultimate goal of "designer" surfaces for particular analyte molecules based on the foundations offered by MO calculations is clearly a daunting task.

A crucial goal in chemical sensor science is the incorporation of molecular recognition into a device-chemistry structure to produce a sensor that is both selective and sensitive (1). Such recognition has generally been based on empirical grounds involving existing chemistry which is known to exhibit some degree of selectivity to certain species or by using adsorptive films to trap gas phase probes. Success using these approaches has been somewhat limited, accordingly, strategies to design chemical sensors rationally is an area of prime importance. A distinct possibility here is the use of molecular modelling techniques based on molecular orbital calculations for structure determination (2). These methods can be employed in two ways, one being calculations of interaction energies for specific analyte-receptor combinations for comparison with experimental response, and the second is the design of receptor sites for particular analytes. The latter, although of course desirable, constitutes a field of considerable difficulty. There has been little or no "designer" chemistry incorporated into chemical sensor technologies.

Our group has made use of the surface acoustic wave (SAW) device in combination with gas chromatography as a tool for studying surface-gas phase probe interactions (3). The results of these experiments have been compared with calculated interaction energies for the same probes with highly simplified "model" surfaces. A successful preliminary study involving this type of approach was that

0097–6156/94/0561–0155$08.00/0

concerning the selective binding of nitrobenzene derivatives to surfaces (sensor) treated with (aminopropyl) triethoxysilane (4). The theoretical calculations by the AM1 semiempirical MO method indicated that differences in hydrogen bond strength and dipole moment of the nitrobenzene-type compound were responsible for the observed selectivity.

In the present paper, we turn our attention to a critical evaluation of MO procedures for the study of "docking" of specific analyte molecules to mythical receptor sites. Particular attention is paid to interactions largely founded on hydrogen-bond interactions.

Problems and Perspectives

Any computational study of a chemical problem begins with a search for possible molecular structures through one of three distinct theoretical methods: molecular mechanics, semiempirical or ab initio methods. The suitability of a computational method for a particular problem is partly determined by the size of the molecular system. Large systems such as biological macromolecules can be treated only with molecular mechanics methods, which are based on a classical-mechanical model of molecular structure. Smaller molecules may be approached with the more rigorous and time consuming quantum mechanical methods including ab initio methods for small molecules and semiempirical methods for molecules too complex to be treated using ab initio methods. Each technique has its uses and limitations, however the semiempirical methods offer the most versatility and for this reason several aspects of their use will be discussed.

Semiempirical methods are readily accessible to the public in the form of a program named MOPAC which is available through the Quantum Chemistry Program Exchange. A general introduction to MOPAC has been published recently by the author of MOPAC (5) in which it has been described as a "general-purpose, semiempirical molecular orbital program for the study of chemical reactions involving molecules, ions, and linear polymers. It implements the semiempirical Hamiltonians MNDO, AM1, MINDO/3, and MNDO-PM3, and combines the calculations of vibrational spectra, thermodynamic quantities, isotopic substitution effects, and force constants in a fully integrated program. Elements parameterized at the MNDO level include H, Li, Be, B, C, N, O, F, Al, Si, P, S, Cl, Ge, Br, Sn, Hg, Pb, and I; at the PM3 level the elements H, C, N, O, F, Al, Si, P, S, Cl, Br, and I are available. Within the electronic part of the calculation, molecular and localized orbitals, excited states up to sextets, chemical bond indices, charges, etc. are computed." Descriptions of molecular mechanics and ab initio methods have also been published (6).

MOPAC calculations can not supersede experimentation. Instead, they offer models of chemical reality that allow the experimentalist to rationalize patterns in observed data, and to gain insight into the chemical phenomena involved in experiments. Semiempirical methods use experimentally determined parameters to evaluate certain integrals approximately, rather than using the ab initio approach of evaluating the integrals numerically. This simplification reduces computational expense but also reduces the universality of the calculations in that the quantitative data obtained from the calculations will only be as good as the data it draws on. The predictive power of semiempirical methods relies on similarities between new molecules and those molecules used as standards within the semiempirical technique. Accordingly, a search of the types of molecules accurately reproduced by a particular method is required before confidently beginning calculations on a new system.

From the perspective of chemical sensor development, computational chemistry would be useful as a tool for investigating the mechanism of interaction between analyte molecules and a sensor surface, and the question that naturally arises is whether or not MOPAC may be used to model these interactions. In applying semiempirical methods to the problem of weak intermolecular interactions on sensor surfaces, the problem is made computationally manageable by extracting elements of interest from the sensor surface. This reduction of the surface-analyte interaction to a hypothetical interaction between two molecules yields insight into the interaction but also presents problems in interpretation. Any calculational method applied to this problem of intermolecular interactions must meet several requirements. The method must accurately reproduce the molecular geometries both for the monomers and for the hydrogen bonded complexes, and must also calculate reasonable values for the relative energies of the monomers and the complexes.

Parameterization

Semiempirical techniques are based on a quantum mechanical framework in which some terms are neglected and approximations are used to evaluate other terms. The approximations involve the use of parameters determined from experiments, giving rise to the name semiempirical. The earlier methods such as MNDO/3 (7) and AM1 (8) used an outdated parameter optimization method which was slow and wieldy, and allowed only a limited number of representative experimental data to be exploited. In PM3 (9), a novel optimization routine uses a pool of 400 to 500 reference data to define the values of the parameters. This vastly improved optimization procedure for PM3 results in an overall 40% decrease in ΔH_f errors relative to AM1 (5). Stewart's review of MOPAC includes a series of tables summarizing heats of formation, geometries, and other properties of molecules obtained with MNDO, AM1 and PM3. Experimental results are also included when available, giving some insight into the merits and weaknesses of these methods. It is true that the PM3 method is a vast improvement over its predecessors in predicting heats of formation, however it is also true that all of the methods including PM3 produce errors in ΔH_f that vary for different classes of compounds. For example, the average difference between the experimental and calculated ΔH_f for the group of compounds containing n-ethane, n-propane, and n-butane is 1.5 kcal/mol, while for the group containing 1,2-dinitroethane, 1,3-dinitropropane, and 1,4-dinitrobutane it is 4.8 kcal/mol (5). The difference between experimental and calculated ΔH_f for individual compounds can be quite large, which indicates that one can only compare compounds of the same functional group class. The calculations can only be useful when these limitations are kept in mind.

Choice of Method

The choice between the various methods available within MOPAC is left entirely up to the investigator. The issue of parameterization has been discussed above and it has been pointed out that the parameterization for PM3 is the most extensive, resulting in the most accurate heats of formation for most organic compounds. The next question concerns the ability of the methods in MOPAC to reproduce hydrogen bonding. Hydrogen bonding interactions are not successfully modelled by all the various methods, and present a serious limitation in the use of MOPAC for studying intermolecular interactions. The MNDO method completely fails to reproduce hydrogen bonding. The AM1 method was developed to remedy this failure, however regular use of the AM1 method in studying hydrogen bonded complexes exposes its deficiencies in this regard. One problem is its prediction of bifurcated hydrogen bonds between two neutral molecules in instances where experiment has indicated a linear hydrogen bond, as is the case for the water dimer (*10*). A bifurcated hydrogen bond contains several H atoms hydrogen bonded to a single heavy atom, and is favoured by AM1 when two or more hydrogen bonding sites are available, see Figure 1. PM3 successfully predicts the expected linear structure of a hydrogen-bonded water dimer, however, the predicted oxygen-oxygen distance of 2.73Å (10) is shorter than the predicted ab initio oxygen-oxygen distance of 3.00Å. It was found that the difference between the hydrogen bonding geometries predicted by AM1 and PM3 arises because the AM1 method does not allow significant charge transfer from donor to acceptor molecules (*10*). The upshot of this work is that none of the methods can accurately model the geometry of hydrogen bonds, although the PM3 method does reproduce linear hydrogen bonds with more success than AM1.

In spite of the bad news about geometry predictions in hydrogen bonds, the calculations can be useful in examining relative interaction energies. The interaction energy for the complex is the difference between the ΔH_f for the hydrogen bonded complex and the sum of ΔH_f for the monomers:

$$E_i = \Delta H_f(\text{hydrogen bonded complex}) - \Sigma[\Delta H_f(\text{monomers})].$$

The reality is that there is no semiempirical technique which is reliable in all situations, and a knowledge of the strong and weak points of the various methods is required in order to decide on the best method for a particular task.

Input Geometry

The final calculated geometry of the hydrogen bonded complex is quite sensitive to the initial geometry used in the input file and the use of a consistent initial geometry for the hydrogen bond is suggested in order to obtain any consistency in the results. An added concern when examining an output file is whether or not the predicted geometry makes sense. The example of bifurcated hydrogen bonds appearing in AM1 results points to a need for care in assessing the output file.

Validity of Model

A final comment concerns the model nature of these calculations. The calculations themselves are based on models of chemical structure and are calculations of properties in the gas phase. In the instance of interactions between a surface and a molecule the model is further reduced by excluding the surface. An example of this is given in Figure 2: on the left, a methanol molecule is hydrogen bonded to part of a polymer. The dimer on the right is the structure that one might use as a model of the system for computational purposes. There is no way of using semiempirical methods to calculate the influence of the surface on the interaction. Some surfaces are designed for and used to bind molecules from a solution, and once again there is a large difference between the gas phase model obtained with MOPAC and the real thing. It appears that a bleak picture is being painted, but the intention is not to dissuade one from using MOPAC rather it is to produce a heightened sense of the pitfalls that exist when doing theoretical calculations.

Application to Chemical Sensors

As mentioned in the introduction, the first application of molecular modelling to sensor response by this group was the use of AMI to model hydrogen bonding interactions between nitoaromatic compounds and surface comprising of amino functional groups (4). In this case the method was applied as an interpretive tool for understanding the principle mechanism involved in determining the relative selectivity obtained for the nitroaromatic compounds over an admittedly limited range of other molecules. Subsequent work (Stone, D.C.; Thompson, M. unpublished results), has extended the range of compounds studied using the SAW-GC system described elsewhere (3). This included measurements of sensor response at both room and elevated (80°C) temperatures. For consistency, the same AMI calculation procedure was employed as previously: only one (acetonitrile) of the twenty-one compounds studied gave rise to a bifurcated hydrogen bonded complex, and so was omitted from the analysis. Interestingly, when a comparison was made between sensor response and calculated E_i values across the twenty-one compounds at both temperatures, a distinct pattern emerged. It was found that the best correlations between sensor response and E_i values were obtained for similar classes of compounds e.g. alcohols, hydrocarbons, nitro-aromatics, etc. In the light of the effects of parameterization noted earlier, this is not a surprising observation. Clearly, better parameterization (as assessed by broader agreement in ΔH_f values between experiment and calculation) would allow a more confident comparison of theoretical results for different classes of compounds. A recent study of the response of polymer-coated SAW sensors (11) helps to illustrate this point further: simple models for the solubility of the analyte in the polymeric film based on a simple boiling point model provided good correlation only when the results were grouped by compound type, while a more complex linear solvation energy relationship (LSER) model incorporating terms for polarizability, dipole moment and hydrogen bond acceptance/donation better modeled the data without grouping.

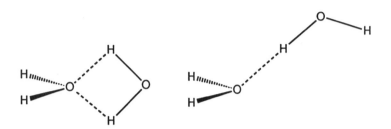

Figure 1.   Structures of water dimer predicted by AM1 (left) and PM3 (right).

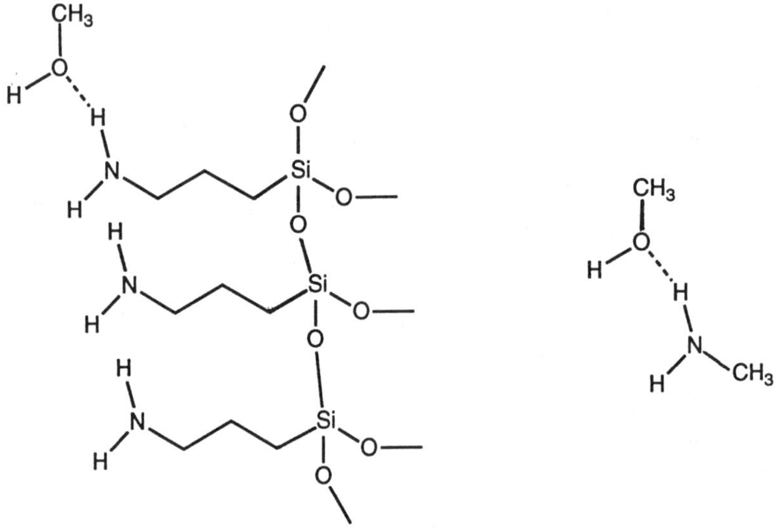

Figure 2.   Left: methanol hydrogen bonded to part of polymer surface.   Right: model of the interaction used in computation

Conclusion

We have described semiempirical MO calculation methods in general terms, with particular emphasis on the effects of parameterization and input geometry as they affect the validity of calculations from the perspective of modelling chemical sensor response. Such techniques can be applied in both an interpretive and predictive manner, provided that the strengths and weaknesses of the chosen method are clearly understood. Correlation with experimental data provides a valuable check for the relevance of a method to a particular situation. Although the application of such methods to the rational design of selective chemical sensors remains to be demonstrated, a better understanding of the issues involved is a step toward this goal.

Acknowledgment

We are grateful to the Federal U.S. Aviation Adminstration for support of this work.

Literature Cited

1. Thompson, M.; Frank, M.D.; Heckl, W.M.; Marassi, F.M.; Vigmond, S.J. In *Chemical Sensor Technology*; Seiyama, T., Ed.; Kodansha, Ltd., Tokyo, 1987, Vol.2; pp 237.
2. Thompson, M.; Stone, D.C. In *Sensors and Sensory Systems for an Electronic Nose*; Gardner, J.W.; Bartlett, P.N., Eds.; Kluwer Academic Publ.; Dordrecht, The Netherlands, 1992, p.25.
3. Thompson, M.; Stone, D.C. *Anal. Chem.* **1990**, *62*, 1895.
4. Heckl, W.M.; Marassi, F.M.; Kallury, M.R.K.,; Stone, D.C.; Thompson, M. *Anal. Chem.* **1990**, *62*, 32.
5. Stewart, J.P. *J. Computer-aided Mol. Design*; 1990, Vo. 4, 1-105.
6. Clark, T. *A Handbook of Computational Chemistry*; John Wiley and Sons, New York, NY, 1985; ISBN: 0-471-88211-9.
7. Bingham, R.C.; Dewar, M.J.S.; Lo, D.H. *J. Am. Chem. Soc.* **1975**, *97*, 1285.
8. Dewar, M.J.S.; Zoebisch, E.G.; Healy, E.F.; Stewart, J.J.P. *J. Am. Chem. Soc.* **1985**, *107*, 3902-3909.
9. Stewart, J.J.P. *J. Comp. Chem.* **1989**, *10*, 209.
10. Jurema, M.W.; Shields, G.C. *J. Comp. Chem.* **1993**, *14(1)*, 89-104.
11. Patrash, S. J.; Zellers E. T. *Anal. Chem.* **1993**, *65*, 2055-2066.

RECEIVED April 5, 1994

Chapter 15

Enhanced Metal Nucleation with Self-Assembled Monolayers of ω-Substituted Silanes Studied by Scanning Force Microscopy

David J. Dunaway and Robin L. McCarley

Department of Chemistry, Louisiana State University, Baton
Rouge, LA 70803

Scanning force microscopy was used to note the chemical
sensitivity of Au nucleation on various silane monolayers. Thin
(1-10 nm) evaporated Au films deposited onto monolayers of
$(CH_3O)_3Si(CH_2)_3X$ (X= -SH, -NH$_2$ and -CH$_3$) on SiO$_2$/Si(111)
display morphologies which are dependent on the strength of the
Au-X interaction. Microscopic adhesion was probed using the
SFM tip under high loads. The electrical resistivity of Au films
deposited on the amine-terminated surface was found to be
substantially lower than that for the thiol-terminated or bare
surface.

Of fundamental interest to many chemists working with thin films on well-defined
surfaces is the nature of the interaction between the surface and the thin film.
This is of particular importance to those involved with organothiols on Au (*1-8*),
alkanoic acids on metal oxides (*9-11*) and silanes on SiO$_2$ surfaces (*2, 12-14*).
Much of the research has focused on understanding the structure of these self-
assembled monolayers on the given substrate and correlating that information
with the macroscopic properties of the thin films (*1-14*). Other areas have been
directed at applications of the monolayer films to microelectronics or molecular
electronics (*2*). Our interest in the nature of the adsorbed layer/substrate
interaction comes from our current research dealing with the fabrication of
nanometer-sized structures using alternating metal and insulator (organic) layers.
We plan on utilizing this layered method for the fabrication of electrodes with
nanometer dimensions. Thus is it imperative that we understand the quality of the
adsorbate bond to the substrate under investigation.

Scanning tunneling (*15*) (STM) and scanning force microscopies (*16*)
(SFM) have had a substantial impact on the thin films area and have been used
most often to obtain information regarding the structure and orientation of
monolayers on substrates. SFM has been used to observe the molecular
arrangement of various Langmuir-Blodgett films on Si surfaces (*17-18*) and
perform nanofabrication of such thin films (*19*). There have been recent reports
of STM and SFM imaging of organothiols on Au(111) with molecular resolution
(*4,7,8*). Fundamental to all of these studies concerned with organic films on
metals is defining the chemical and physical nature of the metal-organic interface.

0097–6156/94/0561–0162$08.00/0

Our approach to this issue, in the case of an adsorbate on a metal, is to reverse the situation and think about how the metal interacts with a thin film of the adsorbate. This can be accomplished by evaporation of a metal onto a monolayer bearing the terminal group of interest; the monolayer is covalently bound via another functionality to a suitable substrate. It is well known that the electrical and morphological properties of thin metal films on insulators are strongly affected by the nature of the insulator (20-21), and a few studies have shown that organic modifiers can enhance metal nucleation (22-27). If little chemical interaction occurs between substrate and metal or metal self-diffusion is too high, the metal will be very mobile and tend to form many large crystallites on the surface. Such films are rough and discontinuous at thicknesses as large as 10 nm, i.e. the Volmer-Weber growth model (21). Thus, information regarding the chemical interaction between a particular functional group and a given metal can be easily assessed from evaluation of the metal morphology using SFM or STM. Such information is important not only to chemists studying adsorbed organic layers on metal surfaces, but to those investigating the physical properties of thin metal films. This is evidenced by the number of publications concerning the applications of thin metal films in areas including catalysis (28), electronic device fabrication (29), microscopy (30), electrochemistry (31) and chemical sensors (32).

The use of molecules which can react with substrates and metals in order to form a well-defined *chemical* mode of attachment of the metal onto the insulator would be useful in vacuum deposition techniques. Organic layers containing two groups capable of forming chemical bonds with metal and inorganic surfaces have recently been implemented as "molecular adhesives" for vacuum-deposited metals (22-27). An example of such a bifunctional molecule is $(CH_3O)_3Si(CH_2)_3SH$, which polymerizes onto oxide containing surfaces such as SiO_2 through the silane end and can then bind metals such as Au, Ag or Cu on the thiol "surface" (22).

Condensation   Metal Deposition   Metal

Allara has demonstrated, using transmission electron microscopy (TEM), that Au films produced with a cystamine adhesive were very smooth in the 5-10 nm range of film thickness (24). Electrical continuity was shown to be a function of crystallite size; the films formed on the thiol surface exhibited smaller crystallites due to the inhibition of adatom diffusion by the anchoring agent. The roughness of the Al_2O_3 substrates was not considered in the discussion of Au film surface roughness (24), but certainly should affect the nature of the Au film. Limitations in the resolution of the electron microscopy used prevented characterization of such films less than 5 nm thick.

The advent of SFM and STM has allowed the study of conductors and insulators in a variety of environments with ultra-high resolution, yielding information not previously available with other microscopies. There has been only one report of SFM studies concerned with chemically enhanced nucleation of metals; that particular investigation (27) dealt with the deposition of Cu onto carboxylic-acid-terminated organothiols on Au(111). With the large variety of molecular adhesives applicable to metals such as Au, Ag, Cu, Al or Pt, investigations using SFM and STM should provide a wealth of information regarding the relationship between substrate surface chemistry and metal film properties.

Experimental

Cleaning SiO₂/Si Substrates. n-type Si(111) and Si(100) wafers (Virginia Semiconductor, Fredricksburg, VA) with a misorientation $\leq 0.25^{\circ}$ and resistivity of 0.06 Ω cm were cleaved into appropriate-sized pieces. The Si was then cleaned in 1:4 30% H_2O_2:98% H_2SO_4 at 75°C for 20 minutes (*Caution: These solutions are highly oxidizing and should be handled with extreme caution!!*), rinsed with copious amounts of 18 MΩ cm water and then blown dry with high purity N_2. Samples were then modified with the desired alkoxysilane by one of several methods. Those samples which were used in electrical measurements were heated at 1000°C for 2-3 h in order to grow a thick layer (60-100 nm) of insulating SiO_2.

Preparation of Silane-Modified SiO₂/Si Substrates. Silane layers were formed on the silicon oxide surfaces by a heated solution preparation, a solution "dipping" or a vapor-phase method. Briefly, the heated solution preparation involved placing the SiO_2/Si in a boiling solution of isopropanol, water and the alkoxysilane for 8 minutes, followed by isopropanol rinses and a 100°C anneal for 10 minutes. This was repeated three times in order to ensure complete coverage of the SiO_2/Si surface (*22*). The solution "dipping" involved placing the freshly cleaned SiO_2/Si in toluene solutions of 10^{-3} M alkoxysilane for 2-5 minutes, followed by toluene rinses and drying in N_2. Vapor-phase depositions were carried out by suspending the SiO_2/Si surfaces 2-3" above a refluxing ~8 % (v/v) silane/toluene solution in a closed system fitted with a condenser for 4 hours (*33*). Samples were then removed, rinsed with toluene and dried in a stream of high purity N_2.

Instrumentation. Metal evaporations were carried out in a cryogenically pumped vacuum system with a base pressure of 5 x 10^{-7} torr (Edwards Auto 306A). Au (99.99%) was thermally evaporated at a rate of 0.1 Å s^{-1} onto the Si substrates which had been affixed to a large aluminum holder with Ag paste (SPI Products). The temperature during the depositions, measured by a thermocouple imbedded in the aluminum holder, did not rise more than 1°C. Thicknesses were monitored with a quartz crystal microbalance. Once evaporations were complete, the chamber was back-filled with high purity N_2 and the samples immediately transferred to the SFM for imaging. Au(111)/mica substrates were prepared as previously described (*30*).
      SFM images were obtained with a Nanoscope III Scanning Probe Microscope (Digital Instruments, Santa Barbara, CA). Cantilevers with pyramidal Si_3N_4 tips with a force constant of 0.58-0.06 N m^{-1} were used throughout the work discussed here. Cantilevers were force calibrated on the sample of interest by measuring the slope of the tip deflection vs. tip-sample separation curves. The minimum amount of force necessary for reproducible imaging was used in order to avoid damage to the sample. In the cases where microscopic adhesion testing was performed, the tip/sample force was increased until substantial motion of crystallites occurred. The radius of curvature for the tips, R_t, was obtained by imaging atomic step edges (of height H) on Au(111)/mica surfaces and measuring the lateral distance traveled by the tip in traversing the step, L (*34*). R_t can be found from Equation (1):

$$R_t = (L^2 + H^2)(2H)^{-1} \tag{1}$$

Images presented here were low pass filtered and if necessary, flattened using the Nanoscope filtering software. All images were obtained in air.

Ellipsometric measurements were made with a Rudolf 437 ellipsometer equipped with a rotating analyzer detector and 488.0 nm (Ar$^+$ laser) light. The angle between the incident beam and surface normal was fixed at 70°. All data were obtained in air and represent the average of five measurements on at least three samples. A film refractive index of 1.45 was used in the calculation of silane film thickness.

Electrical measurements were performed using either a two- or four-point-probe technique. A mask with 3 mm wide slots 3 mm apart was used to define four contact fingers during deposition of 100 nm of Au on the various surfaces. The mask was removed and copper wires were attached to the pads at the end of the fingers using silver paint (Ted Pella, Redding, CA). After curing at room temperature for 30 minutes, the sample was placed in the evaporator on an aluminum stage, and external contact to the copper wires was made by electrical feedthroughs. For the four-point method, a PAR 273 Potentiostat/Galvanostat was used as a current source, and the voltage was monitored with a Fluke 8060A digital voltmeter. Current was passed through the outer fingers, and the voltage drop across the inner fingers was measured. The four-point method was used to measure the contact resistance of thick Au films. We have determined the contact resistance for our apparatus to be approximately 1-5 Ω, using 18 nm Au films. We assumed that this contact resistance is constant throughout the measurement and would only dominate the measured resistance values of fairly thick Au films. Thus, we used a two-point method to measure the resistance of the very thin films. The two-point method involved using the high impedance meter to measure the resistance across two of the Au fingers during deposition of the thin Au films. Resistivities were calculated using the geometric measurements of the fingers and the experimentally determined resistances.

Reagents. All solvents were HPLC grade or better and were used without further purification. The silanes were used as received or vacuum distilled and stored at -20°C until used. Sulfuric acid 98% (Fisher) and hydrogen peroxide 30% (Fisher) were used for preparation of cleaning solutions. All other chemicals were reagent grade or better.

Results and Discussion

SiO$_2$/Si Substrates. Shown in Figure 1 is a 2 μm x 2 μm constant-force SFM image of a clean Si(111) wafer which has approximately 2 nm of native oxide on the surface. Mechanical polish marks approximately 50 nm wide were observed on all samples; these are the "lines" running diagonally in Figure 1. There were no noticeable effects of scanning the surface with the SFM tip at forces as high as 200 nN. We will denote this surface as SiO$_2$/Si from this point on. The surface is extremely flat and devoid of any defects over large scan areas. We have obtained a roughness of ±0.2 nm for these substrates and have observed only slight variations when using wafers from another batch of single crystal Si(111). We have observed a surface roughness of ±1.0 nm for Si(100) surfaces prepared in the same manner and attribute it to the cleaning/polishing procedure used by the manufacturer. The roughness quoted here is a rms value obtained from several scan areas on various wafers. Thus, we opted to use the flatter Si(111) for our studies because the expected changes in topography for the Au films should be less than 1 nm.

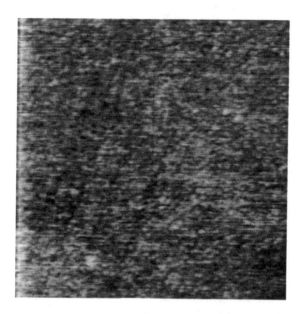

Figure 1. 2 μm x 2 μm constant force SFM image of bare $SiO_2/Si(111)$. Z-range is 5 nm.

Silane-Modified SiO_2/Si Substrates.

Solution Treatments. Our first attempts at forming stable silane monolayers using the solution treatments were unsuccessful, but educational. The heated solution treatments (referred to as "refluxed" from now on) yielded smooth, homogeneous films of the mercaptopropylsilane (MPS) which were measured to be 2.7±0.2 nm, or approximately 3-4 monolayers thick. SFM images of MPS modified silicon surfaces displayed topographies almost identical to bare $SiO_2/Si(111)$. We never observed tip-induced damage during imaging at forces as high as 200 nN. Similar thickness values were obtained for aminopropylsilane (APS) but with larger scatter, indicating the presence of clumps on the surface. This was confirmed by inspection of SFM images of the APS on $SiO_2/Si(111)$, Figure 2. Alkoxysilanes are known to rapidly form polymeric species in the presence of water, so this is not surprising (*35*). In the case of APS, the polymerization reaction is self-catalyzed by the amine group. We have no evidence that would indicate catalysis by the thiol group of MPS.

Dipping the $SiO_2/Si(111)$ in 10^{-3} M MPS/toluene resulted in monolayer thick (0.8±0.1 nm) films of the MPS which were as smooth as the original $SiO_2/Si(111)$ surface. The "dipped" films were not stable under SFM scanning - holes could be cut into the surface using relatively low tip-sample forces (30-50 nN). No damage was observed when the refluxed MPS samples were imaged under high forces (150-20 nN). We suspect that the dipped MPS films had laterally cross linked, but the silanols had not reacted with the $SiO_2/Si(111)$ surface.

Scotch Tape adhesion tests of 50 nm Au films on the MPS dipped samples revealed complete removal of the Au. Similar adhesion studies of the refluxed MPS samples demonstrated no removal of Au. We feel that the multilayers

Figure 2. Constant force SFM image of $(CH_3O)_3Si(CH_2)_3NH_2$ reflux-treated $SiO_2/Si(111)$. Scan size is 900 x 900 nm and Z-range is 5 nm.

formed during the reflux treatment were strongly bonded to the $SiO_2/Si(111)$, due to enhanced surface Si-O-Si bond formation during the heating. With this thought in mind, we annealed the dipped MPS samples for 20 minutes at 100°C before Au deposition. The annealing did not improve the adhesion, as noted by removal of the entire Au film in the tape tests. In addition, exposing the $SiO_2/Si(111)$ to water vapor immediately before dipping in the MPS/toluene did not prevent adhesive failure. Such a pretreatment has been reported to enhance adhesion of silanes to oxide surfaces (*24*). We feel that an insoluble cross-linked monolayer is formed during the MPS dips which is not easily removed with solvent rinses, but can be mechanically dislodged (M. Wirth, University of Delaware, private communication, 1993). With the above information in hand, we turned to a vapor-phase deposition protocol.

Vapor-Phase Depositions. Monolayer films of alkoxysilanes on SiO_2/Si can be prepared by exposing SiO_2/Si surfaces to the hot vapors of the silane in toluene for at least four hours, as long as the silane is sufficiently volatile (*33*). Films of MPS on $SiO_2/Si(111)$ prepared in this manner (referred to as vapor films) were found to be as smooth as the underlying substrate and did not display any polymeric features, Figure 3. We have not observed any microstructure attributable to the MPS monolayer. Ellipsometric analysis of these films gave thicknesses indicative of monolayers (0.6-0.8 nm). This assumes a monolayer thickness of 0.7-0.8 nm. The other ω-substituted silanes deposited from the vapor exhibited similar behavior.

In contrast to the dipped films, there was no damage to vapor-phase films during SFM imaging, even at forces as high as 150-200 nN. In addition, 50 nm Au films deposited on the MPS vapor films showed no loss of Au during Scotch Tape testing. The vapor-phase deposition of alkoxysilanes apparently causes cross linking and strong surface binding of the monolayer.

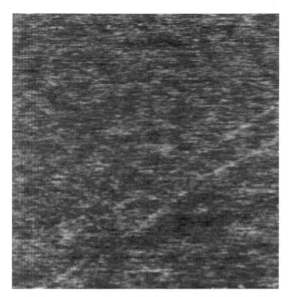

Figure 3. 1 μm x 1 μm constant-force SFM image of $(CH_3O)_3Si(CH_2)_3SH$ vapor-treated $SiO_2/Si(111)$. Z-range is 5 nm.

Characteristics of Au Films on Silane-Modified SiO_2/Si Substrates. In order to probe the interactions between the ω-substituent of the silane and Au, we investigated the morphology of ultra-thin Au films (0.2-10 nm) deposited on the silane-modifed SiO_2/Si substrates. We also probed the microscopic adhesion of Au on the modified surfaces by watching for damage to the thin Au film as the force was increased between tip and sample during SFM imaging experiments. Macroscopic adhesion tests were carried out using Scotch Tape. In addition, electrical resistivity measurements were carried out to note the onset of electrical continuity.

Morphology of Au on Vapor-Phase Silane-Modified SiO_2/Si Substrates. Shown in Figure 4 are SFM constant-force images of 2 nm Au films vacuum-deposited on $SiO_2/Si(111)$ with A: no silane, B: $(CH_3O)_3Si(CH_2)_3SH$, C: $(CH_3O)_3Si(CH_2)_3NH_2$ and D: $(CH_3O)_3Si(CH_2)_3CH_3$. There is a noticeable difference in Au crystallite size on the bare $SiO_2/Si(111)$ in comparison to the amine and thiol modified surfaces. We routinely find 13-16 nm crystallites of Au deposited on freshly cleaned $SiO_2/Si(111)$ surfaces. It is well-known that the difference in surface energies for Au and SiO_2 is quite high, and adhesion of Au to SiO_2 is almost non-existent. Both of these factors lead to fairly rapid Au self-diffusion, which results in "islanding" of the Au crystallites. A similar but larger effect is noted for the $(CH_3O)_3Si(CH_2)_3CH_3$ treated surfaces; crystallites of 20-25 nm diameter were observed, indicating little interaction between the Au and methyl surface. Obtaining clear images of thin Au films (< 2 nm) on the methyl and bare SiO_2 surfaces was difficult due to the forces between tip and Au being large in comparison to those between Au and the terminal methyl group of the silane or the oxide, even at the lowest imaging force of 1-2 nN. No attempts were made at reducing the tip/sample forces by imaging under water, for fear of Au film removal.

A.   B.

C.   D.

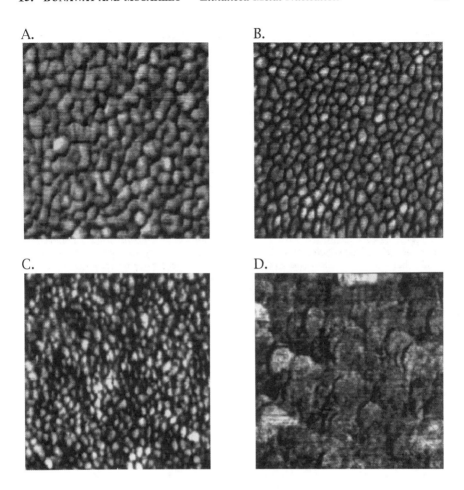

Figure 4. 200 nm x 200 nm constant-force images of 2 nm Au deposited on SiO$_2$/Si(111) with A: no modifier, B: with (CH$_3$O)$_3$Si(CH$_2$)$_3$SH, C: (CH$_3$O)$_3$Si(CH$_2$)$_3$NH$_2$ and D: (CH$_3$O)$_3$Si(CH$_2$)$_3$CH$_3$.

Both the (CH$_3$O)$_3$Si(CH$_2$)$_3$SH and (CH$_3$O)$_3$Si(CH$_2$)$_3$NH$_2$ treated SiO$_2$/Si(111) samples display smaller grain sizes than those on bare SiO$_2$/Si(111), with the grain sizes on the thiol and amine surfaces being 5-9 nm. Such a decrease in grain sizes indicates that surface diffusion of Au on the amine and thiol surfaces is substantially reduced.

There is an abundance of literature concerned with adsorption of thiols and disulfides on Au (*1-8*), and it has been shown that (CH$_3$O)$_3$Si(CH$_2$)$_3$SH can be used as a very effective adhesion promoter for Au (*22*). Organothiols are known to form robust monolayers on Au. Little information regarding the adsorption of amines on Au exists in the literature. Bain has shown that long-chain alkylamines adsorb very poorly on clean Au surfaces (*36*). Ellipsometric and contact angle measurements indicated that the long-chain amines were not well-packed and were easily displaced by alkanethiols. In addition, various amines have been shown to be readily desorbed from Au electrodes (*37*) at potentials near 0 V vs. the sodium chloride calomel electrode (SSCE), whereas thiols were not desorbed (*38*) until at least -1.0 V. Such behavior indicates that amines do not interact strongly with Au. Our results concerning the morphology of Au deposited on amine surfaces, however, points to a fairly strong interaction between Au and the amine. **Microscopic and Macroscopic Adhesion of Au on Silane Monolayers.** In order to compare microscopic and macroscopic adhesion properties of Au on various ω-substituted silane surfaces, we imaged 2 nm Au films at various forces and also used Scotch Tape films on 50 nm Au films to note Au removal. As was discussed earlier, the methyl-terminated and bare SiO_2 surfaces could not be imaged at forces above 1-2 nN. Au films on all of these surfaces were completely removed during the tape tests. In contrast, Au films on both the $(CH_3O)_3Si(CH_2)_3SH$ and $(CH_3O)_3Si(CH_2)_3NH_2$ treated $SiO_2/Si(111)$ samples were unchanged after removal of the tape. Repeated application and removal of tape did not cause any visible damage to the Au. Similar results were obtained when 5 nm of Cr (a commonly used metal adhesion promoter) was used instead of the silanes. It was expected that the Au would adhere very strongly to the thiol surface but not to the amine. A previous study using a polymeric amine as an adhesion promoter indicated that there was substantial interaction between vacuum-deposited Au and the polymer but details concerning the nature of the bonding were not discussed (*23*).

Scanning force microscopy images of 2 nm thick Au films deposited on the amine- and thiol-terminated silane monolayers revealed no difference in microscopic adhesion at high tip-sample loads. We were unable to remove Au deposited on vapor-phase $(CH_3O)_3Si(CH_2)_3SH$ or $(CH_3O)_3Si(CH_2)_3NH_2$ treated $SiO_2/Si(111)$ with forces as high as 150-200 nN but did notice plastic deformation of the Au in the scanned area. We now turn to a calculation which we will use to estimate strengths of Au adhesion on the various surfaces.

Using a simple Hertzian model for a sphere (the tip) pressing a flat surface (the Au film), the radius of the circular contact area can be obtained from (*39*) :

$$R_c^3 = 0.75 \, R_t \, F \left(\frac{1-v_{Au}^2}{E_{Au}} + \frac{1-v_t^2}{E_t} \right) \tag{2}$$

where R_t is the radius of curvature for the tip, F is the tip-sample force, E_{Au} and E_t are the Young's moduli for Au (78.5 GPa) and the tip (200 GPa), and v_{Au} and v_t are the Poisson ratios of the Au (0.42) and tip (0.27) (*40-42*). With the force necessary to remove Au from the bare oxide or methyl-terminated surfaces being 1 nN, a R_c of 0.8 nm is obtained. If one assumes an approximate 3:1 ratio of Au to $(CH_3O)_3Si(CH_2)_3CH_3$ molecules at the interface and knows the number of Au atoms in the (111) plane to be 15 nm^{-2} (*8,43*), then the force needed to remove the Au from the methyl surface can be shown to be approximately 0.1 nN molecule^{-1}. This corresponds to a small Au-methyl interaction strength of roughly 30 kJ mole^{-1}. Similar calculations for the thiol and amine surfaces give a minimum Au-amine and Au-thiol interaction strength of 180 kJ mole^{-1}. With our instrument we are unable to cause adhesive failure of the Au-amine or Au-thiol surface, thus we can only assign an upper limit on these interactions. At this time we are uncertain what effects adhesion has on cohesion and vice-versa. We only quote the numbers here as estimates for the interactions between the various functional groups and Au. The estimated bond energy *(44)* for alkanethiols with Au is 185 kJ mole^{-1}, so it is not surprising that we do not observe adhesive failure. Such a large value for adhesion of Au to the amine surface is unexpected but agrees well with the data from the tape tests and the morphology data. The small Au crystallites on the $(CH_3O)_3Si(CH_2)_3NH_2$ treated $SiO_2/Si(111)$ indicate that there is a very large affinity of Au for the amine surface. We are currently using X-ray photoelectron spectroscopy to characterize the Au-amine interaction.

Electrical Resistivity Studies. Shown in Figure 5 is a plot of resistivity vs. Au film thickness for A: untreated $SiO_2/Si(111)$, the oxide treated with B: $(CH_3O)_3Si(CH_2)_3SH$ and C: $(CH_3O)_3Si(CH_2)_3NH_2$. The untreated surface yields Au films with a high resistivity at Au thicknesses less than 10 nm, but then the resistivity of the film decreases rapidly until it approaches the bulk value of 2 x 10^{-6} Ω cm *(45)*. Such behavior is characteristic of the island-growth mechanism of thin metal films on insulators. Until the grains reach a size which allows contact with other grains, the film will be highly resistive. This is similar to what previous researchers have observed for bare Al_2O_3 surfaces *(24)*.

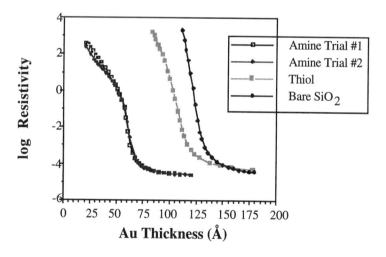

Figure 5. Resistivity plot for various terminated surfaces.

Reduced grain size is confirmed by the decreased onset of electrical continuity for the Au/thiol-terminated surface. We have noted a 3-fold decrease in Au grain size for the thiol terminated surface in comparison to the bare SiO_2. A similar trend was noted for Au evaporated onto Al_2O_3 surfaces treated with 3,3'-dithiodiproprionic acid *(24)*. The Au film thickness at the onset of electrical continuity for the $(CH_3O)_3Si(CH_2)_3SH$ treated $SiO_2/Si(111)$ surfaces is higher than that in the study using adsorbed 3,3'-dithiodiproprionic acid on alumina. We feel that this difference is due to the increased order of the organic disulfide layer in comparison to the organothiol layer investigated here. Such an increase in the order of the film would result in an increased number of nucleation sites. A surface with a large number of nucleation sites would cause a deposited metal film to display a lower resistance than a film formed on a surface with fewer nucleation sites. The more highly ordered disulfide surface (versus the thiol) should then have a lower onset of electrical continuity than the thiol surface.

The Au film thickness at the onset of electrical continuity for $SiO_2/Si(111)$ treated with $(CH_3O)_3Si(CH_2)_3NH_2$ is strikingly lower than that for the bare or thiol-treated surfaces. In fact, the Au films formed on the amine surface are the thinnest, electrically continuous films formed at room temperature that we are aware of. We note a resistivity of 250 Ω cm at 20 Å Au on the amine surface, which is much smaller than the 10^4 Ω cm value for a 30 Å Au film formed on 3,3'-dithiodiproprionic acid-treated alumina (24). It appears that the amine interacts strongly with the Au and reduces surface diffusion, thereby decreasing the Au grain size. But, for such a drastic difference in resistivities to occur, we feel that the number of nucleation sites on the amine surface must be substantially larger. A higher density of nucleation sites would allow growing grains to coalesce and form a continuous Au film at an extremely low Au thickness. As can be seen in Figure 4B, there appears to be a large amount of void space between Au grains for the thiol surface, but there is little for the amine surface. We note a lack of void space for Au on the thiol surface at about 80 Å Au, which is the onset for measurable electrical resistivity. We have performed side-by-side SFM experiments for 20 Å Au films on the various surfaces; *the void space that we observe is not due to tip artifacts.* Thus it appears that the amine-terminated silane is able to cause a decrease in Au surface diffusion and an increase in the number of nucleation sites, which results in a substantially thinner, electrically continuous film. We believe that the increased number of nucleation sites on the amine surface may be due to increased film order brought about by hydrogen bonding between amine groups on the surface. We are currently investigating longer chain thiols and amines to note any differences in nucleation.

The precise nature of the amine-Au interaction is not well understood at this juncture. There is some literature regarding the adsorption of amines on Au, but the observed strength of adsorption was very poor (36, 37, 46). Electron donating substituents on amines have been shown to increase the interaction of the amine with Au surfaces, but the results are somewhat ambiguous due to possible adventitious surface contaminants (46). Au has been shown to act as a catalyst for certain reactions in solution, if and only if the Au is in the form of a very finely divided powder (47). There have been no previous gas-phase measurements of the type of Au-amine interactions described here, and we believe that there may be a reaction between Au atomic vapor and amines which is not found in solution. We are currently investigating the nature of the Au-amine interaction using IR and XPS in order to define the chemical species involved.

Summary

This study has demonstrated that the nucleation and growth of thin Au films can be controlled by tuning the chemical nature of the surface onto which it is deposited. The use of alkoxysilanes terminated with thiol and amine moeities on $SiO_2/Si(111)$ results in small Au crystallite formation during thermal evaporation of Au, but the electrical properties of these films are substantially different at 20 Å film thickness. An increased number of nucleation sites on the amine surface causes onset of electrical continuity at 20 Å, which is attributed to increased accessibility of the terminal groups. Both clean $SiO_2/Si(111)$ and the methyl-terminated silane surface display larger Au crystallites, indicating substantial surface diffusion of the Au. Thus, we have shown that ultra-thin, electrically continuous films of Au can be formed on smooth surfaces with proper control of the surface chemistry of the substrate. Future applications include fabrication of nanoelectrodes, optically transparent electrodes and molecular devices and assemblies.

Acknowledgments

This work was supported by the National Science Foundation (CHE-9221646) and the Louisiana Education Quality Support Fund (LEQSF(1993-96)-RD-A-09). We are grateful to Chris Bell for performing the ellipsometric measurements. Helpful discussions with A. W. Maverick and J. T. McDevitt are acknowledged.

Literature Cited

1.   Whitesides, G. M.; Laibinis, P. E. *Langmuir,* **1990**, *6*, 87 and references therein.
2.   Ulman, A. *An Introduction to Ultra-Thin Films From Langmuir-Blodgett to Self-Assembly*, Academic: San Diego, CA, 1991.
3.   Chidsey, C. E. D.; Liu, G.-Y.; Rowntree, P.; Scoles, G. *J. Chem. Phys.* **1989**, *91*, 4421.
4.   Widrig, C. A.; Alves, C. A.; Porter, M. D. *J. Am. Chem. Soc.* **1991**, *113*, 2805.
5.   Bain, C. D.; Troughton, E. B.; Tao, Y.-T.; Evall, J.; Whitesides, G. M.; Nuzzo, R. G. *J. Am. Chem. Soc.* **1989**, *111*, 321.
6.   Porter, M. D.; Bright, T. B.; Allara, D. L.; Chidsey, C. E. D. *J. Am. Chem. Soc.* **1987**, *109*, 3559.
7.   Kim, Y.-T.; McCarley, R. L.; Bard, A. J. *J. Phys. Chem.* **1992**, *96*, 7416.
8.   Alves, C. A.; Smith, E. L.; Porter, M. D. *J. Am. Chem. Soc.* **1992**, *114*, 1222.
9.   Allara, D. L.; Nuzzo, R. G.; *Langmuir* **1985**, *1*, 45.
10.   Ogawa, H.; Chihera, T.; Taya, K. *J. Am. Chem. Soc.* **1985**, *107*, 1365.
11.   Schlotter, N. E.; Porter, M. D.; Bright, T. B.; Allara, D. L. *Chem. Phys. Lett.* **1986**, *132*, 93.
12.   Sagiv, J. *J. Am. Chem. Soc.* **1980**, *102*, 92.
13.   Moaz, R.; Sagiv, J. *Langmuir* **1987**, *3*, 1045.
14.   Tillman, N.; Ulman, A.; Penner, T. L. *Langmuir* **1989**, *5*, 101.
15.   Binnig, G.; Rohrer, H. *IBM J. Res. Develop.* **1986**, *30*, 355.
16.   Binnig, G.; Quate, C. F.; Gerber, Ch. *Phys. Rev. Lett.* **1986**, *56*, 93.
17.   Smith, D. P. E.; Bryant, A.; Quate, C. F.; Rabe, J. P.; Gerber, Ch.; Swalen, J. D. *Proc. Natl. Acad. Sci. U.S.A.* **1987**, *84*, 969.
18.   Smith, D. P. E.; Hober, J. K. H.; Binnig, G.; Nejoh, H. *Nature* **1990**, *344*, 641.
19.   Albrecht, T. R.; Dovek, M. M.; Lang, C. A.; Crutter, P.; Quate, C. F.; Kuan, S. W. J.; Frank, C. W.; Pease, R. F. W. *J. Appl. Phys.* **1988**, *64*, 1178.
20.   Bauer, E. Z. *Kristallogr.* **1958**, *110*, 373.
21.   Geus, J. W. In *Chemisorption and Reactions on Metallic Films*, Anderson, J. R.; Ed.; Academic: London, 1971, Chapter 3.
22.   Goss, C. A.; Charych, D. H.; Majda, M. *Anal. Chem.* **1991**, *63*, 85.
23.   Chaudhury, M. K.; Plueddeman, E. P.; *J. Adhesion Sci. Tech.* **1987**, *1*, 243.
24.   Allara, D. L.; Hebard, A. F.; Padden, F. J.; Nuzzo, R. G.; Falcone, D. R. *J. Vac. Sci. Technol. A* **1983**, *1*, 376.
25.   Wasserman, S. R.; Biebuyck, H.; Whitesides, G. M. *J. Mater. Res.* **1989**, *4*, 886.

26.   Czanderna, A. W.; King, D. E.; Spaulding, D. *J. Vac. Sci. Tech.* **1991**, *A9*, 2607.

27.   Smith, E. L.; Alves, C. A. ; Anderegg, J. W.; Porter, M. D. *Langmuir* **1992**, *8*, 2707.

28.   Dubois, L. H.; Hansma, P. K.; Somorjai, G. A.; *Appl. Surf. Sci.* **1980**, *6*, 173.

29.   Gupta, D. In *Diffusion Phenomena in Thin Films and Microelectronic Materials*; Gupta, D; Ho, P. S.; Eds.; Noyes: Park Ridge, NJ, 1988.

30.   Chidsey, C. E. D.; Loaicano, D. N.; Sleator, T; Nakahara, S; *Surf. Sci.* **1988**, *200*, 45.

31.   Winograd, N. In *Laboratory Techniques in Electroanalytical Chemistry*; Kissinger, P. T.; Heineman, W. R.; Eds.; Dekker: New York, 1984, Chapter 11.

32.   *Chemical Sensors and Microinstrumentation*, Murray, R. W.; Dessy, R. E.; Heineman, W. R.; Jinata, J. ; Seitz, W. R.; Eds.; ACS: Washington, DC, 1989.

33.   Haller, I. *J. Am. Chem. Soc.* **1978**, *100*, 8050.

34.   Goss, C. A.; Brumfield, J. C.; Irene, E. A.; Murray, R. W. To appear in *Langmuir*, November 1993 issue.

35.   Plueddemann, E. P. *Silane Coupling Agents*; Plenum: New York, NY, 1991.

36.   Bain, C. D.; Evall, J.; Whitesides, G. M. *J. Am. Chem. Soc.* **1989**, *111*, 7155.

37.   Horányi, G.; Orlov, S. B. *J. Electroanal. Chem.* **1991**, *309*, 239.

38.   Widrig, C. A.; Chung, C.; Porter, M. D. *J. Electroanal. Chem.* **1991**, *310*, 335.

39.   Timoshenko, S. P.; Goodier, J. N. *Theory of Elasticity*; McGraw-Hill: New York, NY, 1970.

40.   Goodfellow Catalog, Goodfellow Corporation: Malvern, PA, 1990; p. 202.

41.   *CRC Handbook of Materials Science*; Lynch, C. T., Ed.; CRC Press: Boca Raton, FL, 1975; p.382.

42.   *Preparation and Properties of Silicon Nitride Based Materials*; Bonnell, D. A.; Tien, T. Y., Eds.; Trans Tech Publications: Zurich, 1989; p. 169.

43.   The surface density of Au(111) was used only as an approximate value for the Au surfaces presented here.

44.   Dubois, L. H.; Nuzzo, R. G. *Annu. Rev. Phys. Chem.* **1992**, *112*, 558.

45.   Shackelford, J. F. *Introduction to Materials Science for Engineers*; MacMillan: New York, NY, 1992; p. 543.

46.   Steiner, U. B.; Neuenschwander, P.; Caseri, W. R.; Suter, U. W. *Langmuir* **1992** , *8*, 90.

47.   Puddephatt, R. J. In *The Chemistry of Gold*; Clark, R. J. H., Ed.; Topics in Inorganic and General Chemistry; Elsevier: Amsterdam, 1978, Monograph 16; pp. 253-6.

RECEIVED March 25, 1994

Chapter 16

Scanning Tunneling Microscopy on a Compressible Mercury Sessile Drop

Jiri Janata[1], Cynthia Bruckner-Lea[1], John Conroy[2], Andras Pungor[2], and Karin Caldwell[2]

[1]Molecular Science Research Center, Pacific Northwest Laboratories, Richland, WA 99352
[2]Department of Bioengineering, University of Utah, Salt Lake City, Utah 84112

The concept of "molecular vise" using a compressible mercury sessile drop is described. This drop represents a micro-Langmuir-Blodgett experiment in which thin films of adsorbates can be examined in various stages of surface compression. The macroscopic phase transitions of an octanethiol surface film during the compression/expansion regime are observed visually, and STM images of the mercury/air interface are obtained before and after thiol deposition.

The major motivation for this work has been to realize a system that would allow a molecule to be clamped and studied by scanning microscopy on a liquid substrate. The control over packing of molecules at a liquid/air and liquid/liquid interphase, as offered by the Langmuir-Blodgett (LB) technique, can be translated into a "molecular vise" for clamping molecules within a two-dimensional crystalline array. In such an arrangement, single molecules of the minority component can be effectively immobilized by the surrounding molecules of the majority component in a compressed state of a two-dimensional binary mixture (Figure 1). This concept defines two experimental requirements: (1) the surface area of the interface must be controllably compressible and expandable and (2) in the case of scanning tunneling microscopy (STM) the liquid phase (substrate) must be electronically conducting. A mercury/air or mercury/liquid sessile drop system satisfies both requirements and allows the use of a combination of STM and LB techniques.

For STM the coupling involves tunneling of electrons into a liquid (i.e., stochastic) substrate. Since STM involves extremely small scan areas and signal currents, the instrument and the sample must be rigorously electrically insulated from noise and mechanically stable. The vibrational problem is compounded by the fact that the substrate is a liquid, which is inherently less mechanically stable than a solid. A high density/inertia liquid such as mercury is beneficial because it mitigates the vibration per given energy input. Moreover, it is an excellent electronic conductor with well defined electrochemical behavior.

The second step in this arrangement involves the ability to tunnel through an adsorbed film at varying degrees of compression. There are some obvious difficulties in tunneling through compact, insulating organic layers atop a

0097–6156/94/0561–0175$08.00/0

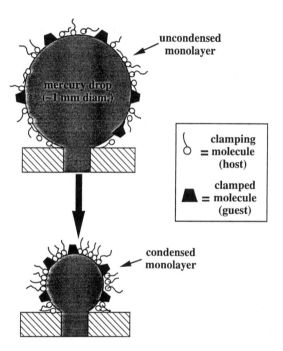

Figure 1. The "molecular vise" concept. Excess of the clamping host molecule is co-deposited on an expanded mercury drop together with a "guest" molecule present in low concentration. Upon compression the "guest" molecule is immobilized in the condensed monolayer formed by the surrounding host molecules.

conducting substrate. While molecular-resolution STM images of n-octadecanethiol have recently been reported (*1*), these images have been difficult to reproduce (*2-4*). From the variability in STM images of similar organic molecules, it is evident that tunneling parameters, tip-sample and sample-substrate interactions, film preparation, and the substrate material are all important factors in interpreting and obtaining STM images. Since it has been shown electrochemically that electron transfer along an eight carbon chain is possible (*5*), and thiols have high affinity for mercury (*6*), n-octanethiol was chosen to adsorb at the interface. Octanethiol-coated gold surfaces have been extensively studied using other surface analysis techniques and are known to form compact monolayers on gold (*7*).

It is known from scanning probe microscopy that one problem specific to imaging individual molecules on surfaces is the tip/substrate interaction (*8-11*). If the tip comes into contact with a molecule, the molecule may transfer from the substrate to the tip. Thus, in effect, the tip becomes enlarged, resulting in decreased resolution. This effect is accentuated by a liquid substrate. Since the substrate surface is not fixed, interfacial molecules adsorbed to both electrodes can follow the tip microelectrode as it travels across the surface. By restraining the guest within a compressed two-dimensional host lattice, high resolution images on a liquid substrate could be possible.

Experimental Section.

The imaging system used was a Digital Instruments Nanoscope II with a 12-μm STM head. The range of this head is 12 by 12 by 4.4 μm in the x,y, and z directions, respectively. This large head allows us to follow waving of the mercury surface. The piezoelectric tube scanner holding the STM tip was positioned over the drop using a three-screw, stand-alone support. This support was fixed to a stage using five parallel springs. The stage was connected to a Plexiglass cube that encased the glass T-valve holding the capillary. The mercury is triply distilled and filtered. In the course of our experiments, we have used both platinum/iridium (80%/20%) and tungsten tips prepared by electrochemical etching (*12*). Although STM images have been obtained using both kinds of tips, best results were obtained with tungsten tips

The first attempts to obtain STM images on mercury were made on a single mercury drop in a conical dimple on a metal sample stage. Such drops are less susceptible to vibration when they are small and well supported by the sample stage. Obviously, in such arrangement the stability of the drop was traded for the ability to manipulate the surface area. Best results were obtained with drop diameters less than 0.5 mm, positioned in a steel cradling cone with an end radius of almost 1 mm. Electrical contact with the mercury was made through the sample plate. The large contact area ensured a good connection. In order to guarantee good contact, the dimple was scraped using a wire brush before each use. The head was then arranged over the sample on the Digital Instruments three-screw sample stage. The head and sample were both enclosed in a Faraday cage and then placed inside a Plexiglass and acoustical foam "deaf box" built upon an optical breadboard. This set-up was then suspended from bungee cords. The images obtained with this system showed surface waving with an amplitude of approximately 600 nm.

The hydraulic system for variable surface area work is shown in Figure 2 (*13*). A 25-μl syringe (Gilmont Instruments, Barrington, IL), operated by a stepper motor, was used to regulate the size of the mercury sessile drop, which was extruded through a 0.2-mm i.d. stainless steel capillary with a conical dimple on the end to support the drop. Also connected to the mercury flow system was a 2.5 ml syringe (Gilmont) designed to provide a relatively inexhaustible supply of mercury, while reducing the amount of handling required. Electrical contact to the drop was made through the metal capillary. If needed, a glass tray (2.8 ml) was placed around the

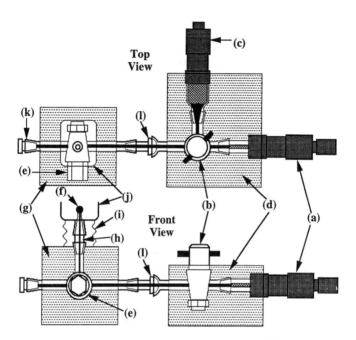

Figure 2. Schematic of the mercury sessile drop flow system (not drawn to scale). The sessile drop size is controlled by a small syringe (25 ml, Gilmont) (a). A glass T-valve (b) connects the small syringe to a large syringe (2.5 ml, Gilmont) (c) which serves as a reservoir for refilling the small syringe. Both syringes and T-valve rest in a Plexiglass mold (d) and are held by screws to the optical breadboard.

Another glass T-valve (e) leads to the sessile drop (f). This valve is permanently epoxied into a transparent Plexiglass holder (g) which is mounted to the optical breadboard. The mercury sessile drop (about 1 mm diameter) sits on a stainless steel capillary (0.1 mm inner diameter) at the top of a male/male joint (h). Springs (i) are used to hold a glass reservoir (j) onto the male/male joint leading to the mercury drop. The third outlet of this T-valve is closed (k), but may be used for an additional mercury reservoir. It was necessary to add a glass ball joint (l) to allow fine positional adjustments between the two rigidly mounted valves. Reproduced with permission from ref. 13. Copyright 1993 ACS.

capillary end to allow depositions from a liquid phase to be made. The surface waving of the variable surface area mercury substrate (~1 μm), while larger than that of the first system was still within a range that could be compensated for by the STM scanner.

The drop area was optically imaged using color CCD camera and a single lens microscope with a magnification of 200X. This optical system has proven indispensable when positioning the STM tip over the surface, and for monitoring the sessile drop during compression experiments.

The initial deposition experiments were done from ethanol solution of thiols. From these experiments we learned the general behavior of the shape of the drop upon compression. However, the uncertainty about the amount of the residual solution near the neck of the drop made us to change to vapor deposition for the set of results reported in this paper. Octanethiol was deposited from the vapor phase onto the mercury surface from a piece of cotton wrapped around a wooden stick wetted with liquid thiol. It was held approximately 1 cm away from the mercury drop, under an inverted beaker. Deposition times before STM imaging varied, ranging from 30 seconds to 2 minutes. STM imaging was performed on a clean mercury drop, on mercury drops after thiol deposition of varying time lengths, and on thiol-deposited mercury drops after compression.

Results.

Optical Observations. Simple optical inspection of the mercury surface after octanethiol vapor deposition shows evidence of the extent of surface coverage. The most dramatic effects can be seen after a long (>30 min) deposition period, at which time highly refractive crystallites form on the mercury creating a nonuniform surface that can be seen with the naked eye. These crystallites appear stationary on the liquid mercury drop. When the drop volume is decreased, the drop immediately begins to flatten. This is evidence that surface compression is occurring, and transfer of thiol into the capillary is minimal. STM images could not be obtained after such long deposition times, due to the insulating nature of the thick thiol film. When attempting to engage the STM tip, the tip penetrates the thiol film and contacts the mercury drop. Consequently, all STM images were obtained after much shorter thiol deposition times (<2 min).

The observation of the shape of the mercury drop during compression/expansion experiments provides qualitative evidence for the presence of surface films. With short deposition times (<5 min) the presence of octanethiol cannot be seen with the naked eye, but its presence is evident when monitoring the mercury drop during contraction. Withdrawal of mercury (compression) causes the mercury drop to change its shape from spherical (Figure 3a), to "flattened" (Figure 3b), to a torroid. This change can be rationalized by considering that in a classical Langmuir-Blodgett compression experiment there is a minimum surface area per molecule below which the surface film collapses. If the surface layer on our variable area sessile drop is a solid-like film that can withstand high surface pressures, the drop surface area to volume ratio will increase as we approach the limiting area of the surface film for a given coverage. Higher surface to volume ratios can be obtained for more complex objects such as ellipsoid and torroid. Thus, in the gas- or liquid-like regions of the L-B pressure curve the drop remains spherical as the volume is decreased. However, flattening of the drop suggests a solid-like surface layer that is near its surface area/molecule minimum.

When octanethiol-coated drops are compressed and then rapidly expanded, the drop surface appears to have smooth cracks showing between larger textured regions. If the STM tip is allowed to penetrate into the thiol film, the tip adheres to the surface film and can be used to lift the surface film from the mercury drop

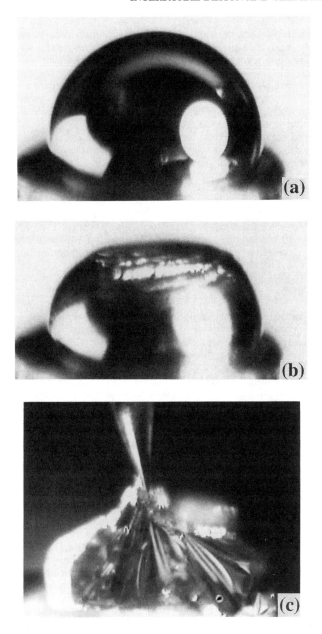

Figure 3. This figure shows a series of photographs of a mercury sessile drop after 40 seconds of octanethiol vapor deposition. Initially, the drop is a 1 mm diameter sphere (a). When the drop volume is decreased, the sphere flattens to maintain constant surface area (b). The surface film sticks to an STM tungsten tip when the tip penetrates the drop surface (c). Reproduced with permission from ref. 13. Copyright 1993 ACS.

(Figure 3c). Analysis of this surface film using atomic absorption spectroscopy indicates that this film is mercury thiolate.

A freshly extruded, uncoated mercury drop remains spherical as the drop size is decreased, and STM tips do not stick to the uncoated mercury surface. However, after exposure to air for 10 minutes, the presence of contamination on the drop surface is indicated by a flattening of the drop shape when the drop size is decreased to about 1/100 of its original volume. This is not surprising since the surface tension of mercury is about 485 dyne/cm, making surface contamination an important consideration (*14*). Consequently, we minimized the time between extruding a clean drop and beginning the thiol deposition and STM imaging procedure. Thiol deposition of 2 minutes or less proceeded immediately after extruding a fresh mercury drop, and STM images were collected within the first few minutes after thiol deposition.

Scanning Tunneling Microscope Results. One consequence of the mechanically unstable mercury substrate is the difficulty in obtaining several consecutive STM images of the same area without collisions between the tip and the waving surface. Figure 4 is a height measurement mode image of an uncoated mercury drop in which the waving of the mercury surface is traced by the STM scanning head (constant current). While these waves are large in amplitude compared to the surface features we were examining, they are low in frequency (3-5 Hz) and, thus, are easily eliminated by re-setting the initial point for the fast scanning in the x-direction. Therefore, the reason that the surface waves are observed only along one axis is that this is the slow scan (y) axis. The waving is too slow to significantly effect the image as the tip scans along the fast scan (x) axis.

Our first attempts at STM imaging of octanethiol-coated mercury drops involved long (> 5 min.) vapor deposition times, resulting in octanethiol layers that could not be imaged. In contrast, after shorter (<2 minutes) vapor deposition times, stable STM images were obtained. For direct comparison of thiol-coated and uncoated mercury sessile drops, Figure 5 shows STM images of a mercury drop that was vapor deposited for 50 seconds with octanethiol (5a), and an uncoated mercury drop (5b). Both are current mode images that are produced by resetting the height baseline for every x-scan in order to remove the y-direction surface waving shown in Figure 4. While the thiol-coated drop shows large surface features (5a), the uncoated mercury appears smooth within the resolution of the system (5b). With only 50 sec thiol deposition, the drop shown in 5a visually appears identical to an uncoated drop (5b). However, the presence of a surface film is evidenced by decreasing the volume of the drop while monitoring the drop shape. Octanethiol layers formed with 50 seconds of vapor deposition remain spherical with small decreases in volume, then flatten with larger volume decreases. This initial spherical drop compression indicates that image 5a was obtained in the liquid-like region of the surface pressure-area curve.

To our knowledge, the set of consecutive scans over an octanethiol contaminated surface, shown in Figure 6, represent the first stable images ever recorded on mercury. The octanethiol vapor deposition time was 45 seconds; and these images were taken in current measurement mode, 20 seconds apart. Figure 6 shows reproducible islands of similar tunneling current on a background of lower tunneling current. The blurring of the images of Figure 6 in the fast scan (x) direction is probably due to tip/sample interactions and the movement of material on the surface. Since initial spherical drop compression suggests that the STM images shown in Figure 5 were obtained in the liquid-like region of the pressure-area curve, it is surprising that the islands do not travel on the liquid mercury substrate during the 2 minutes that elapsed between the start of the first image and the completion of the last.

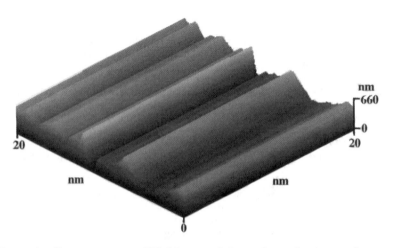

Figure 4. Constant current STM image of the surface of a 1 mm diameter mercury sessile drop with no thiol coating. The bias voltage is 270 mV and scan frequency is 78 Hz. The scanning range is 20 by 20 nm with the fast scanning x-direction shown on the right axis, and the slow scanning y-direction shown on the left axis.

After compression of the mercury drop to a flattened shape, it was easier to obtain STM images of the drop without contacting the mercury when engaging the STM tip, and we sometimes obtained island images as shown in Figure 6. However, we often obtained images as shown in Figure 7, which appear to contain small octanethiol ridges (1-4 nA high). This STM image was obtained after an approximately 50% drop volume reduction beyond the close-packed surface. (Close-packing of the thiol surface is assumed to occur at the minimum drop volume required to maintain a spherical drop). The image shows a series of parallel regions of higher current density on the surface. The direction of the ridges changes with change in scanning rotation, indicating that the ridges are not strongly influenced by the STM scanning process. These ridge images were never seen on the uncompressed (spherical) drops. The streaks in the horizontal direction are likely due to tip/sample interactions.

Summary.

The series of self-consistent visual and STM observations described in this paper confirm the possibility of performing tunneling microscopy on a mercury surface. Both STM images and visual observations of drop shape and the STM tip/drop surface interaction indicated the presence and absence of octanethiol on the mercury drop. In addition, STM images and sessile drop shape are both affected by the compression of an octanethiol coated drop. Expanded drops are spherical in shape and contain island-like STM features. And, compressed drops become flattened in shape and often contain parallel ridge STM features. These island and ridge STM features seen on octanethiol deposited mercury drops as well as "removable films" observed by optical microscopy were never seen on uncoated mercury drops. They are attributed to the formation of Hg thiolate which may possibly take place in the gas phase. Mercury has a significant vapor pressure at room temperature. (0.0017 mm of Hg) and its atoms may react with the incoming molecules of the thiol in the gas phase and subsequently deposit on the surface. STM imaging of the thiolate film

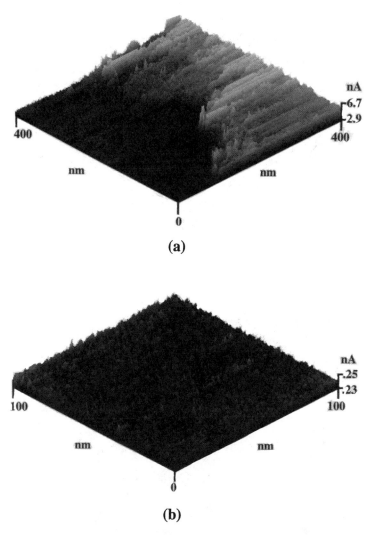

(a)

(b)

Figure 5. STM images showing the difference between octanethiol-coated (a) and uncoated (b) mercury drops. Both are current mode images, with the tip-sample distance reset at the beginning of each x-scan line to remove the y-direction surface waving shown in Figure 4. The imaging conditions for the thiol-coated drop shown in (a) are: vapor deposition time=50 seconds, bias voltage=2.9 V, and scan frequency=39 Hz. And, the imaging conditions for the uncoated mercury drop in (b) are: bias voltage=0.11 V, and scan frequency=11.6 Hz.

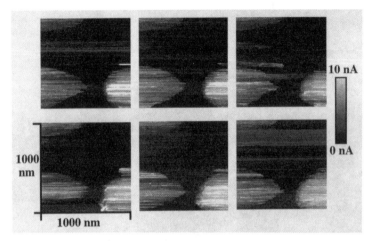

Figure 6. STM images of the same location on an octanethiol-deposited mercury drop. The octanethiol vapor deposition time was 45 seconds, and the time between the beginning of each image is 20 seconds. The fast and slow scanning directions are shown horizontally and vertically, respectively. The imaging conditions are: bias voltage=2.9 V and scan frequency=39 Hz.

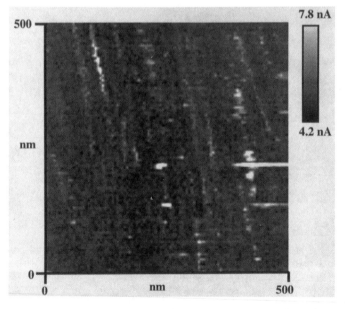

Figure 7. STM image of an octanethiol-coated mercury drop after an approximately 50% reduction in drop volume, relative to the volume at the compaction point. The imaging conditions are: bias voltage=1.6 V and scan frequency=39 Hz.

on the mercury drop was only possible for short thiol deposition times (<2 minutes). After very long thiol deposition times (>30 minutes), crystallites were visible on the drop surface, and STM imaging was not possible.

Although images of single molecules clamped in a Langmuir-Blodgett-like experiment have not yet been obtained, we believe that with some modifications of the apparatus this goal is achievable. The absence of periodicity of the bulk substrate on the time scale of our experiment opens a new possibility for tunneling microscopy. The use of a liquid substrate is particularly important in view of the fact that the interfacial ad-layer can be examined in a compressed (i.e., ordered two-dimensional solid) or an expanded (i.e., disordered two-dimensional liquid) state.

Acknowledgements.

Financial support from the Office of Naval Research is most gratefully acknowledged. The Pacific Northwest Laboratory contributed to this project by making time of the two authors (C. Bruckner-Lea) and (J. Janata) available. Pacific Northwest Laboratory is operated by Battelle for the US Department of Energy under Contract DE-AC06-76RLO 1830.

Literature Cited.

1. Widrig, C. A.; Alves, C. A.; Porter, M. D. *J. Am. Chem. Soc.* **1991**, *113*, 2805-2810.
2. Edinger, K.; Golzhauser, A.; Demota, K.; Woll, C.; Grunze, M. *Langmuir* **1993**, *9*, 4-8.
3. McCarley, R. L.; Kim, Y.-T.; Bard, A. J. *J. Phys. Chem.* **1993**, *97*, 211-215.
4. Kim, Y.-T.; McCarley, R. L.; Bard, A. J. *J. Phys. Chem.* **1992**, *96*, 7416-7421.
5. Miller, C.; Cuendet, P.; Gratzel, M. *J. Phys. Chem.* **1991**, *95*, 877-886.
6. Birke, R. L.; Mazorra, M. *Anal. Chem. Acta* **1980**, *118*, 257-269.
7. Dubois, L. H.; Nuzzo, R. G. *Ann. Rev. Phys. Chem.* **1992**, *43*, 437-463.
8. Lea, A. S.; Pungor, A.; Hlady, V.; Andrade, J. D.; Herron, J. N.; E. W. Voss, J. *Langmuir* **1992**, *8*, 68-73.
9. Salmeron, M.; Beebe, T., Jr.; Odriozola, J.; Wilson, T.; Ogletree, D. F.; Siekhaus, W. *J. Vac. Sci. Technol. A* **1990**, *8*, 635-641.
10. Stroscio, J. A.; Eigler, D. M. *Science* **1991**, *254*, 1319-1326.
11. Frommer, J. *Angew Chem Int Ed* **1992**, *31*, 1298-1328.
12. Heben, M. J.; Dovek, M. M.; Lewis, N. S.; Penner, R. M.; Quate, C. F. *J. Microsc.* **1988**, *152*, 651-661.
13. Bruckner-Lea, C; Janata, J.; Conroy, J.F.; Pungor, A.; Caldwell, K. *Langmuir* **1993**, 9, 3612-3617.
14. Smith, T. *Adv. Colloid Interface Sci.* **1972**, *3*, 161-221.

RECEIVED April 25, 1994

Chapter 17

Nucleation and Growth of Molecular Crystals on Molecular Interfaces

Role of Chemical Functionality and Topography

Phillip W. Carter[1], Lynn M. Frostman, Andrew C. Hillier, and Michael D. Ward

Department of Chemical Engineering and Materials Science, University of Minnesota, 421 Washington Avenue SE, Minneapolis, MN 55455

The nucleation and growth of molecular crystals at interfaces has been investigated in the context of nucleation rates, growth orientation, nanoscale growth modes, and the influence of topographic features on growth characteristics. For instance, real time imaging of crystal growth in solution using the atomic force microscope (AFM) reveals that the nanoscopic surface topography and molecular structure of the crystal-solution interface play an important role in the growth and dissolution characteristics of molecular crystals. This influence can be understood in terms of the strength and orientation of intermolecular bonding at the crystal-solution interfaces. Additionally, the well-defined structure exposed on the surface of molecular single crystals can provide a unique substrate for heterogeneous crystal growth. Nucleation and growth of secondary crystalline phases on molecular crystals is both oriented and facile at certain substrate ledge sites, behavior which is attributed to lowering of the prenucleation aggregate free energy via "ledge directed epitaxy." The nucleation and growth of molecular crystals has also been studied on chemically modified surfaces. For instance, nucleation of malonic acid (HOOC-CH_2-COOH) is significantly more rapid on organosulfur monolayers terminated with carboxylic acid groups than on corresponding monolayers terminated with methyl, methyl ester, or ethyl ester functionalities. In addition, the malonic acid crystals are oriented on these monolayers in a manner which is consistent with interfacial stabilization of malonic acid prenucleation aggregates by interfacial hydrogen bonding and surface polar forces, in contrast to monolayers in which hydrogen bonding functionality is absent. When taken together, these three complementary systems provide much insight into the interfacial forces controlling crystal growth.

Crystalline solids based upon molecular components exhibit numerous properties of fundamental scientific and technological interest, including electrical conductivity, superconductivity, nonlinear optical behavior, and ferromagnetism.(1) The principle

[1]Current address: Corporate Research, Nalco, One Nalco Center, Naperville, IL 60566

advantage of these materials over their inorganic counterparts lies in the potential for control of their bulk properties through molecular design. A sizable body of work exists regarding the application of "crystal engineering" strategies, which utilize thermodynamically preferred intermolecular interactions in the bulk crystal to control crystal packing.(2) However, relatively little effort has been spent examining the formation of molecular crystals in the context of nucleation and crystal growth *processes.*

The formation of a macroscopic crystal is preceded by crystal nucleation, which generally occurs at an interface due to the reduction in surface free energy associated with favorable substrate-nuclei interactions. As a result, the structure of such an interface can play an important role in the crystallization pathway. For example, investigations of the nucleation and growth of amino acids,(*3-4*) inorganic compounds,(*5-6*) and ice(*7*) beneath Langmuir monolayers indicate that enhanced nucleation, oriented crystal growth, and polymorphic selectivity can be achieved by careful design of these interfaces. Further advances in this area would have a significant impact on the ability to control the formation of molecular crystals under circumstances in which crystal morphology, growth rates, growth orientation, defect density, and phase selectivity are problematic. This prompted us to examine strategies for influencing the nucleation and growth of molecular crystals by the *molecular level design of solid substrates.*

Our approach includes examining single crystal substrates as well as monolayer-modified surfaces. Molecular crystals typically exhibit a variety of unique crystallographic planes, with different molecular structures and enantiotopic properties, which can be examined during growth and dissolution or used as chemically and structurally distinct substrates for the nucleation of secondary phases. Understanding the character of a specific crystal surface requires a thorough characterization of the chemical and topographic structure of the exposed crystal face. Indeed, topographic features on crystal interfaces play a critical role in defining the chemical sensitivity of aggregates toward interfaces during nucleation and growth. Studies of the epitaxial growth of secondary phases on single crystal substrates generally have ignored the role of topographic features, such as substrate ledges and kinks, in nucleation, even though these features play an important role in single-phase crystal growth.(*8*) An alternative to molecular crystal substrates, self-assembled monolayers, prepared from the spontaneous adsorption of thiol reagents on polycrystalline gold, allow the fabrication of molecular interfaces with specific chemical functionalities.(*9*) Due to the roughness of the underlying gold substrate, which is prepared by thermal evaporation, the topographic features of these interfaces are typically not as well-defined as freshly cleaved single crystals. However, the value of organosulfur monolayers derives from the virtually unlimited variety of terminal functional groups they can display. This report examines several aspects of the nucleation and growth of molecular crystals on interfaces with well-defined molecular structure. These studies include dynamic *in situ* atomic force microscopy (AFM) studies of molecular single crystals during etching and growth, the nucleation of secondary organic crystalline phases on molecular crystal substrates, and the nucleation of molecular crystals on self-assembled monolayers.

Results and Discussion

Atomic Force Microscopy of Crystal Growth at Organic Crystal Interfaces. The mechanisms of crystal nucleation and growth are most easily surmised from observations of crystal growth and etching upon well-defined single crystal substrates. We have examined the role of hydrogen bonding and charge transfer interactions in the growth and dissolution of α-glycine and $(TMTSF)_2ClO_4$, respectively.(*10*) Both these materials exhibit solid state intermolecular interactions which are dominated by highly anisotropic bonds. These bonding interactions promote the formation of specific surface features, such as oriented kink and ledge structures, during crystal growth and etching.

Glycine is an amino acid which crystallizes in the α-form from aqueous solutions and is dominated by strong hydrogen bonding interactions in the solid state. The mature crystal belongs to the monoclinic space group $P2_1/n$ (a = 5.102 Å, b = 11.971 Å, c = 5.457 Å, and β = 111.7°)(11-12) and is bipyramidal with three principal, chemically inequivalent faces: the {110} and {011} family of planes form the eight side faces of the crystal bipyramid while the {010} planes cap the ends. The x-ray crystal structure reveals glycine molecules packed into hydrogen bonded sheets within the {010} plane (Figure 1). Each {010} sheet is characterized by N-H-O contacts of 2.76 and 2.88 Å between molecules along the crystallographic [001] and [100] axes, respectively. The zwitterionic glycine molecules exhibit strong electrostatic interactions between pairs of {010} sheets resulting in a bilayer structure.

AFM experiments were performed with a Digital Instruments Nanoscope III scanning probe microscope operating in constant force mode. The surface topography of the {010} face of α-glycine crystals, determined by *in situ* AFM studies at room temperature in an aqueous solution near supersaturation conditions, is rich in terraces bounded by ledges along the [001] and [100] directions. We will subsequently refer to ledges along the [hkl] direction as [hkl] ledges. The [001] ledges are predominant in solution over length scales of 1 nm to 1 mm (Figure 2), while the [100] ledge population is low. The smallest observed step height of ~10 Å is in good agreement with the height of a single {010} bilayer (~8.9 Å). The ledge structure of the {010} face is a reflection of the strong hydrogen bonded vectors in the solid state structure. The shorter N-H-O contacts along the [001] direction suggest that this interaction is stronger than that along the [100]. Thus, the predominance of [001] ledges on the {010} crystal surface can be attributed to the stronger solid state bonding interactions between molecules within the step plane of the [001] ledge. The energy required to disrupt the [001] ledge exceeds that of all other ledges exposed on the {010} face. The intersection of the [100] and [001] ledges results in the formation of kink sites on the crystal surface (labeled as region A in Figure 2). The step planes of the [001] and [100] ledges most likely comprise the {110} and {011} planes, respectively, as the observation of these faces in the macroscopic crystal suggests that they are low energy surfaces.

In situ observation of {010} terraces indicates that growth occurs with rapid [100] ledge advancement along the [001] direction, whereas the movement of [001] ledges along the [100] direction is considerably slower. The slowly advancing [001] ledges gradually merge as a result of lowering supersaturation, which reduces both the ledge frequency and step height. Initial rates of [100] ledge advancement on {010} are typically on the order of 200 nm/sec at supersaturations of ~25±5% (4.3M). The rate of [001] ledge advancement is approximately a factor of 5-10 smaller than [100] ledge advancement under similar growth conditions. Differing [001] ledge advancement rates are observed when step heights are less than 50 Å, often resulting in ledge coalescence.

A crystal growth mechanism dominated by oriented topographical features is also observed during growth of the organic charge-transfer salt $(TMTSF)_2ClO_4$ (TMTSF = tetramethyltetraselenafulvalene). This material is a low dimensional organic conductor which is typically grown by electrochemical means. It crystallizes into the triclinic $P\bar{1}$ space group (a = 7.266 Å, b = 7.678 Å, c = 13.275 Å, and α= 84.58°, β = 86.73°, γ = 70.43°),(13-14) with a needle-like morphology in which the [100] corresponds to the needle axis and the (001) plane exhibits the largest area. The solid state structure is dominated by a strong π-π charge transfer interaction between TMTSF molecules along the [100] direction. Weaker interactions along [$\bar{1}$20] and

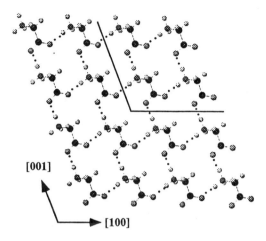

Figure 1. Single crystal x-ray structure of α-glycine, showing the packing of hydrogen bonded glycine molecules in the (010) plane. Hydrogen bonds are indicated with dotted lines. The [001] and [100] ledge directions forming the observed kink sites are shown by solid lines. (Adapted from ref. 10.)

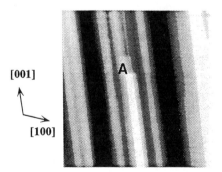

Figure 2. AFM image of {010} face of α–glycine in aqueous solution. The vertical scale is approximately 100 nm. The [001] ledges predominate, although small [100] ledges can be observed, for instance in region A. (Adapted from ref. 10.)

[001], associated with interstack bonding, result in a solid comprising (001) and (210) layers.

In situ AFM studies of $(TMTSF)_2ClO_4$ were performed in an electrochemical fluid cell employing a three-electrode configuration, with the $(TMTSF)_2ClO_4$ crystal as the working electrode. AFM of the (001) plane of a mature single crystal of $(TMTSF)_2ClO_4$ reveals a periodic terraced structure along the [$\bar{1}$20] direction bounded by ledges extending along [100]. The [100] ledges which separate (001) terraces exhibit heights of 20-100 nm, while the distance between ledges is ~200-400 nm. The ledge orientation along [100] corresponds to the stacking axis of the TMTSF molecules in the solid-state (Figure 3) and also corresponds to the orientation of the intrastack π-π charge transfer interaction. The [100] corresponds to the fastest growing crystal direction as it is coincident with the needle axis of the mature crystal. The presence of [100] oriented ledges in the mature crystal suggests that the growth mechanism involves fast incorporation along [100].

The dissolution of $(TMTSF)_2ClO_4$ in ethanol results in a flattening of the terrace structure. The rate of dissolution can be controlled by the electrochemical potential applied to the crystal. At potentials greater than the reversible potential (E > E_0 = 0.42 V versus SCE), the (001) face exhibits large terraces intersected by static [100] ledges with heights of 13.3 Å (Figure 4), in agreement with the c lattice parameter and corresponding to the distance between layers of TMTSF molecules (and accompanying ClO_4^- anions) in the (001) plane. At more cathodic potentials (E < E_0), the relative supersaturation decreases (this is tantamount to electrochemical etching) and [100] ledges recede along the [$\bar{1}$20] direction. The motion of [100] ledges occurs following the formation of small [$\bar{1}$20] ledges and [100]/[$\bar{1}$20] kink sites. These kink sites, with heights of 13.3 Å and widths ranging from 10 to 50 nm, are unstable and quickly advance in the [100] direction. The instability of [$\bar{1}$20] oriented ledges results in the preservation of [100] ledges during the dissolution process. This is a manifestation of the strong charge transfer interaction directed along the [100] axis and the correspondingly high surface energy of the exposed (100) planes on [$\bar{1}$20] ledges compared with the (010) planes that appear upon the step plane of the [100] ledges. The slow etch sequence of Figure 4 illustrates the mechanism of $(TMTSF)_2ClO_4$ dissolution whereby large [100] ledges flow in the [$\bar{1}$20] direction. Dynamic AFM of the growth of $(TMTSF)_2ClO_4$ indicates similar involvement of the ledges and kink sites, with the evolution of unit cell-high (001) terraces via rapid growth at kink sites along the [100] direction. This is manifested by the movement of [100] ledges in the [$\bar{1}$20] direction at longer times.

Crystallization of Secondary Crystalline Phases on Molecular Crystal Substrates. In addition to their role in growth and dissolution, topographical features such as ledges can also play an important role in the nucleation of secondary crystalline phases. Conventional nucleation theories suggest that ledges can provide two surfaces for interaction with the molecules that form a nucleus, in contrast to two-dimensional nucleation modes in which only one interface is available for nucleation. Under conditions where interaction at a ledge site is favorable, nucleation at ledges should occur at lower supersaturations than on sites where ledges are absent.

This has been demonstrated by the nucleation and growth of benzoic acid on succinic acid substrates.(15) Succinic acid crystallizes in the monoclinic space group $P2_1/c$ (a = 5.519 Å, b = 8.862 Å, c = 5.101 Å, and β = 91.6°).(16) The crystal structure of succinic acid reveals hydrogen bonded succinic acid chains arranged in sheets parallel to {010}$_{sa}$, with the chains oriented along the [10$\bar{1}$]$_{sa}$ direction. (Note: the subscript refers to the crystalline material for which the plane or direction is described: sa = succinic acid; ba = benzoic acid; pna = 4-nitroaniline). Single crystals can be cleaved readily by applying a small force with a sharp razor blade or microtome along [001]$_{sa}$ perpendicular to the (100)$_{sa}$ face, providing freshly prepared, clean {010}$_{sa}$

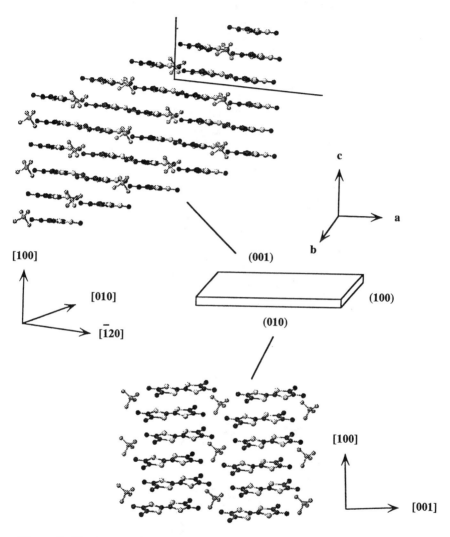

Figure 3. Single crystal x-ray structure of (TMTSF)$_2$ClO$_4$, showing stacks of TMTSF molecules along the [100] direction in the (001) plane. The [100] and [$\bar{1}$20] ledge directions forming the observed kink sites are shown by solid lines. (Adapted from ref. 10.)

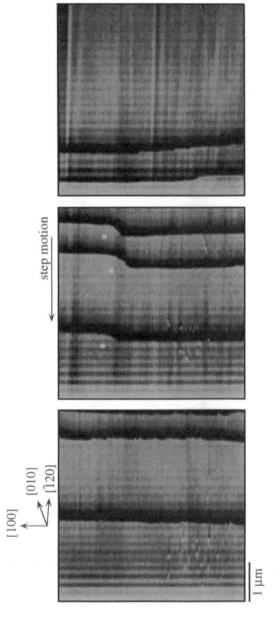

Figure 4. *In situ* AFM of the {001} face of $(TMTSF)_2ClO_4$ showing dissolution under potential control. During dissolution (*middle*), steps oriented along [100] flow in the [$\bar{1}$20] direction. The step flow can be arrested by increasing the electrochemical potential (*left* and *right*). The step heights are 1.3 nm. (Adapted from ref. 10.)

step motion

[100]
[010]
[$\bar{1}$20]

1 μm

surfaces with typical dimensions of 3 mm x 9 mm. These faces possess linear features along $[10\bar{1}]_{sa}$, which are observed readily by optical microscopy. AFM reveals that the features along $[10\bar{1}]_{sa}$ are macroscopic ledges with heights > 5 nm. AFM experiments also indicate that the minimum dihedral angle between $(0\bar{1}0)_{sa}$ terraces and the step planes separating the terraces is 113°, which agrees with the $(0\bar{1}0)_{sa} \cap \{111\}_{sa}$ dihedral angle determined from the crystal structure (112.6°). Both the $\{010\}_{sa}$ and $\{111\}_{sa}$ crystallographic planes have high molecular packing densities, and contain layers of hydrogen bonded chains. These planes can be characterized as low energy surfaces as there is strong hydrogen bonding between succinic acid molecules *parallel* to these planes and minimal molecular corrugation. Strong hydrogen bonding along the succinic acid chains parallel to $[10\bar{1}]_{sa}$ favors cleavage along planes containing this direction. Weak van der Waals forces between $\{010\}_{sa}$ and $\{111\}_{sa}$ layers are responsible for the predominance of the $\{010\}_{sa}$ and $\{111\}_{sa}$ planes, respectively, in the cleaved crystal. The structure and microscopy results therefore support a surface topography principally consisting of $[10\bar{1}]_{sa}$ ledges containing the $(0\bar{1}0)_{sa} \cap \{111\}_{sa}$ planes.

Sublimation of benzoic acid onto a freshly cleaved succinic acid crystal substrate at temperatures between 35-60°C produces *oriented* single crystals of benzoic acid on the $(0\bar{1}0)_{sa}$ face growing from the $[10\bar{1}]_{sa}$ ledges, with nucleation frequencies approaching 50 mm^{-2} (Figure 5a). Crystallographic analysis reveals that the benzoic acid crystals are oriented such that the $(001)_{ba}$ plane is in contact with $(0\bar{1}0)_{sa}$, with the $[010]_{ba}$ axis oriented at an angle of 46°± 1.7° with respect to $[10\bar{1}]_{sa}$. The $[100]_{sa}$ makes a 42° angle with $[10\bar{1}]_{sa}$, therefore, $[010]_{ba}$ is *not* coincident with $[100]_{sa}$. The orientation of benzoic acid on succinic acid involves $[001]_{ba}$ parallel to $[001]_{sa}$ and $[010]_{ba}$ nearly parallel to $[100]_{sa}$. This is contrary to expectations for epitaxially driven two-dimensional nucleation of benzoic acid on succinic acid, which would predict alignment of $[100]_{ba}$ with $[100]_{sa}$ and $[010]_{ba}$ with $[001]_{sa}$ as a result of the agreement between lattice constants of the respective planes. Rather, growth from the succinic acid ledge sites is attributed to lowering of the prenucleation aggregate free energy via "ledge directed epitaxy." This involves a lattice match between the substrate and growing phase along the ledge direction and equivalent dihedral angles of the substrate ledge sites and a pair of aggregate planes, whose identity is assigned on the basis of the structure of the mature crystal. The $[10\bar{1}]_{sa}$ ledge has a 1.0% lattice mismatch with the [110] direction of benzoic acid and a difference of only 0.6° between the ledge dihedral angle and the dihedral angle of the $(001)_{ba} \cap (1\bar{1}2)_{ba}$ planes (Figure 6). Based on the crystal structures, these interfaces consist of "molecularly smooth" low energy planes, favoring stabilization of the prenucleation aggregates by dispersive interactions. As a consequence of these epitaxial effects and the crystallographic symmetry of the monoclinic space groups of the substrates and benzoic acid, benzoic acid growth is highly oriented. It is noteworthy that only one of the possible orientations is observed (Figure 5a). This is a consequence of the crystallographic inequivalence of these two orientations because of the monoclinic symmetry of the succinic acid substrate. This behavior also has been observed for other substrate-crystal systems.(*15*)

The crystallization of 4-nitroaniline on succinic acid is similar to that observed for benzoic acid, as exhibited by the formation of oriented, rectangular single crystals of 4-nitroaniline at the $[10\bar{1}]_{sa}$ ledges (Figure 5b). Under mild saturation conditions (nucleation frequencies between 100-1000 mm^{-2}) 4-nitroaniline crystals grow *exclusively* on the $[10\bar{1}]_{sa}$ ledges. Analysis of the resulting 4-nitroaniline crystals reveals that they belong to the centrosymmetric monoclinic space group $P2_1/n$ (a =12.34 Å, b = 6.07 Å, c = 8.59 Å, and β = 91.45°), consistent with the only known phase of 4-nitroaniline.(*17*) The solid state structure of this compound contains chains along the $[10\bar{1}]_{pna}$ axis, which are arranged head-to-tail by hydrogen bonding between nitro and amine groups. These chains are assembled into densely packed, polar

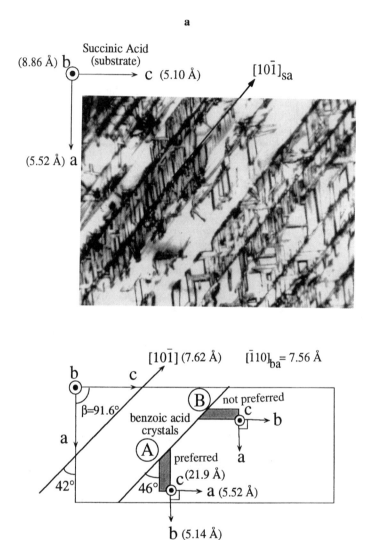

Figure 5. *Top,* preferred orientation exhibited by a) benzoic acid crystals and b) 4-nitroaniline crystals grown from the gas phase onto $(0\bar{1}0)_{sa}$. Nucleation is initiated at the $[1\bar{0}\bar{1}]_{sa}$ ledges. *Bottom,* schematic representation of the orientations of benzoic acid and 4-nitroaniline on $(0\bar{1}0)_{sa}$. The lattice parameters and directions for succinic acid, benzoic acid and 4-nitroaniline are also depicted. (Adapted from ref. 15.)

b

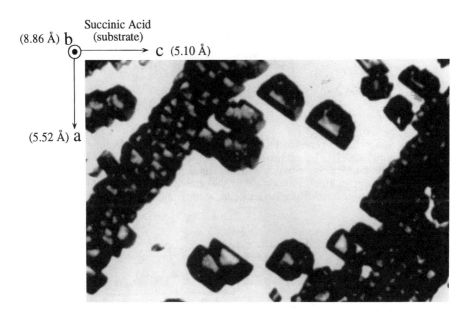

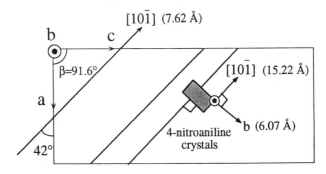

Figure 5. Continued.

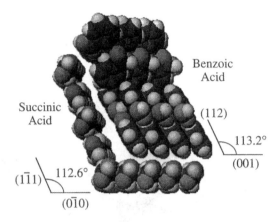

Figure 6. Space-filling representation of a benzoic acid prenucleation aggregate and a succinic acid $[10\bar{1}]_{sa}$ ledge site containing the $(0\bar{1}0)_{sa} \cap \{111\}_{sa}$ planes, illustrating the dihedral angle match with the contacting pair of $(001)_{ba} \cap (112)_{ba}$ planes of the aggregate. Epitaxy along the shared directions of $[10\bar{1}]_{sa}$ and $[110]_{ba}$ is also suggested from the small lattice mismatch. (Adapted from ref. 15.)

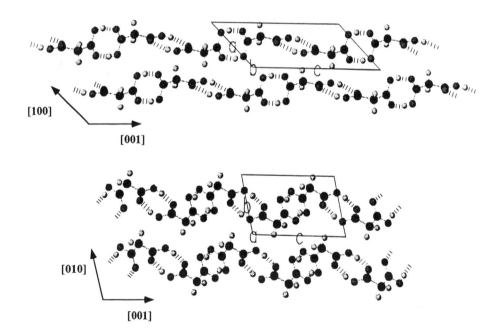

Figure 7. Single crystal x-ray structure of malonic acid, showing the packing of hydrogen bonded malonic acid chains on the (010) (*top*) and (100) (*bottom*) faces. Hydrogen bonding between carboxylic acid groups, indicated by the dashed lines, is truncated at the (001) face. (Adapted from ref. 19.)

$(101)_{pna}$ layers. Large $(101)_{pna}$ faces contact $(0\overline{1}0)_{sa}$, and $[10\overline{1}]_{pna}$ is aligned with $[10\overline{1}]_{sa}$. The orientation of 4-nitroaniline grown on succinic acid can be attributed to negligible lattice mismatch between $[10\overline{1}]_{sa}$ and $[10\overline{1}]_{pna}$, and agreement of the dihedral angles of $(101)_{pna} \cap (121)_{pna}$ and the $[10\overline{1}]_{sa}$ ledges to within 0.9°. Since the hydrogen bonded chains of the substrate and aggregate are aligned, hydrogen bonding along the ledge direction may also be important.

Nucleation and Growth on Self-Assembled Monolayers. The ability to modify surfaces with specific chemical functional groups provides an opportunity for creating designer surfaces for the enhanced nucleation and growth of molecular materials. Unfortunately, single crystal substrates, while providing well-defined topography, are limited to primarily low energy surfaces. Hence, we investigated the use of self-assembled monolayers as substrates. For instance, nucleation and crystal growth of malonic acid, $HOOC-CH_2-COOH$, on self-assembled organosulfur monolayers was examined in order to elucidate the contribution of hydrogen bonding in these processes.(19) The single crystal x-ray structure of malonic acid ($P\overline{1}$ space group, $a = 5.33$ Å, $b = 5.14$ Å, $c = 11.25$ Å, $\alpha = 102.70°$, $\beta = 135.17°$, $\gamma = 85.17°$) reveals hydrogen bonded chains linked by cyclic -COOH dimers along the c axis.(18) These -COOH hydrogen bonding functionalities are oriented parallel to the (100) and (010) crystallographic planes (Figure 7) and protrude from the (001) face. The resulting (001) surface of malonic acid is capable of forming hydrogen bonding interactions with other molecules and surfaces. Therefore, it is reasonable to suggest that nucleation of malonic acid would be favored on monolayers capable of interacting via hydrogen bonding or polar forces, compared to monolayers capable of providing only dispersive forces.

Monolayers were formed by overnight immersion of the fresh gold substrates in ethanolic solutions of the thiols. Comparative experiments were performed by attaching three substrates to a glass stage in an inverted position above the malonic acid in a custom made sublimation chamber. Analyses of the frequency and size distributions of the crystals were obtained with Image Analyst software. Crystal growth by sublimation of malonic acid onto organosulfur monolayers is indeed consistent with more rapid nucleation on monolayers terminated with -COOH groups (Figure 8). The number of malonic acid crystals per unit area grown on [Au]-$S(CH_2)_{11}COOH$ and [Au]-$S(CH_2)_{15}COOH$ is typically an order of magnitude greater than on monolayers terminated with methyl, methyl ester, or ethyl ester groups, or on bare gold. Crystals grown on carboxylic acid-terminated monolayers are also much smaller than on other monolayers or bare gold, consistent with significantly higher nucleation frequencies on carboxylic acid-terminated surfaces. This suggests that the carboxylic acid-functionalized surface serves to stabilize the prenucleation aggregates of malonic acid during nucleation.

The measured contact angles of monolayers with water are as follows: [Au]-$S(CH_2)_{15}COOH$ = 24°; [Au]-$S(CH_2)_{11}COOH$ = 48°; [Au]-$S(CH_2)_{11}COOCH_3$ = 61°; [Au]-$S(CH_2)_{11}COOCH_2CH_3$ = 90°;.[Au]-$S(CH_2)_{15}CH_3$ = 112°; [Au]-$S(CH_2)_{11}CH_3$ = 109°. The rapid nucleation rates of malonic acid on carboxylic acid-terminated monolayers are consistent with the low sessile contact angles measured on these monolayers, which are a consequence of strong polar and hydrogen bonding interactions between the monolayer and water. It is unlikely that the methyl ester-terminated monolayer would participate in hydrogen bonding with malonic acid. Nevertheless, the contact angle on this monolayer indicates that it is a relatively polar surface. The methyl ester monolayers exhibit malonic acid nucleation frequencies nearly identical to those on the methyl-terminated monolayer, suggesting that the observed preference for nucleation on carboxylic acid monolayers is due to interfacial stabilization of malonic acid aggregates via interfacial hydrogen bonding.

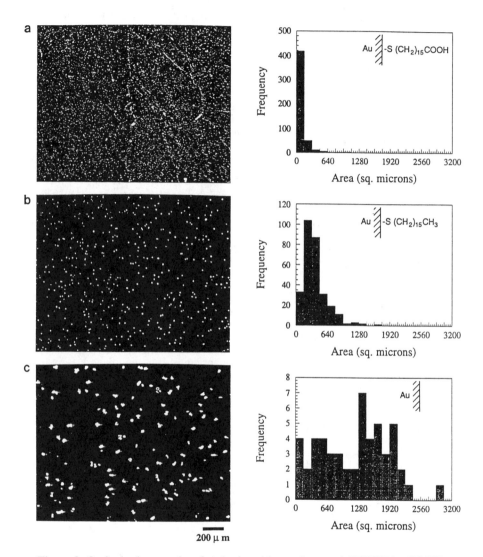

Figure 8. Optical micrographs of malonic acid growing on a) HS(CH$_2$)$_{15}$COOH, b) HS(CH$_2$)$_{15}$CH$_3$, and c) bare gold. Frequency distributions of crystal size for each monolayer are depicted on the right of each micrograph. Note that the frequency scale is different for each histogram. (Reproduced from ref. 19.)

X-ray diffraction of crystalline films of malonic acid were obtained with a Rigaku D-Max II diffractometer equipped with a thin film sample stage for grazing angle measurements. Data were collected at a grazing angle of 4°. Diffraction data of malonic acid crystals grown on the [Au]-$S(CH_2)_{11}COOH$ monolayer exhibit a relatively strong (001) peak (Figure 9), consistent with orientation of this plane parallel to the monolayer interface. In contrast, the (001) peak is noticeably absent for crystalline films on [Au]-$S(CH_2)_{11}COOCH_3$ or [Au]-$S(CH_2)_{11}CH_3$, and is weak in powder samples of malonic acid. It is noteworthy that the (001) plane contains truncated hydrogen bonded chains, which implies that the [Au]-$S(CH_2)_{11}COOH$ interface stabilizes malonic acid aggregates via hydrogen bonding during nucleation, in agreement with the more rapid nucleation on this monolayer. It is likely that this interaction involves the formation of cyclic hydrogen bonded carboxylic acid heterodimers of malonic acid molecules and the [Au]-$S(CH_2)_{11}COOH$ monolayer, as this is a common hydrogen bonding motif for these moieties. The absence of the (001) peak on the other monolayers reflects the lack of hydrogen bonding functionalities on these interfaces.

The (010) peak is strong for malonic acid crystals grown on the [Au]-$S(CH_2)_{11}COOH$ and [Au]-$S(CH_2)_{11}CH_3$ monolayers, but absent on [Au]-$S(CH_2)_{11}COOCH_3$. The (100) peak is also strong on [Au]-$S(CH_2)_{11}COOH$, weaker on [Au]-$S(CH_2)_{11}CH_3$ and negligible on [Au]-$S(CH_2)_{11}COOCH_3$. Finally, crystalline films grown on [Au]-$S(CH_2)_{11}COOCH_3$ exhibit strong ($\bar{1}01$), ($0\bar{1}1$), (011), and {111} peaks, which are substantially weaker on the other monolayers. The additional observation of (100) and (010) orientation on the carboxylic acid-terminated monolayer suggests hydrogen bonding with the hydrogen bonded chains oriented parallel to these faces, or directional polar interactions. The observation that the (010) plane is more highly oriented, relative to (100), on the more hydrophobic [Au]-$S(CH_2)_{11}CH_3$ monolayer is qualitatively consistent with the greater protrusion of hydrophobic methylene groups from the (010) plane compared to (100). The chemical basis for orientations observed on the [Au]-$S(CH_2)_{11}COOCH_3$ is less obvious. However, the contrast between the orientations and nucleation behavior on the [Au]-$S(CH_2)_{11}COOCH_3$ and [Au]-$S(CH_2)_{11}COOH$ monolayers strongly suggests that hydrogen bonding plays an important role in the latter.

While the orientations evident from x-ray diffraction are not singularly specific for each monolayer, the x-ray data clearly indicate that the orientation of the crystals depends upon the chemical nature of the terminal functional group of the monolayer. This result was also confirmed using grazing angle infrared reflection adsorption spectroscopy (IRRAS).(*19*) It is reasonable to suggest that this behavior is a consequence of interfacial interactions during the nucleation process.

Conclusions

These studies demonstrate that the nature of an interface can have a marked effect on crystal nucleation and growth. The advantage of single crystal and monolayer substrates is that these effects can be understood at a molecular level. For molecular crystals, the growth and etching behavior as well as the macroscopic morphology is controlled by the nanoscale ledge structure, which is solely a manifestation of the underlying intermolecular bonding in the absence of solvent and impurity influences. Ledge structures on molecular crystals can be employed to direct the nucleation and growth of secondary crystalline phases via a "ledge directed epitaxy" mechanism. In addition, the growth of molecular crystals on organosulfur monolayers demonstrates the ability of interfacial forces, such as hydrogen bonding, to dictate nucleation frequencies and crystal orientations. These observations suggest several potential methods for creating designer substrates for crystal growth. Further studies, including extension of these principles to nucleation and growth in solutions, should provide

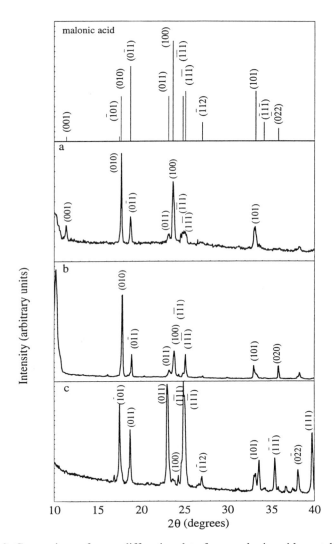

Figure 9. Comparison of x-ray diffraction data from malonic acid crystals grown on a) $HS(CH_2)_{11}COOH$, b) $HS(CH_2)_{11}CH_3$, and c) $HS(CH_2)_{11}COOCH_3$ monolayers. The diffraction intensities of the significant peaks of a powder sample of malonic acid, in which the crystallographic planes are randomly oriented, is depicted at the top of the figure. The Miller indices for the significant diffraction peaks are indicated. (Reproduced from ref. 19.)

substantial insight into the understanding and control of topics concerning crystal seeding, morphology control, and polymorph selectivity.

Acknowledgments

The authors gratefully acknowledge the support of the National Science Foundation (NSF/DMR-9107179) and the Center for Interfacial Engineering (NSF Engineering Research Centers Program, CDR 8721551). LMF thanks the National Science Foundation for a Graduate Fellowship. ACH also thanks the Upjohn Company for a Graduate Fellowship.

Literature Cited

(1) *Extended Linear Chain Compounds,* Miller, J. S., Ed.; Plenum Press: New York, NY, 1980-1983; Vols. 1-3.

(2) For instance, see: Desiraju, G. R. *Crystal Engineering-the Design of Organic Solids*, Elsevier: New York, NY, 1989.

(3) Landau, E. M.; Grayer Wolf, S.; Levanon, M.; Leiserowitz, L.; Lahav, M.; Sagiv, J. *J. Am. Chem. Soc.* **1989**, *111*, 1436.

(4) Weissbuch, I.; Frolow, F.; Addadi, L.; Lahav, M.; Leiserowitz, L. *J. Am. Chem. Soc.* **1990**, *112*, 7718.

(5) Heywood, B. R.; Mann, S. *Adv. Mater.* **1992**, *4*, 278.

(6) Zhao, X. K.; McCormick, L. D.; Fendler, J. H. *Adv. Mater.* **1992**, *4*, 93.

(7) Popovitz-Biro, R.; Lahav, M.; Leiserowitz, L. *J. Am. Chem. Soc.* **1991**, *113*, 8943.

(8) Tiller, W. A. *The Science of Crystallization: Microscopic Interfacial Phenomena*, Cambridge University Press: New York, NY, 1991; pp 327-381.

(9) Ulman, A. *An Introduction to Ultrathin Organic Thin Films from Langmuir-Blodgett to Self-Assembly;* Academic Press: Boston, MA, 1991.

(10) Carter, P. W.; Hillier, A. C.; Ward, M. D. *J. Am. Chem. Soc.,* in press.

(11) Albrecht, G.; Corey, R. B. *J. Am. Chem. Soc.,* **1939**, *61*, 1087.

(12) Power, L. F.; Turner, K. E.; Moore, F. H. *Acta. Cryst. B* **1976**, *32*, 11.

(13) Bechgaard, K.; Cowan, D. O.; Bloch, A. N. *J. Chem. Soc., Chem. Commun.* **1974**, 937.

(14) Bechgaard, K.; Carneiro, K.; Rasmussen, F. B.; Olsen, M.; Rindorf, G.; Jacobsen, C. S.; Pedersen, H. J.; Scott, J. C. *J. Am. Chem. Soc.* **1981**, *103*, 2440.

(15) Carter, P. W.; Ward, M. D. *J. Am. Chem. Soc.,* **1993**, *115*, 11521.

(16) Leviel, J.-L.; Auvert, G.; Savariault, J.-M. *Acta Cryst. B* **1981**, *37*, 2185.

(17) Trueblood, K. N.; Goldish, E.; Donahue, J. *Acta. Cryst.* **1961**, *14*, 1009.

(18) Goedkoop, J. A.; MacGillavry, C. H. *Acta Cryst.* **1957**, *10*, 125.

(19) Frostman, L. M.; Bader, M. M.; Ward, M. D. *Langmuir,* in press.

RECEIVED March 25, 1994

Chapter 18

Organic–Inorganic Molecular Beam Epitaxy

Ordered Monolayers of Phthalocyanines, Naphthalocyanines, and Coronene on Cu (100), SnS_2 (0001), and MoS_2 (0001)

C. D. England[1], G. E. Collins[2], T. J. Schuerlein, and N. R. Armstrong[3]

Department of Chemistry, University of Arizona, Tucson, AZ 85721

Packing structures are described for monolayers of various large aromatic molecules on metal and layered semiconductor surfaces, formed by the process of organic/inorganic molecular beam epitaxy. These ultrathin films are of interest as the starting point for a new generation of photonic and chemical sensor materials. Commensurate or coincident lattices are formed, depending upon the symmetry of the unit cells of the substrate and overlayer, and the strength of their interaction. Modeling studies show that even when the interactions are weak (i.e. van der Waals forces only), their accumulation over large surface lattice areas can affect the orientation of the domains of at least the first monolayer.

Molecular beam epitaxy (vacuum deposition) has recently emerged as a possible compliment to Langmuir-Blodgett (LB) or self-assembly (SA) strategies in the formation of ordered ultrathin film dye assemblies (1,2). In organic-MBE the component molecules are generally unsubstituted (crystalline) and compatible with ultrahigh vacuum deposition conditions [e.g. phthalocyanines (Pc), naphthalocyanines (NPc), perylenes, C_{60}, etc.](1,2). Highly ordered thin films can be produced by deposition at low rates (e.g. less than 10 monolayers per hour), with appropriate choice of substrate material and temperature (1,2). Most of the substrates can themselves be created by a molecular beam epitaxy process, in the same MBE deposition chambers, hence the name O/I-MBE (1a,b). Since the side chains which control molecular architecture in LB and SA thin films are missing in the MBE-deposited materials, van der Waals contact of the molecular units can occur in three dimensions, leading to

[1]Current address: Digital Equipment Corporation, 77 Reed Road, Mail Stop HL02–3/L12, Hudson, MA 01749
[2]Current address: Geo-Centers, Inc., 10903 Indian Head Highway, Fort Washington, MD 20744
[3]Corresponding author

unique optical, electrical and photoelectrochemical properties (1;2,d,e). These ordered thin films are of interest for the development of new photonic materials (3), the study of energy and charge transfer processes relating to energy conversion and thin film photoconductivity (1;3,c,d,e), and for the study of chemisorption processes on crystalline organic dye thin films such as the phthalocyanines and perylenes (4). Our previous studies of gas sensing using thin films of polycrystalline phthalocyanines have shown the need for the development of ordered ultrathin organic films which minimize the energetic distribution of chemisorption and charge-trapping sites (4,5). Molecular architecture of thin films of crystalline organic molecules will play a major role in such chemisorption processes. Based upon the successes recently realized for chemical sensors produced from epitaxial layers of metal oxides (6), control of molecular architecture in crystalline organic thin films promises to improve sensitivity, selectivity and response time of sensors based on optical and electrical property changes which occur during analyte chemisorption.

Substrates for most of the organic-MBE studies to date include the basal (0001) planes of SnS_2, MoS_2 and HOPG, cleaved single crystals of halide salts such as KCl and KBr, hydrogen-terminated silicon [Si(111)-H] and Cu(100) (1,2,7). Commensurate lattices are often difficult to form, owing to the mismatch in size and symmetry of the overlayer and substrate surface unit cells. Coincident lattices are formed in many cases, however, with unit cells which may be quite close to those in the bulk structures of the molecule (1,2). The strength of interaction between the substrate and first closest packed monolayer is critical to the final molecular architecture in the thin film. In the studies reported to date the Cu(100) surface represents one of the most strongly interacting substrates, while SnS_2, MoS_2, HOPG and Si(111)-H presumably interact with these aromatic hydrocarbons mainly through van der Waals forces. The packing structures of such monolayers are derived mainly from reciprocal lattice data, obtained from low energy electron diffraction (LEED) and reflection high energy electron diffraction (RHEED) studies of these systems, with methodologies described elsewhere (1,2). These techniques produce high quality diffraction data for surface lattices which are ordered in domains with dimensions of at least 100-10,000Å. Where applicable, visible absorbance and/or reflection/absorption spectra of these thin films have also been obtained, to further characterize the local molecular architecture (1e-g;3c-e). The coupling of transition dipoles in some of these dyes is quite strong for epitaxial thin films, and produces spectral shifts and bandwidths which are sensitive to molecular ordering over distance scales of less than 100Å.

In this account we review the packing structures for monolayers of divalent and trivalent metallated phthalocyanines (e.g. CuPc, InPc-Cl) and coronene on the basal planes of the weakly interacting metal dichalcogenides, and for demetallated and divalent metal phthalocyanines and naphthalocyanines on the Cu(100) surface. These ordered ultrathin films are likely to be critical building blocks in new photonic and sensor materials. For the simplest monolayer, coronene on MoS_2, modeling studies show that the accumulation of van der Waals (VDW) interactions between substrate and overlayer, over a large enough coronene surface lattice, plays a decisive role in the orientation of the coronene unit cell. Our preliminary studies suggest that this approach also describes the interaction of Pc's on similar substrates.

Phthalocyanine and Naphthalocyanine Monolayers on   Copper: Pc/Cu(100) and
NPc/Cu(100)

The LEED data and equivalent real space schematic lattices are shown in Figure 1 for
the divalent metallated Pc's and for the demetallated and copper NPc's. For LEED
characterization of these large molecule overlayers electron beam excitation energies
of 15-20 eV are needed to image the reciprocal lattices, with the result that more than
one Laue zone is generally detectable in these studies. The additional Laue zones are
often essential, however, to unambiguously determine the structures of these multiple
domain overlayers.

The four-fold symmetric Cu(100) surface allows for strong interactions with the
organic overlayer, and the formation of epitaxial Pc and NPc thin films with a minimal
number of equivalent domains (2f,g;7,8). Minimizing the number of equivalent
domains is relevant to the exploitation of unique optical and/or chemical sensor
properties anticipated for these thin films.   Other examples of the use of four-fold
symmetric substrates include the deposition of phthalocyanines and naphthalocyanines
on KCl and KBr single crystal surfaces, which produce one or two equivalent domains,
resulting from slight variations in unit cell spacing of the two substrates (2b). It is
assumed that ionic defects in these substrates act as nucleation sites for the Pc
overlayer, and that the spacing of these sites controls the packing structure of the Pc
monolayer.

In the case of the Cu(100) surface our studies and those of previous
investigators (2f,g,8) have shown that a variety of divalent metal Pc's (CuPc, FePc) and
the demetallated (H_2Pc) will form commensurate lattices, with Pc-Pc spacings close to
14Å. These unit cells are rotated by ±21.8° to accommodate the spacing requirements
of the Cu(100) unit cell, and the distances of closest approach for the Pc's. Thin films
with two equivalent domains are therefore formed from this MBE process. These Pc
thin monolayers are difficult to desorb as intact molecular species from the Cu(100)
surface (8). It has been presumed that the strong interaction of the Pc with the copper
surface is due to interactions with bridging nitrogens in the Pc ring (2f). The packing
structure in Figure 1a is drawn using that assumption, with the center of the Pc ring
over Cu a-top sites, and the bridging nitrogens proximal to other a-top Cu sites.

Figure 1b shows LEED (reciprocal lattice) images that reflect the fact that the
naphthalocyanines CuNPc and H_2NPc can produce either single domain overlayers,
with a uniaxial (R = 45°) surface unit cell, or a two-domain lattice (R = ± 38.7°). In
studies to date it appears that whether one uniaxial NPc domain, or two equivalent
domains are formed, is controlled by the substrate temperature and deposition rate,
suggesting that surface migration of the NPc to an energetically favorable adsorption
site is important. This issue is currently under additional study. The LEED data for
the single domain NPc/Cu(100) monolayer suggest a unit cell dimension $b_1=b_2=18.0$Å,
which is larger than the unit cells seen for epitaxial NPc layers on single crystal salts
$b_1=b_2=16$-17Å (9). It can be seen, however, that this spacing is the smallest which
provides for adsorption of the center of the molecule over a Cu a-top site, and
placement of the four bridging nitrogens in proximity to the same types of Cu a-top
sites seen in the Pc/Cu(100) structure. The two-domain NPc monolayer has unit cell
dimensions, $b_1=b_2= 16.4$Å. The schematic in Figure 1b again places the center of the
NPc over an atop copper site, but with slightly different orientations of the bridging
nitrogens, with respect to the copper lattice, than observed for the uniaxial NPc surface
lattice. The NPc's in Figure 1b have also been internally rotated 12.5° with respect to

the main axis of the molecule to accommodate a packing structure which minimizes the interaction between adjacent hydrogens.   Further optical characterization, and modeling studies similar to those described below will be necessary to discriminate other possible unit cell structures.

It is notable that these types of architectures appear to be sustainable over multiple Pc and NPc layers, and that they accommodate unique packing structures of other dissimilar organic dyes (e.g. perylenes) during the formation of organic/organic' multilayers and superlattices (1h, 8).   For example, we have recently shown that the CuPc/Cu(100) monolayer supports epitaxial deposition of PTCDA thin layers, with a packing structure which is completely different than seen for a PTCDA/Cu(100) monolayer, and the structures observed for monolayers of PTCDA on HOPG and MoS$_2$ (1h,8). The prospect is good, therefore, for unique optical and electronic heterojunction devices from these kind of deposition strategies.

Phthalocyanines and Coronene on Layered Semiconductors: InPc-Cl on MoS$_2$(0001) and SnS$_2$(0001) and Coronene on MoS$_2$(0001)

Figure 2 shows the LEED diffraction data, schematics of this data, and proposed packing structures for the trivalent metal Pc's (e.g. InPc-Cl), and coronene on MoS$_2$ and SnS$_2$ surfaces (1g;2j,k,l;3d,e;7).   Comparable LEED studies of modified perylene monolayers on both the Cu(100) and MoS$_2$ surfaces will be reported elsewhere (1h,8). Of the various large molecule adsorbates studied to date, only coronene has a symmetry in the plane of the overlayer which matches the six-fold symmetry of the substrates, and all of the epitaxial layers formed consist of multiple equivalent domains, starting in the first monolayers deposited (7).   The LEED data and packing structures for Pc's on both MoS$_2$ and SnS$_2$ have been discussed previously (1). 3 x 2 coincident lattices are formed, R = 0° (MoS$_2$) and R = ±4° (SnS$_2$) (Figure 2a). As discussed below, this difference in packing of the Pc monolayer apparently arises primarily from the difference in sulfur-sulfur spacing on the two basal planes (3.1Å - MoS$_2$, 3.5Å - SnS$_2$). Three equivalent domains are formed on the MoS$_2$ surface and six on SnS$_2$, the intersection of which are sites for propagation of disorder in epitaxial Pc thin films (1). Nevertheless, these substrates have supported the formation of ordered Pc layers with thicknesses up to ca. 40 - 50 equivalent monolayers.

Figure 2b shows one of the six-fold symmetric coronene domains formed on six-fold symmetry surfaces like MoS$_2$ (7,11). The unit cell dimensions suggested from the LEED data (b$_1$ = b$_2$ = 11.4Å) are consistent with closest packing of coronene in the first monolayer, at near the expected van der Waals distances, but rotated off the principle axes of MoS$_2$ by ±13.9°. The primary packing structure shown in Figure 2b is slightly different from a previously proposed structure (2j,k), in that the coronenes have experienced rotation (ϕ = 6.5°) internal to the unit cell to minimize repulsive interactions between adjacent hydrogens.   The factors controlling the primary orientation of the Pc, perylene and coronene unit cells, and those which control the possible origin sites for nucleation of these epitaxial layers, are discussed further below.

It is clear that the metal dichalcogenides are amenable to the formation of a variety of epitaxial organic and inorganic overlayers, with formation of coincident lattices facilitated by the weak interactions of the basal planes of these materials with

LEED Reciprocal lattice images

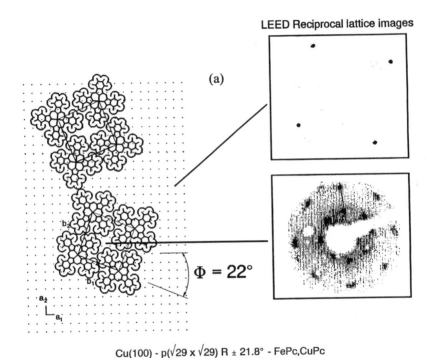

(a)

$\Phi = 22°$

Cu(100) - p($\sqrt{29}$ x $\sqrt{29}$) R ± 21.8° - Fepc,CuPc

Figure 1 --   (a) LEED data and proposed packing structures for the Cu(100) surface (E_{beam} = 80 eV), and that surface covered with an equivalent monolayer of either FePc or CuPc (E_{beam} = 15.5 eV);

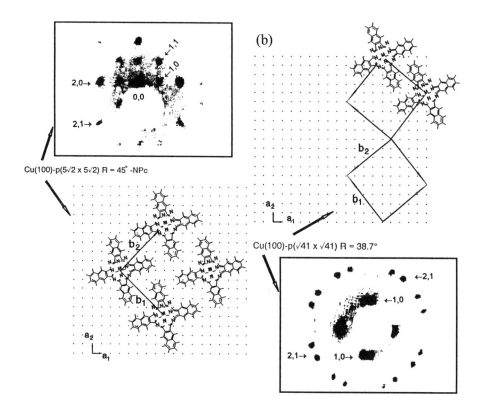

Figure 1 -- (b) the Cu(100) surface covered with an equivalent monolayer of CuNPc or H_2NPc (E_{beam} = 15.5 eV). For the naphthalocyanine overlayers, two different packing structures have been observed, a uniaxial single domain monolayer (left) and a two domain rotated monolayer (right). For both these structures the low index diffraction spots are indicated for the multiple Laue zones typically observed.

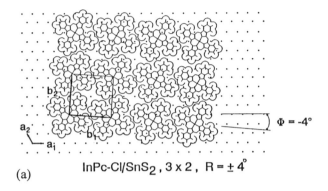

(a)   InPc-Cl/SnS$_2$, 3 x 2 , R = ± 4°

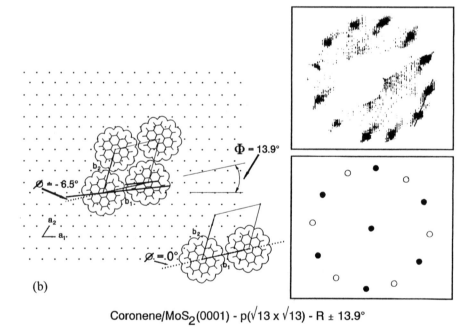

(b)

Coronene/MoS$_2$(0001) - p($\sqrt{13}$ x $\sqrt{13}$) - R ± 13.9°

Figure 2 --   A packing structure schematic for (a) the InPc-Cl/SnS$_2$ monolayer
(References 1c and 11) , (b) LEED data, LEED schematic and
proposed packing structures for an equivalent monolayer of
coronene on MoS$_2$ (Reference 11).

organic overlayers (1,2,12). Their electrical and optical properties appear to be tunable over a wide range by the simple mixing of chalcogens during the formation of the substrate material -- a process which is relatively straightforward to implement by MBE (1,12,13). They will remain a substrate of choice for a variety of emerging technologies, despite the difficulty in forming single domain organic overlayers. The influences they exert on overlayer packing structures are worth further exploration.

Cumulative van der Waals interactions between Coronene and the MoS_2 basal plane

Recent studies have demonstrated that the accumulation of van der Waals interactions over monolayer lattices of long chain hydrocarbons may determine the packing structures of those layers on substrates like HOPG and MoS_2 (14). In recent studies we have also assumed that the packing structures of simple adsorbates like coronene/MoS_2 could also be described on the basis of VDW forces alone (11).

Bulk structures for both coronene and MoS_2 were imported from the Cambridge Crystallographic data base into Sybyl, on a Silicon Graphics Iris workstation. Adjustments were made to create small rhombic arrays of the uppermost S-Mo-S unit cells, and single, flat lying coronene molecules, or small (seven molecule) closest packed lattices of flat lying coronene, with structures consistent with the surface electron diffraction data above. Calculations followed using the Tripos 5.2 force field to estimate the total VDW energy of the system by means of a simple Leonard-Jones 6-12 potential model, which effectively reaches a zero value when atom separations exceed 8.0Å. These modeling processes first involved determining the total van der Waals energy of a flat, seven molecule coronene lattice, as a function of internal rotation angle (ϕ), with no interaction with a substrate. Fluctuations of less than 1% in VDW energy were seen as ϕ was changed. The most stable lattice was found for $\phi = 6.5°$ (Figure 2b).

The MoS_2 substrate and the overlayer molecule(s) were next defined as separate aggregates and the overlayer and substrate were oriented into some desired configuration (e.g. some particular azimuthal angle of the overlayer with respect to the substrate). These structures were then merged into a new molecular area and a new VDW energy was calculated, considering interactions between each atom in the coronene lattice and all of the sulfur and molybdenum atoms within 8Å of those atoms. This process was first carried out for overlayers of one coronene molecule and then coronene overlayer sizes of up to seven molecular units. Our calculations were limited to such overlayer aggregate sizes due to availability of disk space and processor capability. A single coronene molecule, or the point of origin for a full coronene lattice, can be centered above four different surface sites on the MoS_2 surface (S-atop, S-Mo, S-hollow, and S-bridge), such that the molecular plane of the coronene is parallel to the basal plane. The S-atop site places the center aromatic ring directly over a sulfur surface atom, the S-Mo site is directly above the Mo atom in the metal cation plane directly below the surface, a S-hollow site is a three-fold sulfur site, not directly over a Mo atom, and a S-bridging site represents the placement of the point of origin across two adjacent S atoms (1g,7,11b).

Rotation of a single coronene molecule on the MoS_2 basal plane produces the energy fluctuations as a function of ϕ, shown in Figure 3a. The fluctuation in interaction energy with ϕ for the S-atop and S-bridge sites exhibits a sixfold rotational symmetry, $E[\phi]=E[\phi + (n60°)]$. The interaction energy for the S-Mo and S-hollow sites exhibits a fluctuation with ϕ with threefold rotational symmetry, $E[\phi]=E[\phi + n(120°)]$. There are obviously only small absolute and relative fluctuations in van der Waals energy for a single coronene molecule regardless of the type of site chosen for adsorption. From these plots, however, we would predict that there is a slight preference for the S-atop site at the center of the molecule.

The fluctuation in total VDW energy for a two dimensional lattice of seven coronene molecules placed on a 24×24 (S-Mo-S atom pairs) rhombic $MoS_2(0001)$ slab is shown in Figure 3b, as a function of the major angle of rotation Φ between the principal axes of the coronene overlayer and the MoS_2 substrate. These data represent a coronene lattice with an internal rotation angle of $\phi = 6.5°$. Similar data (not shown) have been obtained for $\phi = 0°$ (11). The deepest energy minima (a total VDW energy reduced by ca. 1% of the average energy) occurred when the coronene lattice was centered over a S-atop site at $\Phi \approx +16.0°$ and $\Phi \approx -10.0°$. For $\phi = 0°$ (a slightly more repulsive coronene lattice configuration), the deepest energy minima occurred for the lattice centered over the S-hollow site, $\Phi\approx -13.9°$, which is closer to the orientation deduced from the LEED data.

Comparable studies have been conducted for the InPc-Cl system on both MoS_2 and SnS_2 surfaces, again assuming only VDW interactions (11). As in the case for coronene, small energy minima (ca. 1% fluctuations) were observed for a 9-molecule Pc lattice centered on the S-atop site on SnS_2, at unit cell rotation angles of ca. $\Phi \approx -4°$, close to those predicted from the LEED studies. On the MoS_2 surface, no minima were observed for rotation around the S-atop site. The $\Phi \approx 0°$ rotation, however, had a VDW energy close to the average for all angles, was bracketed by sharp repulsive (1%) maxima, and still represented the lowest possible VDW energy for a Pc lattice like that shown in Figure 2a. For both surfaces the S-atop sites always produced the largest energy fluctuations as the Pc lattice was rotated, as was observed for the coronene lattices. From these calculations it appears that the S-S spacing on the (0001) planes of MoS_2 and SnS_2 helps to control the rotation of these coincident lattices and hence the number of equivalent domains found on these Pc thin films.

These calculations clearly require extension to larger surface lattices before far reaching conclusions can be drawn. We predict that the size of the energy minima will increase in an absolute sense as these lattices are made larger, however the fluctuations may never exceed a few percent of the total VDW energy of the system. This is consistent with the observed difficulty in optimizing ordered layer growth of such aromatic molecules on these van der Waals surfaces. It is likely that nucleation of these thin films starts with small clusters, probably centered over S-atop sites. As the cluster dimensions increase, the accumulated energy savings in a particular orientation will exceed a few kcal/mole, which may be sufficient to guide the growth of the rest of the monolayer. For some systems other intermolecular forces may be significant (e.g. electrostatic forces) and may dominate the packing structure from the earliest stages of film growth. These systems will require a more sophisticated modeling approach to fully understand.

(a)

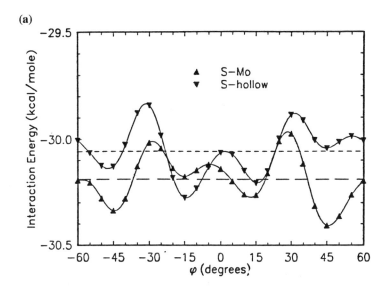

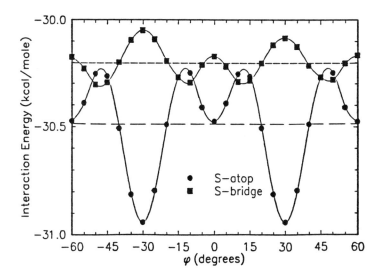

Figure 3a -- Total van der Waals energies for a single coronene molecule on the MoS$_2$ lattice.

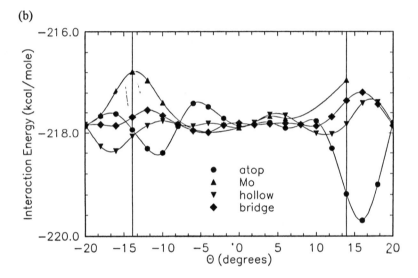

Figure 3b --   Total van der Waals energies for a group of seven coronene
              molecules on the same lattice, as a function of rotation angle.

Conclusions

Vacuum deposition technologies can now be used to create a wide variety of ordered organic thin films whose optical and electronic properties may be as unique as those of the best LB and SA thin films. The ordering created in the thin Pc and NPc overlayers on the Cu(100) surface is now being exploited by us to study the molecular details of small molecule chemisorption on such surfaces, as means of understanding and controlling such phenomena in gas sensors. These surfaces are also intriguing as starting points for multilayers and/or superlattices, where electrical contact is desired to the molecular semiconductor thin film (e.g. in electroluminescent and photoconductive assemblies). We have recently found that thin films of the metal dichalcogenides like SnS_2, which has a bandgap of 2.2 eV, and is transparent in the Pc Q-band spectral region, act as unique substrates for the formation of C_{60}/Pc and PTCDA/Pc multilayers, where epitaxial layer growth can be extended in both organic materials for several lattice periods (1e,g). In the same way that the accumulation of van der Waals forces appears to dictate unit cell orientations of first monolayers, these VDW forces at the interfaces between each organic layer are also anticipated to dictate ordering in organic superlattices (15). This has made the production of ordered superlattices with more than ca. 5 highly ordered lattice periods very challenging, as is often the case in the buildup of LB and SA thin films. Exploitation of molecules with the potential for interactions which exceed the strengths of van der Waals forces alone appears to be a reasonable strategy in the development of the next generation of these materials.

Acknowledgments

This research was supported by grants from the National Science Foundation (Chemistry and Small Grants for Exploratory Research), the Air Force Office of Scientific Research, and the Materials Characterization Program, State of Arizona.

Literature Cited

[1] a) Nebesny, K.W.; Collins, G.E.; Lee, P.A.; Chau, L.-K.; Danziger, J.L.; Osburn, E.; Armstrong, N.R. *Chemistry of Materials*, **1991**, *3*, 829.

   b) Armstrong, N.R.; Collins, G.E.; Nebesny, K.W.; England, C.D.; Chau, L.-K.; Lee, P.A.; Parkinson, B.A. *J. Vac. Sci. Technol A.*, **1992**, *10*, 2902.

   c) Armstrong, N.R.; Arbour, C.; Chau, L.-K.; Collins, G.E.; Nebesny, K.W.; Lee, P.A.; England, C.D.; Parkinson, B.A. *J. Phys. Chem.*, **1993**, *97*, 2690.

   d) Armstrong, N.R.; Chau, L.-K.; England, C.D.; Chen, S. *J. Phys. Chem.* **1993**, *97*, 2699.

   e) Armstrong, N.R.; Collins, G.E.; Williams, V.S.; Chau, L.-K.; Nebesny, K.W.; England, C.D.; Lee, P.A.; Lowe, T.; Fernando, Q. *Synthetic Metals*, **1993**, *54*, 351.

f) Chau, L.-K.; Chen, S.-Y.; Armstrong, N.R.; Collins, G.E.; England, C.E.; Williams, V.S.; Anderson, M.L.; Schuerlein, T.J.; Lee, P.A.; Nebesny, K.W. *Mol. Cryst. Liq. Cryst.*, in press.

g) Anderson, M.L.; Collins, G.E.; Williams, V.S.; England, C.D.; Chau, L.-K.; Schuerlein, T.J.; Lee, P.A.; Nebesny, K.W.; Armstrong, N.R. *Surf. Sci.* in press.

h) Schmidt, A.; Schuerlein, T.J.; England, C.D.; Collins, G.E.; Armstrong, N.R., manuscript in preparation.

[2] a) Hara, M.; Sasabe, H.; Yamada, A.; Garito, A.F. *Jap. Journ. Appl. Phys.* **1989**, *1306*, 28.

b) Tada, H.; Saiki, K.; Koma, A. *Jap. Journ. Appl. Phys.* **1991**, *L306*, 30.

c) Dann, A.J.; Hoshi, H.; Maruyama, Y. *J. Appl. Phys.*, **1990**, *67*, 1371, 1845.

d) Yanagi, H.; Ashida, M.; Elbe, J.; Wörhle, D. *J. Phys. Chem.*, **1990**, *94*, 7056; Yanagi, H.; Kouzeki, T.; Ashida, M. *J. Appl. Phys.*, **1993**, *73*, 3812.

e) Yanagi, H.; Dauke, S.; Ueda, Y.; Ashida, M.; Wörhle, D. *J. Phys. Chem.* **1992**, *96*, 1366.

f) Buchholz, J.C.; Somorjai, G.A. *J. Chem. Phys.* **1977**, *66*, 573.

g) Lippel, P.H.; Wilson, R.J.; Miller, M.D.; Wöll, Ch.; Chiang, S. *Phys. Rev. Lett.* **1989**, *62*, 171.

h) Fryer, J.R.; Kenney, M.E. *Macromolecules* **1988**, *21*, 259 and Fryer, J.R. *Mol. Cryst. Liq. Cryst.* **1986**, *137*, 49.

i) Ashida, M. In *Electron Crystallography of Organic Molecules*; Freyer, J.R.; Dorset, D.L., Eds,; Fluwer Academic Publishers: Netherlands, **1990**, pp. 227-240.

j) Zimmerman, U.; Karl, N. *Surf. Sci.* **1992**, *2168*, 296.

k) Ludwig, C.; Gompf, B.; Glatz, W.; Petersen, J.; Eisenmenger, W.; Möbus, M.; Zimmerman, U.; Karl, N. *Z. Phys. B* **1992**, *86*, 397.

l) Haskal, E.I.; So, F.F.; Burrows, P.E.; Forest, S.R. *Appl. Phys. Lett.* **1992**, *60*, 3223.

m) Weaver, J.H. *Acct. Chem. Res.* **1992**, *143*, 25; and references therein.

n) Sakurai, M.; Tada, H.; Saiki, K.; Koma, A. *Japanese Jour. of Applied Phys.* **1991**, *30*, 1892.

[3] a) Williams, V.S.; Sokoloff, J.P.; Ho, Z.Z.; Arbour, C.; Armstrong, N.R.; Peyghambarian, N. *Chem. Phys. Lett.*, **1992**. *193*, 317.

b) Williams, V.S.; Mazumdar, S.; Armstrong, N.R. Ho, Z.Z.; Peyghambarian, N. *J. Phys. Chem.*, **1992**, *96*, 4500.

c) Terasaki, A.; Hosoda, M.; Wada, T.; Tada, H.; Koma, A.; Yamada, A.; Sasabe, H.; Garito, A.F.; Kobayashi, T. *J. Phys. Chem.*, **1992**, *96*, 10534.

d) So, F.F.; Forrest, S.F. *Phys. Rev. Lett.*, **1991**, *66*, 2649.

e) So, F.F.; Forrest, S.F.; Shi, Y.Q.; Steier, W.H. *Appl. Phys. Lett.*, **1992**, *56*, 674.

[4] Collins, G.E.; Pankow, J.W.; Odeon, C.; Brina, R.; Arbour, C.; Dodelet, J.-P.; Armstrong, N.R. *J. Vac. Sci. Technol.*, **1993**, *11(4)*, 1383.

[5] a) Brina, R.; Collins, G.E.; Lee, P.A. Armstrong, N.R. *Anal. Chem.* **1990**, *62*, 2357.
    b) Waite, S.; Pankow, J.; Collins, G.E.; Lee, P.A.; Armstrong, N.R., *Langmuir*, **1989**, *5*, 797.
    c) Pankow, J.W.; Arbour, C.; Dodelet, J.P.; Collins, G.E. *J. Phys. Chem.*, **1993**, *97*, 8485.
[6] a) Semancik, S.; Cavicchi, R.E. *Thin Solid Films*, **1991**, *206*, 81.
    b) Cavicchi, R.; Semancik, S. *Surf. Sci.* **1991**, *56*, 70.
[7] Collins, G.E. Ph.D. Dissertation, University of Arizona, 1992.
[8] Schuerlein, T.J.; England, C.D.; Armstrong, N.R. *J. Vac. Sci. Technol.* (submitted).
[9] Yanagi, H.; Kouzeki, T.; Ashida, M. *J. Appl. Phys.* **1993**, *8*, 3812.
[10] a) Dahlgren, D.; Hemminger, J.C. *Surf. Sci.* **1981**, *109*, L513.
    b) Prince, K.C. *J. Electron Spec Related Phen.* **1987**, *42*, 217.
    c) Holland, B.W.; Woodruff, D.P. *Surf. Sci.* **1973**, *36*, 488.
[11] England, C.; Collins, G.E.; Schuerlein, T.J.; Armstrong, N.R., *Langmuir*, (submitted).
[12] a) Saiki, K.; Ueno, K.; Shimada, T.; Koma, A. *J. Cryst. Growth* **1989**, *95*, 603.
    b) Ueno, K.; Koichiro, S.; Toshihiro, S.; Koma, A. *J. Voc. Sci. Technol.* **1990**, *A8*, 60.
    c) Ohuchi, F.S.; Parkinson, B.A.; Ueno, K.; Koma, A. *J. Appl. Phys.* **1990**, *68*, 2168.
[13] a) Tributsch, H. *Structure and Bonding* **1982**, *49*, 127.
    b) Physics and Chemistry of Materials with Layered Sturctures, Vols. 1-6; Mooser, E., Ed.; Reidel: Dordrecht, 1976-79.
    c) Electronic Structures and Electronic Transitions in Layered Materials; Groasso, V., Ed., Reidel: Dordrecht, 1986.
[14] a) Rabe, J.P.; Buchholz, S.; Askadskaya, L. *Synthetic Metals*, **1993**, *54*, 339.
    b) Askadskaya, L.; Boeffel, C.; Rabe, J.P. *Ber. Bunsenges, Phys. Chem.* **1993**, *97*, 517.
    c) Rabe, J.P.; Buchholz, S. *Science*, **1991**, *253*, 424.
[15] Zhang, Y.; Forrest, S.R.; *Phys. Rev. Lett.*, **1993**, *71*, 2765.

RECEIVED April 5, 1994

Chapter 19

Surface Sensitivity of Electron Energy Loss Spectroscopy and Secondary-Ion Mass Spectrometry of Organic Films

M. Pomerantz[1,4], R. J. Purtell[2], R. J. Twieg[3], S.-F. Chuang[2,5], W. Reuter[2,6], B. N. Eldridge[2], and F. P. Novak[2]

[1]Chemical Physics Department, Weizmann Institute of Science, 76100 Rehovot, Israel
[2]Thomas J. Watson Research Center, IBM Research Division, Yorktown Heights, NY 10598
[3]IBM Almaden Research Center, Almaden, CA 95120

Electron energy loss spectroscopy (EELS) and secondary-ion mass spectrometry (SIMS) are particularly useful for the study of surfaces because they give information about a very shallow depth (of order nm) of the sample. Precisely what these depths are in organic solids has been difficult to measure because the surface needs to be highly regular. We used Langmuir-Blodgett layers on silicon wafers. The films proved to be well packed and smooth. The chemical nature of the films was arranged to place "markers" at specific depths, which aided in the analysis of the data. We derive remarkably small information depths: less than 0.4 nm for EELS and 2 nm for SIMS.

In order to understand effects at an interface it is necessary to determine the chemical composition of the surface. Fortunately, there are a number of analytical techniques that are sensitive, to a greater or lesser degree, to the molecules at the surface and which do not see into the bulk. In this paper we report our measurements and interpretations of the depth sensitivities of two techniques of chemical analysis that have high surface sensitivities.

   Electron energy loss spectroscopy (EELS) measures the frequencies of the molecular vibrations, and thus may be used to identify molecules. Its surface sensitivity arises from the fact that electrons of about 5 eV in energy are used. These have a short mean free path in matter. The scattering lengths have been extensively studied in inorganic solids (1), but the information depths reported for organic solids show wide variations. Values as different as about 2.5 nm (2) and 0.4 nm (3) have been reported. We shall attempt to understand this discrepancy.

[4]Current address: Cedar Lane Apartments, Ossining, NY 10562
[5]No longer at this address. Current address unavailable.
[6]Current address: Brookdale Road, Mahopac, NY 10541

0097–6156/94/0561–0216$08.00/0
© 1994 American Chemical Society

Secondary-ion mass spectrometry (SIMS) reveals the composition of the surface by measuring the masses of the ions ejected from the surface under bombardment by ions of keV energies. The damage that is done by the incident ions can be made tolerable by the use of the highly sensitive time-of-flight analysis of the mass spectra (4, 5). This is called "static SIMS" because the damage is to less than a monolayer and the sample is effectively constant. Here, also, the reported depth sensitivities have varied widely, between 1 nm (6) and 12 nm (7, 8). We wish to understand this as well.

Part of the problem in making measurements of the surface of an organic solid is in preparing a surface that is uniform, i.e., in both depth and along the surface. With organic solids this is particularly difficult because cleaving or polishing techniques are usually not applicable. As model systems, Langmuir-Blodgett films (9) have many advantages. A single molecular layer is spread and compacted on the surface of water. It is then transferred to a solid substrate. The chemical identity of this monolayer is thus guaranteed.

To produce a smooth film it is necessary to use a well-compacted film on a smooth support. This is where the previous works have had the widest variations. The film should also be free of holes in order that an intrinsic depth can be measured. This may be achieved if the films are of rather straight molecules and sufficient pressure can be applied to pack the molecules tightly without collapsing the films. This gives the desired uniformity along the surface. The smoothest commonly available substrates that we know are electronic grade Si wafers. Langmuir-Blodgett films on Si have been shown to be very smooth (10) and the films seem to bridge over the roughness and thus are smoother than the substrates.

Langmuir-Blodgett films offer a natural means to achieve controlled chemical variations with depth of the film. Molecules of different types may be deposited sequentially, thus making a profile of known chemical composition on a length scale of the thickness of a molecular layer. To make a chemical profile on a smaller scale of depth, one can make substitutions within the molecule of a single layer. We utilize both of these approaches in the present work.

Experimental methods

In order to have layers with distinctive chemical entities at known positions, one family of molecules we chose was the fluorine substituted alkoxybenzoic acids, of the general form $F-(C-F_2)_n-(C-H_2)_m-O-Ph-COOH$ where "Ph" represents a benzene ring connected to the para substituents. They can be prepared with varying numbers of carbons that are fluorinated, n, or hydrogenated, m. We refer to the compounds as FnHm. (The number of fluorines is 2n+1.) When deposited as Langmuir-Blodgett films, the fluorinated end is on the outside. For the EELS measurements we are interested in the extent to which the fluorinated segment acts as a shield between the incoming electrons and the C-H groups further down the chain. As a comparison molecule we also employed the unfluorinated analogs denoted by F0Hm, i.e., $H-(CH_2)_m-O-Ph-COOH$.

For the SIMS experiments, the mass of the F is a distinctive marker. In addition, for the SIMS we used perdeuterated stearic acid layers. The mass of the deuteron is another distinctive marker in the mass spectrum. The details of the layers will be given below.

The semifluorinated alkoxybenzoic acids were prepared by the following general sequence of reactions: First, ethyl 4-hydroxybenzoate was reacted with an ω functionalized (I, Br, or OH) 1-alkene to give the terminal olefin alkylated benzoate. In the case of the halogen terminated olefins, the ether linkage was formed by alkylation using potassium carbonate in N-methylpyrollidinone. In the case of the alcohol terminated olefins, the ether linkage was formed by a Mitsunobu coupling reaction with diethylazodicarboxylate and triphenylphosphine in tetrahydrofuran. Second, the olefinic material was reacted with the perfluoroalkyl iodide catalyzed by AIBN, 2,2'-azobisisobutyronitrile, or triethylborane. Third, the resulting material was dehalogenated with zinc and HCl in ethanol or Bu_3SnH. Last, the ester was saponified with alcoholic base to give the final semifluorinated alkoxy benzoic acid. Details of the preparation will be given elsewhere (11).

Orientation of the molecules as a surface film was produced by the Langmuir-Blodgett method (9). Amphiphilic molecules are spread on the surface of water. The molecules naturally align with the acid (-COOH) ends into the surface of the water. A movable barrier on the surface of the water is used to apply pressure sufficient to obtain the closest packing possible, without collapsing the film. Under continuous pressure the film is then transferred from the water surface to the substrate by smoothly inserting and removing the substrate from the water. The direction of the meniscus confirms that the ionic ends of the molecules attach to the Si and the hydrophobic ends face away from the surface, as usual. We believe it is an important point of technique that we used polished Si wafers (electronic grade, 0.1 Ω-cm, p-type). The smoothness of Langmuir-Blodgett films on these substrates is about 0.3 nm and seems to improve when multilayers of Langmuir-Blodgett films are deposited on them (10). The wafers were cleaned by scrubbing with detergent, thorough washing with water and spin drying. (We remark that no traces of the cleaning solution or ions from the bath, such as Cl, were found by secondary ion mass spectrometry, as can be seen in the SIMS spectra shown below.)

The Langmuir-Blodgett films of the alkoxybenzoic acids were all made in the same way: each of the FnHm molecules were spread in hexane solution onto a water bath containing about 10^{-4} M $AlCl_3$, held at a pH of about 5.5. The Al^{+3} reacts with the COOH to form an ionic compound (12, 13). We used Al as the counterion here because these fluorinated molecules collapsed at low surface pressures when the usual divalent ions were used in the bath. With trivalent Al they are remarkably strong; the films were capable of withstanding much more than the 30 dyne/cm. pressure applied during deposition. This is a sign that good packing may be achieved, which is important in avoiding pinhole artifacts.

The stearic acid and perdeuterated stearic acid (D-$(C$-$D_2)_{17}$ -COOH) are the purest commercially available and were used without further purification. These molecules were spread in a concentration of 1 mg per ml hexane solution on a bath containing about 10^{-3} M $MnCl_2$ at a pH of about 6.6. The Mn replaces the carboxylic acid hydrogen. Thus the perdeuterated layer contains no hydrogen. The deuterium thus serves as a marker isotope for the SIMS. These films were deposited at a surface pressure of about 20 dynes/cm. We shall refer to them as "MnDSt".

Information Depth of EELS

In our experiments, the electron energy loss spectra were measured in a Leybold-Heraeus ELS 22 spectrometer, with incident electron energy of 5 eV. We report data taken under specular conditions, at an incident angle of 60° from the normal. The current density was about 5×10^{-11} amp/mm^2, and the pressure was about 10^{-10} Torr. The energies of the reflected electrons were analyzed with a resolution of about 7 meV. The energy lost by an incident electron, ΔE, is transferred to an excitation of a molecular vibration. The EELS spectrum measures the number of electrons that have a specified energy loss.

We wish to measure the "information depth" to which a particular molecular vibrator can be detected, by the characteristic energy loss it creates. It should be understood that an electron may lose energy to a vibration, but not contribute to an EELS spectrum if it suffers additional scatterings that knock it out of the detection windows set for the instrument. The information depth thus depends on both the energy resolution and angular resolution of the instrument.

An experimental approach to the measurement of the information depth, ID, is as follows. The molecular vibrator, whose frequency is known, is placed at a known distance below the surface. The fraction of electrons that lose energy to this, and only this, vibrator is measured as a function of the number of covering atoms. The covering thickness that reduces the relative intensity of this energy loss by a factor 1/e is a measure of the information depth to which this vibration can be detected.

Interestingly, the information depth depends on the mode of excitation of the marker. If the marker is excited by the dipolar mechanism, which is long-range, the depth is predicted to be about 3 nm (14). If the excitation is by an impact mechanism, the information depth will be influenced by the electron mean free path in the solid. There is abundant evidence (15, 16), and our experiment reported here confirms, that the common C-H vibration of organics is excited by impact. The C-H vibrators are an important example for obtaining the ID in an impact scattering process.

For these experiments we used the semi-fluorinated alkoxybenzoic acids as the Langmuir-Blodgett film. The entire 1 inch wafer was covered with one monolayer of film. Ellipsometry gave values of about 1.6 nm for the thickness of the monolayer, which is reasonable for molecules of this length standing on the surface. We use the excitations of the buried C-H groups as the indicator vibrations. The information depth of the scattered electrons is determined from the number of fluorinated carbons at the ends of the molecule, i.e., the coverage, which reduces the signal from the C-H group. The C-F stretch vibration excitation is at an energy of about 140 meV, which is much different from the energy of the C-H stretch, 360 meV, and so cannot be confused with it. (See Figure 1).

The data in Figure 1 are for single monolayers of FnHm molecules on Si wafers. The electron energy loss spectra in this series of samples are normalized to their respective elastic peaks. Thus we are measuring the information depth by the changes in the CH stretch losses relative to the elastic peak. By using this procedure, we have neglected changes in the overall spectral intensities due to such things as roughness and ordering variations from sample to sample. These effects may change the information depth we deduce but we presume them to be small because the packing is determined by the bulky benzene ring, which is the same for all the molecules.

Curve a is for the unfluorinated molecule F0H16. It shows a strong loss at about 360 meV, characteristic of the C-H stretch (17); this will be our marker vibration. There are also broad losses centered at about 150 meV due to a variety of C-H and C-C vibrations. When the Langmuir-Blodgett film is of the molecule F2H8, the two outer carbons are fluorinated. The EELS spectrum, curve b, shows a relatively smaller C-H stretch loss compared to curve a. This occurs because the C-H groups are now shielded by the overlying C-F groups, but perhaps also because there are fewer C-H groups. This uncertainty is decided by the measurements on films of molecules F4H6 (with 4 fluorinated C's on the outside), F6H6 (with 6 fluorinated C's on the outside) and F8H8 (with 8 fluorinated C's), c.f., curves c, d, and e of Figure 1. The numbers of C-H groups are not much changed among the last three molecules, but the intensities of the C-H losses decrease noticeably. This is shown in Figure 2, in which the relative intensities of the C-H stretch losses are plotted vs. the number of fluorinated C's. From this figure we deduce that there is an attenuation of the signal from the C-H stretch by about a factor of 1/3 (about e^{-1}) for each pair of $C-F_2$ groups between the C-H groups and the electron beam. This immediately gives a remarkably short information depth of about 0.3 nm, the length of two C-C bonds. This is conclusive evidence that the excitation of the C-H groups is by impact scattering. In the dipolar process, the range of the interaction is ten times larger, about 3 nm (14). If the scattering were dipolar, the interposing of even eight $C-F_2$ groups would not much reduce the interaction with the underlying $C-H_2$ groups.

The result for the information depth is to be regarded as an upper limit because there may be extrinsic effects, such as holes in the film or poor packing, etc., that could lead to the uncovering of some C-H groups. But since our result is that the excitation depth is only about 2 atoms, the extrinsic effects must be virtually nil. Variations in overall signal level in the specular direction from sample to sample would also change our result. However, as noted above, within this group of samples we do not expect a large variation in surface ordering or roughness, which would change the relative elastic specular scattering intensity.

There is evidence from our computer models of the packing (18) that some C-H is exposed even with near optimal packing for the case of 2 fluorinated carbons. This may explain why there is less effect of the two fluorinated C in F2H8 in reducing the C-H signal than expected by comparison with longer fluorinated chains. (See Figure 2).

Our result, that the excitation of the C-H vibration is reduced e-fold by two $(C-F_2)_2$ groups, or ID $\leq$ 0.3 nm, is in reasonable agreement with the result of Schreck, et al (3). It is remarkable that, despite being contained in a Van der Waals solid, vibrators beneath a mere two atomic layers begin to be hidden from an EELS measurement.

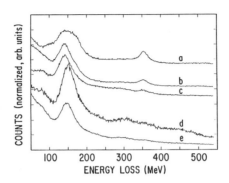

Figure 1. EELS spectra of single Langmuir-Blodgett films of the Al salts of FnHm molecules on Si wafers.   Curve a is F0H16, b is F2H8, c is F4H6, d is F6H6, e is F8H8, which is in the order of increasing coverage of the C-H groups by C-F.   The ordinate is the number of counts normalized to the number of counts at the peak of the elastic scattering, for n = 0. The rationale for the normalization is given in the text. The abscissa is the energy loss by the electrons in the specularly scattered beam.

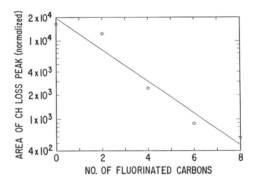

Figure 2. The relative number of electrons which lose energy to a C-H vibration, vs. the number of fluorinated carbons above the C-H groups.   For the normalization, see the caption of Figure 1.

Information Depth in SIMS

In the SIMS experiments we use a similar strategy to determine the depth from which secondary ions are detected. We cover a marker, this time a distinctive mass, with another distinctive layer of material. In SIMS, the length scale is larger than that in EELS; it is of the order of several nm rather than fractions of a nm. Thus the marker and covering masses are not chosen to be in the same molecule, but are placed in different molecular layers. One sample was constructed using the molecule F4H6, F-$(C-F_2)_4$-$(C-H_2)_6$ -O-benz-COOH. Here fluorine is a distinctive mass marker. The covering film was per-deuterated manganese stearate, $(D-(C-D_2)_{17}$ -COO$)_2$ Mn. The deuterium serves as a marker isotope. Three layers of the F4H6 films were deposited over the entire Si surface, and two layers of MnDSt films were coated over the F4H6, but only over half of the surface (see inset of Figure 3). This small innovation in sample design allows us to avoid one of the problems with SIMS measurements, that of comparing samples made at different times and measured in different runs. With this dual sample, we measure the SIMS of the three layer part, and then with no change in the SIMS parameters, we move the sample a small distance. We can then observe the effect of the additional two covering layers on the SIMS of the continuation of the same three layer sample.

Our static SIMS system (19) operated with an Ar^+ energy of 10 KeV. UHV conditions were maintained. The resolution of the instrument is about 1 part in 17,000.

In Figure 3 is shown part of the mass spectrum of positive ions ejected from the sample described above and illustrated in the inset. The right side is taken from the area covered only with F4H6. It shows a very strong peak (>24,000 counts) for a large fragment of F4H6 with no Al but a H^+ added. On the left side, the same mass region is shown, as observed from the area in which the F4H6 is covered with two monolayers of MnDSt. The peak due to the fluorinated ion is only about 1000 counts, for the same counting time. Thus, the two MnDSt layers reduce the emission by a factor of > 24. The other peaks on this part of the sample correspond to fragments from the MnDSt films.

One might ask whether the large attenuation of the ejection of the entire F4H6 molecule is also found for smaller fragments. In fact, the attenuation ratio is similar for lighter fragments as well, as illustrated in Figure 4. This shows the spatial variation of the $(C_2F_4)^+$ ion as the sample is moved to direct the beam on the covered and the uncovered F4H6 film. Noting the log scale of intensity, it is found that the $(C_2F_4)^+$ ions decrease by a factor of 18. This gives an attenuation of a factor of 4 per layer, rather similar to the factor of 5 per layer for a full F4H6 molecule which has about a four fold larger molecular weight. In the region of the MnDSt film we observe, of course, the appearance of deuterated ions, such as the $(C_2D_5)^+$.

The attenuation of even smaller fragments, such as the Si^+ from the substrate, are also strongly attenuated by as few as two monolayers. In Figure 5 is shown the low mass part of the spectrum, from an area covered by one monolayer of (hydrogenated) MnSt, (data on the left), and an adjacent area covered in addition by two layers of MnDSt, or a total of three layers. With the single layer, the signal of Si^+ at mass = 28 daltons saturates the detectors and is offscale at 8×10^6 counts;

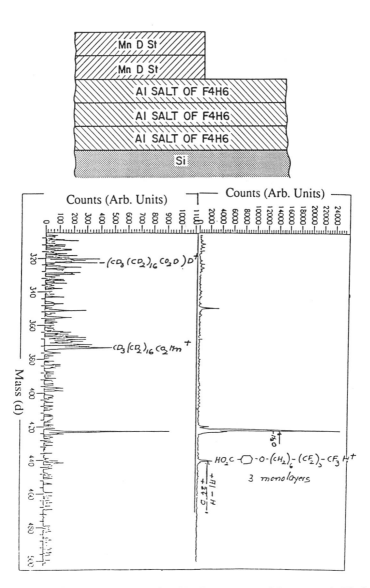

Figure 3. The heavy mass part of a SIMS spectrum of 3 Langmuir-Blodgett films of the Al salt of F4H6, on a Si wafer, is shown to the right side. The spectrum on the left is of the same range of masses, from the same films, but covered with two Langmuir-Blodgett films of Mn deutero stearate.

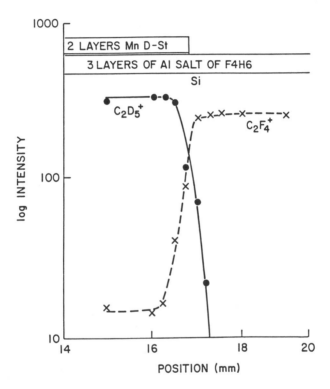

Figure 4. The same sample as Figure 3, but the dependence of emission of light mass fragments is depicted as a function of position as the edge of the Mn deutero stearate film is moved under the ion beam.

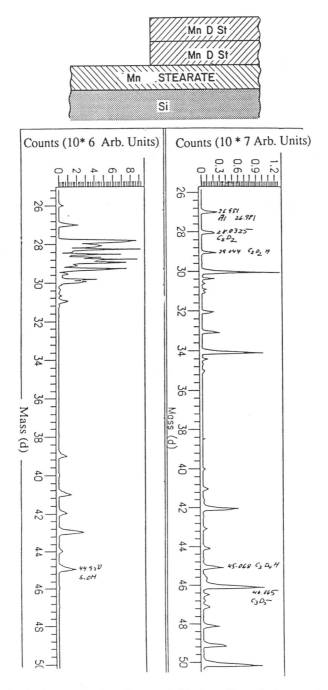

Figure 5. SIMS spectra of one Langmuir-Blodgett film of Mn stearate (on the left) and the same film when covered with two Langmuir-Blodgett films of Mn deutero stearate (on the right). The emission of Si⁺ from the substrate is much reduced by the addition of two monolayers.

one is observing the ringing due to the overload. With the addition of 2 monolayers of MnDSt we observe practically no Si. The peak at 28.0325 daltons is due to $C_2D_2^+$, which can be resolved from Si in this apparatus. These data yield an attenuation of the Si^+ monatomic ion signal by at least a factor of 20 due to the coverage by two layers of MnDSt. Thus, even for ions consisting of single atoms, one monolayer attenuates the emission by at least a factor of 4.

Conclusions

The general result of our measurements is that the information depths we find are at the lower end of the values reported previously in the literature. For EELS, the depth had been stated to be as much as 2.5 nm (2), and as small as 0.4 nm (3). Our value of 0.3 nm is close to the smallest reported value. The reasons for the differences among reported values has likely to do with the choice of layer material and the surface roughness of the substrate. The large depth was observed in a sample in which a polymer, PMMA, was the covering layer. One might question whether a polymer deposited by the Langmuir-Blodgett technique would be as uniform as simple molecules of uniform size.

The same comments about the necessity of smooth substrates and uniform films apply to the SIMS measurements. In published experiments in which the substrate was Ag or Au films on glass (8), the covering Langmuir-Blodgett films did not attenuate the Ag^+ ions as thoroughly as did our films on Si substrates. The surface roughness reported there was less than 1 nm, and the films were stearic acid deposited at rather low surface pressure. The uniformity of such layers is questionable. For these films, at least 6 layers were needed to reduce the monatomic ions from the substrate by a factor of ten. In another report (20) in which the sample was Ba stearate on Ag foil, the ratio of Ag^+ to CH_3^+, from Langmuir-Blodgett overlayers, decreased by a factor of 2 per layer of Ba stearate coverage. This overestimates the attenuation, since the intensity of the CH_3^+ will increase as the coverage is increased, so it gives an estimate of attenuation/layer of less than a factor of two per layer. Our measurements give a much greater attenuation of emitted ions by our well packed Langmuir-Blodgett films on a smooth surface. As few as two layers reduce the substrate ions by a factor more than twenty. This yields an attenuation of Si^+ by a factor greater than four per layer, which is at least twice as large as reported in reference 20. Our result is closer to that reported for Mg and Cd arachidate on Au-covered glass, in which a resolution of about 1 nm was obtained (6).

Our general conclusion is that the preparation and nature of the films and the substrate have important influences on what has been called (8) the "information depth". This is the depth from which signals are observed, and depends on the sample and the instrument. One must then enquire about the details of the experiment in order to interpret such results. The information depths are likely to depend on the binding and structure of the organic solid. Thus careful measurements on a variety of materials are needed before any wide generalizations can be made.

It should perhaps be reiterated that these results are an upper limit of the information depths. Defects in the films, such as pinholes, would tend to expose atoms further away from the surface to the electron beam, thus giving an apparently larger information depth in EELS. Defects are expected to allow underlying secondary ions to escape more readily, thus increasing the apparent SIMS information depth. It would be desirable to have an independent method of measuring the defects in the films, but in such a polycrystalline layer there seems to be none. (Atomic force microscopy might eventually provide a means, but the interpretation of the images is still uncertain.) As mentioned above, the fact that our measured upper limit for the EELS information depth is only two atoms is very good evidence that pinholes are not significantly affecting our results, to the accuracy we can measure. It suggests, in fact, that EELS may be an excellent method for detecting holes in very thin films.

Literature Cited

(1)   Seah, M. P. In *Practical Surface Analysis*; D. Briggs and M. P. Seah Eds., Wiley: NY, 1983. p. 181 - 216.

(2)   Pireaux, J. J., Gregoire, C., Caudano, R., ReiVilar, M., Brinkhuis, R. and Schouten, A. J. *Langmuir* **1991**, *7*, 2433 - 2437.

(3)   Schreck, M., Abraham, M., Lehmann, A., Schier, H. and Göpel, W. *Surface Sci.* **1992**, *262*, 128 - 140.

(4)   Benninghoven, A. *Surf. Sci.* **1973**, *35*, 427.

(5)   Briggs, D., Brown, A. and Vickerman, J. C. *Handbook of Static SIMS*; Wiley and Sons: N. Y., 1989.

(6)   Laxhuber, L., Mohwald, H. and Hashnii, M. *Int. Jour. Mass Spectrom. and Ion Phys.* **1983**, *51*, 93.

(7)   Wandess, J. H. and Gardella Jr., J. A. *Surf. Sci.* **1985**, *150*, L107 - L114.

(8)   Hagenhoff, B., Deimel, M., Benninghoven, A., Siegmund, H. and Holtkamp, D. *J. Phys. D* **1992**, *25*, 818-832.

(9)   Ulman, A. *An Introduction to Ultrathin Organic Films*; Academic Press: N. Y., 1991.

(10) Pomerantz, M. and Segmüller, A. *Thin Solid Films* **1980**, *68*, 38.

(11) Twieg, R. J., Betterton, K. and Nguyen, H. T. unpublished results.

(12) Wolstenholme, G. A. and Schulman, J. H. *Proc. Farad. Soc.* **1950**, *46*, 475.

(13) Aveyard, R., Binks, B. P., Carr, N. and Cross, A. W. *Thin Sol. Films* **1990**, *188*, 361 - 373.

(14) Gadzuk, J. W. In *Vibrational Spectroscopy of Molecules on Surfaces*; J. T. Yates, Jr. and T. E. Madey Eds., Plenum: New York, 1987. p. 49 - 103.

(15) Demuth, J. E., Avouris, P. and Schmeisser, D. *J. of Electron Spectroscopy and Related Phenomena* **1983**, *29*, 163 - 174.

(16) Dai, Q. and Gellman, A. J. *Surface Sci.* **1991**, *257*, 103 - 112.

(17) Avram, M. and Mateescu, G. D. *Infrared Spectroscopy: Applications in Organic Chemistry*; Wiley: New York, 1972.

(18) Unpublished results of T. Jackman.

(19) Eldridge, B. N. *Rev. Sci. Instrum.* **1989**, *60*, 3160.

(20) Wandass, J. H., Schmitt, R. L. and Gardella,Jr., J. A. *Appl. Surf. Sci.* **1989**, *40*, 85 - 96.

RECEIVED March 25, 1994

CHEMICAL SENSOR DESIGNS

Chapter 20

Electroanalytical Strategies with Chemically Modified Interfaces

H. D. Abruña[1], F. Pariente[1], J. L. Alonso[2], E. Lorenzo[1], K. Trible, and S. K. Cha[3]

Department of Chemistry, Baker Laboratory, Cornell University, Ithaca, NY 14853-1301

Various analytical strategies based on the use of chemically modified electrodes are presented. We describe the use of conventionally sized as well as microelectrodes in the determination of mercury and copper ions in solution and point to the potential utility of modified electrodes in carrying out speciation studies. Also discussed is the use of carbon fiber microelectrodes modified with alkaline phosphatase as amperometric biosensors. Finally, we present some preliminary findings on the development of a modified platinum electrode for the determination of nitric oxide.

The application of chemically modified electrodes (CME's) in novel analytical strategies and sensors represents a very active area of investigation [1-8]. This is due in part to the inherent advantages that derive from their use which include high sensitivity and specificity. In addition, there are now numerous materials and methodologies that can be employed in electrode modification. Furthermore, many of these methodologies can be coupled to the use of microelectrodes, so that analytical studies on very small samples including single cell specimens, may be carried out [9]. As a result, numerous analytical applications of chemically modified electrodes have been reported. [10-17]

We have sought to exploit the advantages of polymer modified electrodes for the determination of transition metal ions and organic functionalities [18-23]. Our methods are based on the preconcentration of the analyte (metal ion or organic functionality) at the electrode surface by modifying the same with functionalized polymers that carry reagents for the selective and sensitive determination of the species of interest.

[1]Current address: Departamento de Química Analítica y Análisis Instrumental, Universidad Autónoma de Madrid, 28049 Madrid, Spain
[2] Current address: Biosensores Sociedad Limitada, Centro Europeo de Empresas Innovación, Paterna, Valencia, Spain
[3]Current address: Department of Chemistry, Kyungnam University, Masan, 630-701, Korea

The analysis is based on the electrochemical determination of the amount of immobilized analyte/reagent complex. This serves as the analytical signal which is then related to the concentration of the species of interest. We have demonstrated the applicability of this approach to the determination of metal ions such as iron, copper, cobalt, nickel, calcium and silver and to organic functionalities such as aromatic amines and aldehydes. [18-23].

In addition, we have previously demonstrated that chemically modified electrodes may allow for a new approach to metal ion speciation studies by employing a family of ligands whose formation constants for the metal ion of interest span a broad range. [24,25]

Most recently, some of us have expanded our analytical interests to the development of amperometric biosensors [26-28].

Clearly, the use of chemically modified electrodes lends itself to a multitude of analytical applications. In this manuscript we present various analytical strategies based on CME with emphasis on:

1. determination of mercury with surface immobilized amino acids
2. determination of copper at surface modified microelectrodes
3. amperometric biosensor based on carbon fiber microelectrodes modified with alkaline phosphatase
4. sensor for the determination of nitric oxide (NO)

Experimental:

Materials and general procedures were as previously described [18-28]. Platinum microelectrodes were prepared as previously described [29]. In the studies on the determination of mercury or copper, electrodes were modified with poly-[Ru(v-bpy)3]$^{2+}$ by electroinitiated polymerization [30] of the monomer complex (typically at 0.5 mM concentration) from acetonitrile/0.1 M TBAP solution by scanning the potential (typically at 100mV/sec) between 0.0 and -1.40 V for a prescribed amount of time depending on the desired coverage. The exact coverage was determined by measuring the charge under the voltammetric wave for the Ru(III/II) process at about +1.25 V vs SSCE. Typical coverages employed were in the 2-6 equivalent monolayers range.

The electrodes modified with a polymeric film of [Ru(v-bpy)3]$^{2+}$ were immersed in an aqueous solution of the desired ligand (typically 5-10 millimolar depending on solubility) for 15 minutes while stirring. The electrodes were subsequently rinsed with water and placed in stirred aqueous solutions of the metal ion of interest at various concentrations for typically 5 min. after which the electrodes were rinsed with water and acetone.

The electrochemical response of the metal/ligand complex was used as the analytical signal and was determined by differential pulse voltammetry in CH3CN containing 0.1M TBAP as supporting electrolyte. Acetonitrile (rather than an aqueous medium) was employed since it consistently gave lower background currents.

In the determination of mercury [31], an initial potential of 0.0V was applied for a period of five minutes resulting in the reduction of the preconcentrated mercury ions to metallic mercury. The potential was then scanned in the positive direction where oxidation of the mercury took place as a stripping peak which was employed as the analytical signal. In the case of copper [32], the metal based oxidation at about +0.50V was employed as the analytical signal.

In the determination of mercury and copper, the coverage of polymer on the electrode surface was controlled to be about 10 monolayer equivalents since this gives rise to easily measured responses while minimizing transport limitations.

In the work on surface immobilized alkaline phosphatase [33], carbon fiber electrodes (Donnay; Belgium) (7μm diameter) were sealed with epoxy resin into pasteur pipets and electrical contact was made with a mercury drop and a copper wire. Freshly polished electrodes were modified by immersion (30 min) into a solution (at 4°C) containing 2.5% (v/v) glutaraldehyde and 1% (w/v) bovine serum albumin (BSA). The electrode was subsequently washed with cold 0.1M glycine buffer (pH 9.5) for 1-2 min. The electrodes were then dried in air at 25°C. A 2μl volume containing 0.05-2.0IU of alkaline phosphatase with 1% (w/v) of BSA in glycine buffer was placed over the electrode and allowed to dry. In order to remove any weakly bonded or physically entrapped enzyme, the electrode was immersed in cold glycine buffer for 30 min. prior to use. Enzyme activity was monitored by the oxidation (at +0.30V) of 4-aminophenol which is the product of the enzymatic reaction (hydrolysis) of alkaline phosphatase on 4-amino phenyl phosphate which is employed as the substrate.

Finally on studies on the determination of nitric oxide, platinum electrodes sealed in glass were modified with Nafion and cellulose acetate. Modification with Nafion was achieved by placing over a freshly polished electrode a 10μl volume of a 0.5% (w/v) solution of Nafion in ethanol and allowing the solvent to dry in air. Modification with cellulose acetate was carried out in a similar fashion with an acetone/cyclohexanone solution (2%) of cellulose acetate and again employing a 10μl volume. Prior to use, the electrode was contacted with a phosphate buffer solution for 30 min. to allow the modifying layers to equilibrate and swell. The determination of NO was carried at +0.90V vs SSCE to ensure that the response was mass transport limited.

Results and Discussion:
1. Determination of mercury with surface immobilized amino acids:
    In this study we employed electrodes modified with tryptophan, 6-amino hexanoic acid, 4-amino butyric acid, picolinic acid and nicotinic acid as ligands (incorporated by ion exchange into a polycationic film of electropolymerized [Ru(v-bpy)3]$^{2+}$; see experimental section) for the determination of mercury. Our interest stems from a desire to extend our methodologies to the determination of this most important pollutant and, in addition, to study the interaction of mercury ions with these amino acid-modified electrodes and to assess their possible biological relevance.

    Initially we measured the stripping response for electrodes modified with the above mentioned ligands and in all cases a very well behaved response was obtained. The peak potential showed minor variations with the values ranging from +0.40 to +0.50 V. These vales are not only in a region where background currents are small, but in addition, they are also well removed from the redox response due to the surface immobilized poly-[Ru(v-bpy)3]$^{+2}$ so that no interference was anticipated.

    As mentioned in the experimental section, an initial potential of 0.0V was applied for a period of five minutes resulting in the reduction of the preconcentrated mercury ions to metallic mercury. The potential was subsequently scanned in the positive direction where oxidation of the mercury took place as a stripping peak which was employed as the analytical signal. Although, in principle, the reduction of the mercury ions could be employed as the analytical response, the kinetics of such a process were found to be slow and dependent on the coordination environment (i.e. incorporated ligand) of the mercury. This could be due to transport limitations of the metal complex within the polymer film or to slow charge transfer kinetics. Since the polymeric films employed were relatively thin (thickness of the polymer films was controlled to be about 2-6 monolayer equivalents; vide-supra) we speculate that the latter effect is more likely to be responsible for the observed behavior. However, we have no direct evidence to support this. Much more reproducible and better defined responses were obtained in the stripping modality and thus this approach was followed in all determinations. A similar behavior has been noted previously in the determination of silver with electrodes modified with quaternized poly-vinylpyridine with various polyanionic transition metal complexes incorporated by ion exchange [23].

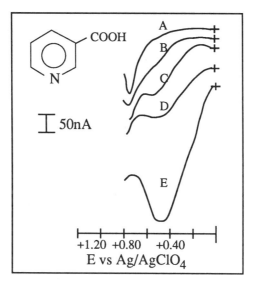

Figure 1. Differential pulse voltammograms for electrodes modified with nicotinic acid after having been contacted with solutions containing no mercury (A) and mercury at concentrations of 1.05×10^{-8}M (B), 1.07×10^{-7}M (C) and 1.06×10^{-5}M (D) and 1.04×10^{-3}M (E). Inset: structure of nicotinic acid. Figure adapted from reference 31.

Figure 1 presents differential pulse voltammograms for electrodes modified with nicotinic acid (see inset for structure) after having been contacted with solutions containing no mercury (A) and mercury at concentrations of 1.05×10^{-8}M (B), 1.07×10^{-7}M (C) and 1.06×10^{-5}M (D) and 1.04×10^{-3}M (E); respectively. In all cases a stripping response was observed at a potential in the vicinity of +0.46 which is very close to the anticipated value of +0.48V. Although the stripping response was readily quantifiable it was somewhat broad. Although we are not certain as to the origin of this broadening, it might be due to interactions between the ions and the polymer film. This could be due, in part, to differences in swelling due to the medium change from aqueous to non-aqueous (acetonitrile). A similar behavior (although with varying degrees of broadening) was observed for all the ligands employed in this work. Based largely on this observation, we believe that, as mentioned above, interactions with the polymer film might be responsible for the observed behavior.

By measuring the peak current (or the area under the voltammetric wave) for the stripping response of the preconcentrated mercury, (normalized to the surface coverage of polymer) calibration curves (Log i/Γ vs. Log [Hg]) for the determination of mercury were constructed. Figure 2 shows representative calibration curves employing electrodes modified with tryptophan, amino hexanoic acid and picolinic acid. In all cases, excellent correlations (r>0.98) were obtained over the range of mercury concentrations from 1×10^{-8} to 1×10^{-2} M. This is illustrative of not only the sensitivity of the method, but also of its wide dynamic range.

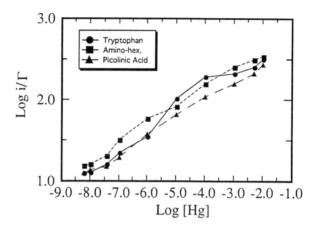

Figure 2. Calibration curves for the determination of mercury with electrodes modified with tryptophan, 6-aminohexanoic acid and picolinic acid.

At the higher concentrations, there appears to be some evidence of saturation as the observed response begins to level off. This is particularly evident in the calibration curves for 6-amino hexanoic acid and tryptophan. That such a leveling-off in the response was due to saturation, was corroborated by the fact that the stripping response for mercury did not increase with further increases in its solution concentration. In addition, the measured charges in the stripping peaks (at saturation) correlated very well with our estimates for a film where all the ligand sites were coordinated by the mercury ions. This was calculated from the experimentally determined surface coverage of the polymer on the electrode surface and assuming complete neutralization of the charge due to the pendant $[Ru(v\text{-}bpy)_3]^{2+}$ groups by the ligands.

Although saturation behavior was generally observed for all of the ligands employed, the mercury concentration in solution at which it became manifested showed some variation with the nature of the ligand as can be observed in a qualitative sense from the various calibration curves in Figure 2.

At the low concentration end, the stripping response began to approach background levels at mercury concentrations around 1×10^{-8} M which we estimate to be the limit of detection.

Some comments on the curves in Figure 2 are warranted. First of all, the very wide dynamic range was rather surprising and puzzling since if the overall process was dictated by thermodynamics, a narrower range would have been anticipated; analogous to a titration curve. The reasons for such behavior are not clear at this time. However, a plausible explanation might be that there are repulsive interactions as the film becomes increasingly metalated; analogous to a Frumkin type isotherm. However, we have no direct experimental evidence in support of this.

We were also interested in determining if, at a fixed solution concentration of mercury, the observed response was dependent on the value of the formation constant for mercury of each of the various ligands employed. We carried out such a study at a solution concentration level of mercury of $1x10^{-4}$ M. This rather high concentration was selected so that a large stripping response was obtained and, accordingly, a high signal-to-noise ratio was attained. Furthermore it was below the level where any saturation response was observed. One difficulty that we encountered was the limited data available on the formation constants of mercury complexes with the ligands studied. In fact, only three values could be found in the literature. [34-36] When the log of the normalized current response (at a solution concentration of mercury of $1x10^{-4}$ M) for the various incorporated ligands is plotted against the log of the formation constant (in solution) for the corresponding ligands a very good correlation is obtained (r=0.98). However, it should also be noted that the slope of such a line is significantly different from unity, pointing to the presence of other effects.

The fact that at constant concentration of mercury the magnitude of the analytical response for different ligands correlates well with the magnitude of the formation constant implies that the relative strength of coordination exhibited in solution is also maintained at the surface. Thus, the affinity of the interface for mercury ions can be varied and controlled over a wide range in a predictable fashion by the appropriate choice of ligand. It is also important to note that the above correlation between the normalized response (at constant concentration) and the formation constant implies that the system reaches equilibrium. Since chemical speciation involves, at a fundamental level, competitive equilibria, the approach outlined here could be employed in such studies.

Thus, by monitoring the electrochemical response of the immobilized metal/ligand complex as a function of the coordinating strength of the surface immobilized ligand, and from a knowledge of the concentration of the competing ligands present in solution, one could make an assessment as to the forms (chemical environment) in which the analyte ion is found. Such studies are currently being carried out.

There are a few other observations that can be made. First of all, the slopes of the calibration curves for all of the ligands studied were quite similar. However, the magnitude of the response varied with the nature of the ligand as was previously mentioned. The smallest signals were obtained for 4-amino butyric acid and this might reflect the difficulty, due to steric constraints, of this ligand in binding the mercury through both the amino and carboxylate groups. The fact that the signals are significantly larger for 6-amino hexanoic acid is consistent with this argument since the binding is identical but in this case the ligand can easily coordinate through both groups. Similarly, the responses for picolinic acid are larger than for nicotinic acid and this may again reflect the difficulty that the later would have in binding through both the pyridine nitrogen and the carboxylate group. It is also possible, although we do not have any direct evidence to support this, that the hydrophobic nature of nicotinic and picolinic acid as well as of tryptophan may enhance the partitioning of the ligand within the polymeric film, making its effective concentration higher and thus giving rise to a larger response. This is mentioned since such hydrophobic interactions would make the resulting complexes more lipid soluble thus enhancing the bioavailability and hence the toxicity.

Thus, this study demonstrates not only the feasibility of performing mercury determinations at very low levels, but in addition points to the possibility of employing this analytical approach to speciation studies.

2.Determination of copper with chemically modified microelectrodes:

Of the analytical applications of CME being currently explored, the use of ultramicroelectrodes is of particular relevance since it allows for determinations that are difficult, if not impossible, to carry out with conventionally sized electrodes. This is due, in part, to the ability of carrying out measurements in unusual and in low ionic strength media, as well as in spatially restricted environments [37]. This last application, however, requires electrodes that not only have a small area, but that are, in addition, structurally small.

The ability to carry out measurements in low ionic strength media is particularly valuable in metal ion speciation studies in water since the addition of an electrolyte (especially one with coordinating anions) can perturb some of the equilibria present in solution. In addition, the very low capacity (for reagent incorporation) of such electrodes would minimally affect the concentration and distribution of species.

We have carried out investigations on the use of CME for the determination of copper in solution with emphasis on the effects of electrode size (from conventional to ultramicro), nature of coordinating group and competitive binding effects by other metals and ligands [32]. Electrodes were modified with electropolymerized films of $[Ru(v-bpy)_3]^{+2}$ into which Eriochrome Cyanine R, Xylenol Orange, Nitroso R salt or Bathocuproine sulfonate were incorporated by ion exchange. Determinations in the submicromolar range could be carried out and calibration curves with a wide linear range were obtained. The effects of pH, competitive binding by other metals (iron, cobalt and nickel) or the presence of other ligands (chloride, bromide, oxalate and phosphate) on the analytical determinations have also been investigated.

Here we will focus on the determination of copper with microelectrodes modified with an elctropolymerized film of $[Ru(v-bpy)_3]^{+2}$ into which bathocuproine sulfonate (BCS) or Xylenol Orange (XO) were incorporated by ion exchange. We had previously demonstrated the utility of surface immobilized BCS in this type of determination using conventionally sized electrodes [19, 24] and the interest here was to determine if such an approach could be transposed to a microelectrode and to determine if there were variations in the signal to noise ratio or in sensitivity.

The copper complex of BCS and XO exhibit metal based (Cu(I/II)) oxidation waves at +0.52 and +0.43V, respectively and these were employed as the analytical signal.

Electrodes of two different sizes were employed and these were of 75 and 5μm in diameter, respectively. The first electrode was made by sealing a piece of platinum wire in 5 mm soft glass tubing as described in the experimental section. The latter was made as previously described [29] and the total tip diameter of the electrode was 11 microns as determined by optical microscopy.

For the 75 μm diameter electrode, the experimental procedures (including polishing) were identical to those employed for the conventionally sized electrodes.

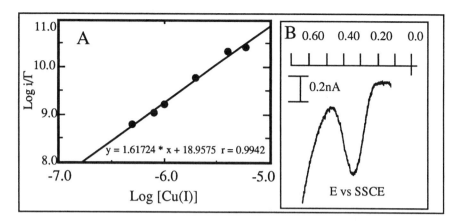

Figure 3. A: Calibration curve for the determination of copper using a 75μm diameter platinum electrode modified with an electropolymerized film of [Ru(v-bpy)3]$^{2+}$ with Xylenol Orange incorporated by ion exchange. B: Differential pulse voltammogram for the determination of copper from a 5×10^{-7} M solution. Figure adapted from reference 32.

The electrode was modified by electropolymerization of [Ru(v-by)3]$^{2+}$ and subsequent incorporation, by ion exchange, of either XO or BCS. The modified electrodes were then employed in the determination of copper in solution. Differential pulse voltammetry (10 mV/s scan rate and a pulse amplitude of 75 mV) was employed in the determination. Figure 3A shows a calibration curve using a 75μm diameter electrode modified with an electropolymerized film of [Ru(v-bpy)3]$^{2+}$ with Xylenol Orange incorporated by ion exchange and Figure 3B presents a differential pulse voltammogram for the determination of copper from a 5×10^{-7} M solution. As can be seen from the voltammogram, the peak is very well defined, the background is flat and featureless and an excellent signal to noise ratio is achieved even at this low concentration. The calibration curve exhibits excellent linearity and an enhanced limit of detection relative to conventionally sized electrodes. Although from analysis of the data the estimated limit of detection would be of the order of 2.0×10^{-7} M, we believe that the real value is lower. We believe this to be the case since the background level of copper in the reagents used was close to the value mentioned above. In any case, the behavior of these electrodes is clearly enhanced over that for conventionally sized electrodes. One can also notice in the calibration curve that the onset of a saturation response is reached at a concentration of about 6×10^{-6}M which is significantly lower than the corresponding value for a conventionally sized electrode. This might be a reflection of the lower capacity of the electrode. Finally, it needs to be pointed out that the slope of the calibration curve is significantly higher than 1. Although, the reason(s) for this behavior are not clear, it represents an improvement for low level determinations.

In addition to the 75μm diameter electrode described above, we also carried out determinations with a 5μm diameter electrode whose total tip diameter was of the order of 11 microns. Figure 4 shows the response for such an electrode

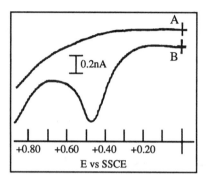

Figure 4. Differential pulse voltammogram for the determination of copper from a 5×10^{-6}M solution with a 5μm diameter platinum disk electrode modified with an electropolymerized film of $[Ru(v\text{-}bpy)_3]^{2+}$ with Bathocuproine sulfonate incorporated by ion exchange (B). A second scan was obtained after holding the potential at +0.60 V for 5 min. (A). Figure adapted from reference 32.

modified with an electropolymerized film of $[Ru(v\text{-}bpy)_3]^{+2}$ with Bathocuproine sulfonate incorporated by ion exchange. Curve B shows the differential pulse voltammogram obtained after copper incorporation from a 5×10^{-6}M solution. Again, a very well defined response is obtained. Curve A shows a second scan following that shown in B and after holding the potential at +0.60V for five minutes. In this case, only a flat and featureless voltammogram is obtained. However, the Bathocuproine sulfonate is still retained within the polymer film as evidenced by the fact that a differential pulse voltammogram run after re-immersion of the electrode into the copper solution gave a response virtually identical (within experimental error) to that shown in figure 4B. In fact in 6 replicate determinations the deviations were of the order of ±9%; comparable to results with conventionally sized electrodes. We have previously documented such (anticipated) behavior and the importance of that fact within the present context is that it allows for reuse of the electrode without the need for polishing or remodification. This is a very valuable feature since the mechanical fragility of these electrodes precludes polishing. Thus, the use of Bathocuproine sulfonate allows for the use dimensionally small ultramicroelectrodes in multiple determinations and with enhanced sensitivity.

3. Amperometric biosensor based on carbon fiber microelectrodes modified with alkaline phosphatase:
We have recently begun studies geared at the analytical application of enzymes immobilized on microelectrodes. We had previously demonstrated that 4-amino phenyl phosphate could be employed as a substrate for the determination of the enzymatic activity of alkaline phosphatase by following the oxidation of 4-amino-phenol which is the product of the enzymatic reaction [28]. In the present case, the interest was in determining whether such an approach could be transposed to a microelectrode and to determine if the immobilization process gave rise to an enzymatically active interface.

The preparation and modification of the carbon fiber microelectrodes was as described in the experimental section. The oxidation of 4-aminophenol

($E^{o\prime}$= -0.05V) which is the product of the enzymatic reaction, was used as the analytical signal.

When a carbon fiber microelectrode modified with alkaline phosphatase was immersed in a solution containing 4-aminophenyl phosphate, a steady-state current (at an applied potential of +0.30V; this potential was employed in order to ensure mass transport control) due to the oxidation of 4-aminophenol was reached within 3-4 sec. indicating a rapid response. Moreover, the magnitude of the steady state current was dependent on the solution concentration of 4-aminophenyl phosphate. The response was linear at the lower concentrations (Figure 5) and (asymptotically) reached a saturation limit (as expected) for substrate concentrations above 20mM. From the linear portion of the calibration curve, a sensitivity of 1.4×10^3 nA M^{-1} and a limit of detection of 5×10^{-7}M were determined. These values are, in fact, superior to those that we had previously obtained at conventionally sized electrodes.

Thus, these results confirm that with the immobilization strategy employed here, enzyme modified carbon fiber microelectrodes of excellent performance characteristics can be prepared.

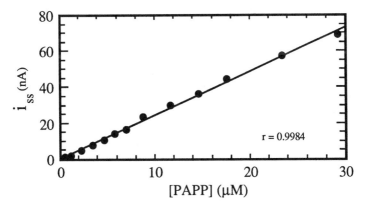

Figure 5.   Calibration curve for the determination of 4-aminophenyl phosphate with a carbon fiber microelectrode modified with alkaline phosphatase.

4.Sensor for the determination of Nitric Oxide (NO).

Since its identification as the endothelial derived relaxation factor (EDRF) there has been an explosive growth in the study of NO (nitric oxide) in biological systems [38-42]. One of the difficulties associated with its study and determination is the fact that it has a very short lifetime. Thus, methodologies for the determination of NO have been largely indirect. There is, however, a growing interest in the direct and in-vivo determination of NO. Because of their inherent sensitivity, electrochemical techniques are especially well suited for the development of analytical approaches for NO and the use of microelectrodes can allow, as was recently demonstrated, for in-vivo determinations [9c]. In addition, NO can be oxidized at about +0.80 V (vs SCE) at a platinum electrode (as well as at other electrode materials) although nitrite, ascorbate and other species can seriously interfere with the determination. In addition, in the case of in-vivo determinations, protein adsorption can also present difficulties. However, one can take advantage, in the development of analytical strategies for the selective determination of NO, of the fact that it is neutral and a gas.

Keeping these requirements and restrictions in mind, we have employed a platinum electrode modified with Nafion and cellulose acetate (hydrolyzed or unhydrolyzed) for the direct determination of NO based on its oxidation at +0.90V.

The purpose of the Nafion film is to exclude anionic species such as NO_2^- which is a severe interferent whereas the cellulose acetate can discriminate on the basis of size. For this particular application, hydrolyzed or unhydrolyzed cellulose acetate could be employed since NO would be anticipated to have a high permeability through either one.

The oxidation of NO at platinum surfaces has been found to depend strongly on the crystallographic face exposed with the (111) exhibiting much higher activity than the other low index faces [43]. Fortunately, a polycrystalline platinum electrode has a significant fraction of its surface with a (111) orientation.

When a bare platinum electrode is exposed to a deaerated solution of NO in phosphate buffer (pH=7.5), a well behaved though irreversible wave associated with oxidation of NO is observed with a peak potential of about +0.85V. The product of this oxidation is nitrite ion which can also be oxidized at these potentials. Thus, nitrite represents a severe interferent in the determination of NO. In addition, ascorbate, which is a species commonly found in biological media can also be a severe interferent. One way to eliminate interferents is to modify the surface of the electrode. Since the two most important interferents are anions or are negatively charged at physiological pH, the use of a poly-anionic modifier, capable of excluding anions by Donnan effects would be desirable and/or optimal. Thus we employed Nafion as a modifying layer. In fact, when a platinum electrode was modified with Nafion as described in the experimental section, determinations of NO could be made even in the presence of a very high concentration (10mM) of nitrite ion. Figure 6 presents the current response at +0.90V of an electrode

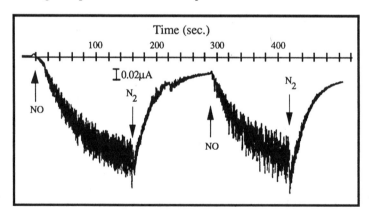

Figure 6. Current vs. time trace for a platinum electrode modified with Nafion and in contact with a pH 7.5 phosphate buffer containing 10 mM sodium nitrite and after bubbling the solution with either N_2 or NO.

modified with Nafion and in contact with a phosphate buffer solution (pH =7.5) containing 10mM nitrite when the solution was bubbled with either NO or nitrogen. It is clear that the electrode responds rapidly and reversibly to NO. Similar results were obtained in the presence of nitrate or ascorbate or to a mixture of any of these anions demonstrating that the Nafion layer acts as a very effective barrier against anions.

Electrodes were also modified with cellulose acetate and the response to NO in the presence of bovine serum albumin was ascertained. Again the electrode responded rapidly and reversibly to NO. When both modification procedures were employed there was a diminution in the magnitude of the response (as would be anticipated), however, the response was again reversible. This is important since it establishes that such an electrode could be employed in in-vivo applications.

By monitoring the current at +0.90V we have made a calibration curve, shown in Figure 7 which establishes the high sensitivity in addition to the linearity and dynamic range of this approach.

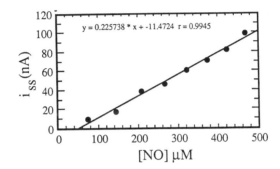

Figure 7. Calibration curve for the determination of NO.

Finally, we have carried out preliminary in-vivo experiments. In here promonocytic cells from the cell line HL-60 were differentiated and activated with phorbol 12-myristitic 13-acetate and calcium ionophore A-23187. Such a procedure activates the cells and induces NO production which could be followed with the above described modified electrode by monitoring the current at +0.90V. Upon injection of methyl-L-arginine acetate there was a decrease in the current consistent with the known fact that methyl-L-arginine acetate inhibits NO production in these cells. The physiologically relevant concentration of NO varies according to function. However, it is estimated that a cell secretes about 200 attomoles of NO/picoliter; amounting to a concentration of 100μM, well within our limits of detection. Although these are preliminary results, they demonstrate the feasibility of employing such a modified electrodes for the in-vivo determination of NO.

Conclusions:

It is clear that the use of chemically modified electrodes lends itself to numerous and varied analytical applications. By the judicious choice of electrode modifier the properties of the interface can be exquisitely tuned in a controlled fashion to achieve discrimination based on various chemical interactions including coordination, Donnan exclusion, enzyme activity, size exclusion and others. In addition, the use of microelectrodes has extended the range of applications. We feel confident that as new analytical challenges arise and as our ability to control interfacial reactivity down to the molecular scale increases, analytical schemes with unprecedented sensitivity and selectivity, based on modified electrodes, will be developed.

Acknowledgments:
This work was supported in part by the National Science Foundation, by a NATO Collaborative Research Grant (91-0047) and by the Spanish Ministry of Education and Science. E.L. acknowledges support by the Precompetitive Award Program, and the CICYT for a Visiting Scientist Fellowship at Cornell University.

Literature cited:
1. Murray, R.W.; in Electroanalytical Chemistry, A.J. Bard (Ed.), Vol. 13, Marcel Dekker, New York, **1984**, p. 191.
2. Murray, R.W.; Ann. Rev. Mat. Sci., **1984**, 14, 145.
3. Murray, R.W.; Accts. Chem. Res., **1980**, 13, 135.
4. Faulkner, L.R.; Chem. Eng. News, February 27, **1982**, p. 28.
5. Abruña, H.D.; Coord. Chem. Rev., **1988**, 86, 135.
6. Abruña, H.D.; in Electroresponsive Molecular and Polymeric Systems, T.A. Skotheim (Ed.), Marcel Dekker, New York, **1988**, p. 92.
7. Fujihira, M.; in Topics in Organic Chemistry, A.J. Fry and W.R. Britton (Eds.), Plenum, New York, **1986**, p. 255.
8. Murray, R.W.; Ewing, A.G.; Durst, R.A.; Anal. Chem., **1987**, 59, 379A.
9. a. Bailey, F.; Malinski, T.; Kiechle, F.; Anal. Chem., **1991**, 63, 395.
   b. Chen, T.K.; Lau, Y.Y.; Wong, D.K.Y.; Ewing, A. G; Anal, Chem. **1992**, 64, 1264
   c. Malinski, T.; Taha, Z.; Nature, **1992**, 358, 676
10. Kalcher, K.; Electroanalysis, **1990**, 2, 419.
11. Gao, Z.; Li, P.; Zhao, Z.; Talanta, **1991**, 38, 1177.
12. Downard, A. J.; Powell, H. K. J.; Xu, S., Anal. Chim. Acta, **1991**, 251, 157.
13. Prabhu, S. V.; Baldwin, R. P.; Kryger, L.; Electroanalysis, **1989**, 1, 13
14. Gao, Z.; Li, P.; Zhao, Z.; Fresenius. J. Anal. Chem., **1991**, 339, 137.
15. Dong, K.; Kryger, L.; Christensen, J. K.; Thomsen, K. N.; Talanta, **1991**, 38, 101.
16. Gardea-Torresdey, J.; Darnall, D.; Wang, J.; J. Electroanal. Chem., **1988**, 252, 197.
17. Ikaniyama, Y.; Heineman, W.R.; Anal. Chem., **1986**, 58, 1803.
18. Guadalupe, A.R.; Abruña, H.D.; Anal. Chem., **1985**, 57, 142.
19. Wier, L.M.; Guadalupe, A.R.; Abruña, H.D.; Anal. Chem., **1985**, 57, 2009.
20. Hurrell, H.C.; Abruña, H.D.; Anal. Chem., **1988**, 60, 254.
21. Kasem, K.K.; Abruña, H.D.; J. Electroanal. Chem., **1988**, 242, 87.
22. Liu, K. E.; Abruña, H. D.; Anal. Chem., **1989**, 61, 2599.
23. Lorenzo, E.; Abruña, H. D.; J. Electroanal. Chem. **1992**, 328, 111
24. Cha, S. K.; Abruña, H. D.; Anal. Chem., **1990**, 62, 274.
25. Cha, S. K.; Kasem, K. K.; Abruña, H. D.; Talanta, **1991**, 38, 89.
26. González, E.; Pariente, F.; Lorenzo, E.; Hernández, L.; Anal. Chim. Acta. **1991**, 242, 267.
27. Lorenzo, E.; González, E.; Pariente, F.; Hernández, L.; Electroanalysis; **1991**, 3, 319.
28. Pariente, F.; Hernández, L.; Lorenzo, E.; Biochem. Bioenergy.; **1992**, 27, 73
29. Pendley, B. D.; Abruña, H. D.; Anal. Chem. **1990**, 62, 782
30. Abruña, H. D.; Denisevich, P.; Umaña, M.; Meyer, T. J.; Murray, R. W.; J. Am. Chem. Soc., **1981**, 103, 1.

31. Cha, S. K.; Ahn, B. K.; Hwang, J-U.; Abruña, H. D.; Anal. Chem. **1993**, 65, 1564.
32. Lorenzo, E.; Fernández, J.; Pariente, F.; Trible, K.; Pendley, B.; Abruña, H.; J. Electroanal. Chem. **1993**, 356, 43
33. Pariente, F.; Hernández, L.; Abruña, H. D.; Lorenzo, E.; Anal. Chim. Acta **1993**, 273, 187.
34. Sillén, L. G.; Martell, A. E.; "Stability Constants of Metal-Ion Complexes", The Chemical Society Special Publication No. 17, The Chemical Society, London, 1964.
35. Sillén, L. G.; Martell, A. E.; "Stability Constants of Metal-Ion Complexes: Supplement No. 1", The Chemical Society Special Publication No. 25, The Chemical Society, London, 1971.
36. Maeda, M; Tsunoda, M.; Kinjo, Y.; J. Inorg. Biochem. **1992**, 48, 227
37. Wightman, R. M., Wipf, D. O.; Electroanalytical Chemistry, **1989**, 15, 267
38. Hibbs, J. B.; Taintor, R. R.; Vavrin, Z.; Rachlin, E. M. Biochem. Biophys. Res. Commun. **1988**, 157, 87
39. Marletta, M. A.; Yoon, P. S.; Iyengar, R.; Leaf, C. D.; Wishnok, J. S. Biochemistry **1988**, 27, 8706
40. Ignarro, J. L.; Buga, J. M.; Wood, K. S.; Byrns, R. E.; Chaudituri, G. Proc. Natl. Acad, Sci. U.S.A. **1987**, 84, 9265
41. Radomski, M.W.; Palmer, R. M. J.; Moncada, S. Proc. Natl. Acad. Sci. U.S.A. **1990**, 87, 10043.
42. Snyder, S. H.; Bredt, D. S.; Sci. Am. **1992**, 68.
43. Ye, S.; Kita, H.; J. Electroanal. Chem. **1993**, 346, 489

RECEIVED April 18, 1994

Chapter 21

Silicon-Based Chemical Microsensors and Microsystems

Elisabeth M. J. Verpoorte[1], Bart H. van der Schoot[2], Sylvain Jeanneret[2], Andreas Manz[1], and Nico F. de Rooij[2]

[1]Central Analytical Research, Ciba-Geigy Ltd., CH—4002 Basel, Switzerland
[2]Institute of Microtechnology, University of Neuchâtel, Rue A.-L. Breguet 2, CH—2000 Neuchâtel, Switzerland

A three-dimensional modular concept for miniaturization of flow systems for chemical analysis is presented. The system uses silicon micromachined flow manifolds and pumps in combination with electrochemical sensors or optical detection. Applications range from simple ion concentration measurements with ISFETs to a multi-step chemical analysis of phosphate. Miniaturization of the flow manifolds leads to simply constructed, compact systems in which reagent consumption is substantially reduced.

The continuous monitoring of chemical parameters is becoming increasingly important in biotechnology, process control, environmental analysis and medicine. In a limited number of cases, chemical sensors are available that show a sufficient specificity for the components to be measured. More often, however, measurement with a chemical sensor can only be carried out after the sample has undergone some form of pretreatment. This is necessary to eliminate the influence of interfering species on sensor response, to obtain the selectivity and sensitivity required for an accurate determination.

A standard approach to the automation of chemical analyses is to incorporate the sample handling, pretreatment and detection procedures involved into a flow system (see, for example, refs. (1-3)). Liquid handling within the manifold is controlled by pumps and valves. Depending on the application, an injected sample can undergo dilution and mixing, a separation step of some kind, or reaction with selected reagents, as it is transported ultimately to the detector. While such flow systems have been successfully employed to simplify and improve the reproducibility of many types of chemical analyses, they tend to consume significant quantities of reagents and carrier solutions. In addition, large numbers of tubing connections mean significant amounts of time spent on system maintenance. Miniaturization of flow systems for analysis is thus a logical "next step" in addressing these and other problems associated with conventional systems. Moreover, it has been shown that the performance of analytical systems can be improved through miniaturization (1, 4-6). In terms of time, this means significantly faster analyses are possible in systems whose performance is not limited by factors such as reaction kinetics. For flow systems incorporating a separation step, reduced channel dimensions also mean improved separation efficiencies. Thus, miniaturized systems would be excellent candidates for real-time monitoring in the medical, environmental and biotechnological fields, to name but a few. Microconduits imprinted in 7.0 × 4.5 × 1.0 cm plastic blocks have been reported for a variety of applications (7). However, channel dimensions were comparable to the tubing used in conventional systems, which generally has an inner diameter ranging from 0.3 to 0.8 mm. The degree of miniaturization that can be achieved with conventional methods of machining has its limitations. Photolithographic techniques permit the design and fabrication of flow manifolds with much smaller dimensions, as suggested in (7) and (Fettinger, J. C.; Manz, A.; Lüdi, H.; Widmer, H. M. *Sensors and Actuators*, in press, 1993.). Microsystem technology also makes possible the fabrication of silicon elements such as pumps, valves, and detector cells. These could in turn be combined to construct truly miniaturized systems for chemical analysis.

In a previous publication, the implementation of a three-dimensional modular strategy for miniaturization of a liquid handling system was described (Fettinger, J.; et al. *Sensors and Actuators*, in press, 1993.). This involved the design of a series of planar elements with simple channel structures and holes, which, when stacked on top of one another in some particular configuration, yielded an interconnected network of channels. Initial flow systems developed according to this concept were made of Plexiglas, and were used to carry out valveless injection of samples. One limitation in this work was the fact that although the flow manifold had been made quite compact, conventional pumps were used to control solution flow. Since flowrates of about 1mL/min were used here, these pumps were well suited for the measurements. However, many applications with miniaturized systems will require flow rates in the μ L/min range, which can no longer be reliably delivered by most conventional pumps. One alternative is to use silicon-based micro membrane pumps as described in (8). These pumps are small planar elements which can easily be incorporated into a stacked flow system. In addition to satisfying the necessary performance criteria, the use of such pumps can drastically reduce the amount of bench space taken up by the analysis system.

Hardware

The micropumps used in the construction of the analysis systems described here are piezoelectrically driven membrane pumps. Their design is similar to that of Van Lintel *et al.* (8). It consists of a micromachined silicon part anodically bonded between two Pyrex glass plates. The thicker of the glass plates (1.5 mm) serves as the base plate on which the valves close. The thinner glass plate (0.3 mm) forms the pump membrane, driven by a ceramic piezo disc (Philips PXE5, 10 mm diameter, 0.2 mm thickness) which is glued onto the glass using a conductive epoxy. The silicon part forms two passive valves on the glass base plate. A 1 μm layer of silicon dioxide on the valve seats prevents bonding to the glass when the pumps are assembled. The operating principle of the pump is presented in Figure 1. The pumps are operated with driving voltages up to 300 Volt at frequencies up to several hundred Hz. Although pump rates of 1mL/min are achievable, the applications described typically use flow rates in the order of 20-300 μL/min. The operation of the pumps at elevated frequencies (>20 Hz) ensures that the influence of remaining flow pulsations on the detector signal can easily be removed by appropriate filtering.

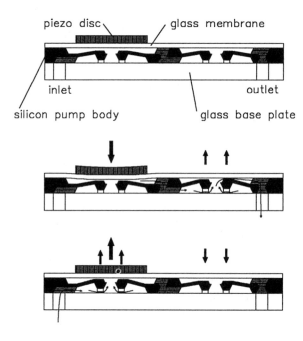

Figure 1. Operation of the piezo-electric micropump. (Reproduced with permission from reference 10.)

As discussed, flow manifolds for the miniaturized analysis systems are constructed by stacking of planar elements. Next to the pumps, a number of different elements with liquid channels and through holes have been realized by anisotropic etching of silicon. All elements measure 22 × 22 mm², with channels that are 600 μm wide across the top and 200 μm deep, and through holes that are 1 × 1 mm² at their widest points. Fluid connections to the pumps are made at the corners. The channel structures have eight holes around the circumference and often a ninth hole in the center, thus allowing a maximum flexibility in the setup of the complete flow manifold.
A practical limitation in the applied micropumps is the limited opening of their valves. All solutions that enter the pumps have therefore to be filtered to prevent clogging. For this reason, and also to ensure a minimum dead volume, the systems are basically

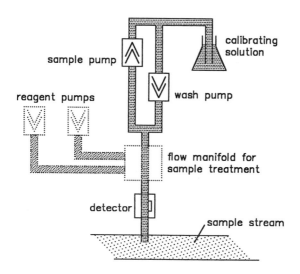

Figure 2   Basic setup for a miniaturized analysis system. (Adapted from ref. 10)

constructed as depicted in Figure 2. A detector is placed at the input of the system and the solution flow is controlled by two pumps, working in opposite directions. The choice of the detection principle is not limited by the system and can, for instance, be electrochemical or optical. To minimize the detection volume, the ideal detector is incorporated into a micromachined flow channel. The control of the fluid flow is as follows. The first pump (wash pump) is used to flush the detector cell with filtered calibrant solution to set the baseline signal for the sensor. The second pump (sample pump) is intermittently used to draw a sample, large enough to fill the detector but small enough not to reach the pump. This injection scheme resembles time-based injection procedures described in (2), since sample volume is a direct function of the length of time the sample pump is operational. Although two pumps are used here, the system does not need an active sample injection valve and is therefore much simpler than the traditional flow-injection setup. Optionally, a flow manifold can be added to dilute or to carry out chemical reactions in the sample. In that case, additional pumps supply reagents to the sample and a true multi-step chemical analysis can be performed. When the reaction is completed, the products are flushed across the detector and the concentration can be measured.

Ion concentration measurements with ion sensitive field effect transistors (ISFETs)

In its most elementary version as depicted in Figure 2, a miniaturized analysis system has been used for the measurement of ion concentrations with ion sensitive field effect transistors (ISFET). The system comprises two pumps and two multi-ISFET detector cells, one of which acts as a reference. The detector cell is built around a 22×5.5 mm^2

chip containing four ISFETs. The chip is mounted on a 22 ×26 mm² carrier to fit in the stack, as can be seen in Figure 3. On the chip a polysiloxane sealing ring is deposited, forming a flow channel around the sensors. The flow channel is closed at the top by placing it against the next element in the stack.

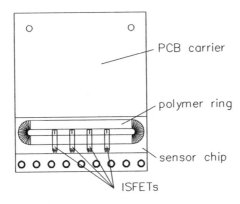

Figure 3   ISFET-based flow-through cell for ion measurements. (Adapted from ref. 9)

As the detector is placed at the input of the system, samples of a few microliters are sufficient to fill the sensor and to perform the measurement. Because of the rapid response of the ISFETs, the sampling frequency can be in the order of several samples per minute.

Preliminary measurements (9, 10) have been performed using potassium sensitive ISFETs, as shown in Figure 4. Samples of various concentrations are injected in a 10^{-2} M K^+ baseline solution. The sample size is approximately 7 μL, and with this injection volume, the peakheight does not reach the full Nernstian response but only about 52 mV/decade. Note that the figures in Figure 4 indicate concentrations, not activities. If the response is corrected for activity, it is linear within ± 5 %. At a measurement rate of 4 samples per minute, the consumption of calibrant solution is less than 3 mL/h.

Multi-step Analysis of Phosphate

Phosphate in water may be determined according to a procedure outlined in (11), known as the "molybdenum blue method". It involves the complexation of phosphate with molybdate, with subsequent reduction of the complex with ascorbic acid. The result is a complex having an intense blue color, which exhibits maxima in its absorbance spectrum at 882 nm and 744 nm. The overall reaction rate is limited by the complexation step, with maximum conversion of phosphate to the reduced complex requiring about 10 minutes. This analytical procedure has been adapted by many groups for phosphate analysis in flow systems (see, for instance, 2, 12, 13). In one instance, a

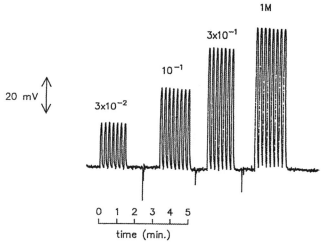

Figure 4   Measurement of various potassium concentrations injected in 10^{-2} K^+ baseline solution. (Reproduced with permission from reference 10.)

system was developed to monitor phosphate concentrations in fermentation broths (12). The flow manifold employed in that application is the model for the phosphate analysis using a stacked system described in this paper.

A schematic representation of the micro flow system employed is given in Figure 5. Included in the diagram are the flow rates for the various reagents used; these rates have been set at values which are about one-tenth of those in (12). To start an analysis, a fixed volume of sample is drawn into the main channel by the sample pump. This volume is adequate to fill the system up to a point between the molybdate line and the sample and wash pumps. Once this has been done, the sample pump is turned off, and the wash pump almost simultaneously turned on. The sample solution is now pushed in the opposite direction with filtered water. Filtered reagent solutions from two side branches are merged with the sample solution at slow and constant flow rates as it is carried to the detector. It is clear that only a portion of the sample will be analyzed. In effect, then, this portion serves as the "injected" sample plug. Subsequent reaction is facilitated by the inclusion of a mixing coil before the detector. Detection of the blue phosphate complex is carried out at 660 nm, a wavelength at which absorbance is strong, as evidenced by a molar extinction coefficient, ε, of 14,300 L/(mole·cm) (11).

Figure 6 depicts a stacked modular version of the flow system shown in Figure 5, realized with 22×22 mm^2 planar silicon elements and micro pumps. The three solution streams are indicated by solid, dashed and dotted lines. The pumps have holes at the corners, some of which are blocked on one side. Through holes are symbolized by $\begin{smallmatrix} 0 \\ 0 \end{smallmatrix}$, whereas blocked holes are shown as $\begin{smallmatrix} x \\ 0 \end{smallmatrix}$ or $\begin{smallmatrix} 0 \\ x \end{smallmatrix}$. The symmetrical spacing of inter-chip fluid

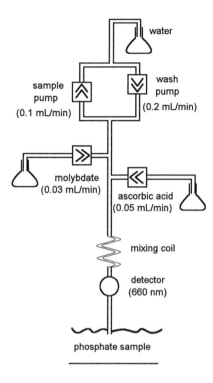

Figure 5   Micro flow system for the measurement of phosphate concentrations

connections allows reorientation of the elements with respect to one another, both through rotation and vertical repositioning within the stack. Thus, this stacked module approach has an inherent versatility which other approaches to flow analysis do not possess.

Shown in Figure 7 are results for the analysis of several aqueous phosphate containing solutions obtained using a stack of silicon elements having the same configuration as that shown in Figure 6. For these initial experiments, the silicon pumps were connected externally to the stack, rather than being incorporated directly within it. The pumps were operated at frequencies between 3 and 30 Hz and a voltage of 240 V. The pumping sequence was controlled using a simple Pascal computer program to switch the various pump power supplies on and off at set times. Data was recorded on a chart recorder. A capillary detector cell having a 1 mm pathlength (14), interfaced to a UV-visible spectrometer (Linear model 206 PHD, Reno, Nevada, USA) using optical fibers, served as the absorbance detector.

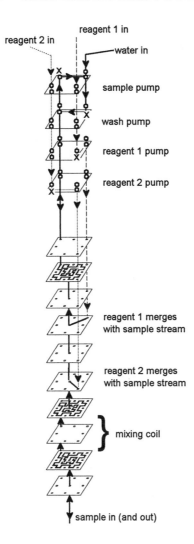

reagent 1 in
reagent 2 in
water in
sample pump
wash pump
reagent 1 pump
reagent 2 pump
reagent 1 merges
with sample stream
reagent 2 merges
with sample stream
mixing coil
sample in (and out)

Figure 6   Stacked modular analogue of the flow manifold in Figure 5 for the detection of phosphate. Three chips make up the mixing coil, which has an overall volume of about 30 μL.

Since the analysis is being carried out under non-equilibrium conditions, measured peak intensity should show a strong dependence on the amount of time the sample spends in the system. The longer the sample residence time, the greater the extent of the reaction on which the detection of the component of interest depends. This in turn means an increase in the sensitivity of the analysis. In this set of experiments, peak intensity was significantly increased by stopping the flow of solution just before the zone of sample to which reagents were added entered the detector cell. The data shown in Figure 7 was obtained in this way, by allowing the wash and two reagent pumps to run simultaneously for 7 seconds after the end of the sampling step, before stopping flow within the system for 40 seconds. The overall residence of the sample plug is then about 50 seconds, when other analysis steps are included. The volume of sample to which reagent is added is determined to a large extent by the length of time the wash pump drives sample past the two reagent lines before flow is stopped. In this case, the sample plug has a volume of about 25 μL, since the wash pump pumps at a flow rate of about 3.3 μL/s (200 μL/min) over a period of 7 seconds. An entire cycle can be carried out in about 4 minutes, and includes, in addition to the reaction period: 1) an initial 60 second period during which the system was flushed with doubly distilled, filtered (> 0.20 μm) water 2) a 55 second injection period during which approximately 90 μL of sample is aspirated into the system 3) a 75 second period during which the reacted sample plug is carried out of the system through the detector.

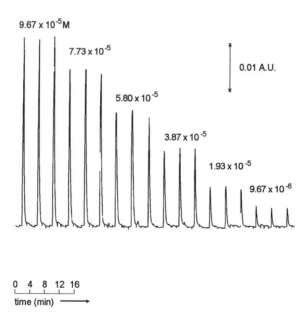

Figure 7   Analysis results for the determination of phosphate

Response to phosphate was found to be linear over the range of concentrations given in Figure 7, with a correlation coefficient, r, of 0.9991. At concentrations higher than these, the extent of phosphate complexation over a given period of time is no longer a linear function of phosphate concentration, as dictated by the reaction kinetics. The deviation from Beer's Law response at concentrations in the 10^{-4} M range is in agreement with other authors (2, 11-13). The sensitivity for phosphate obtained using this stop-flow scheme falls in the range of expected values for the molybdenum blue method when used in conjunction with continuous flow techniques (2, 12, 13).

The phosphate analysis described here is characterized by an analysis time of four minutes, when system flushing is included. This is in contrast to the conventional flow injection analysis of phosphate using the molybdenum blue method reported in (2) and (13). These two studies indicate analysis times of 30 and 80 seconds, respectively. Several factors play a role in the longer analysis times for the micromachined system. First is the incorporation of a time-based injection technique in the system, as opposed to the volume-defined methods using valves described in (2) and (13). While elimination of the need for an injection valve simplifies the system, time-based injection schemes by definition contribute to longer analysis times. A second factor is the design of the stacked manifold itself, which requires that sample both enter and exit it at the same location. This differs from conventional systems, where a sample is introduced at one end and detected at the other. Thorough flushing of the stack is therefore required between samples to avoid carryover from one sample to the next. This precludes injection of a second sample while the first is still being detected, a time-saving measure implemented in (2). Optimization of the stacked system's design, taking these factors into consideration, as well as a further reduction in overall system volume, will lead to improved analysis times.

A comparison of the flow manifold reported in (12) and the µ-TAS reveal that dramatic decreases in reagent and solution consumption can be achieved upon implementation of the µ-TAS. Whereas the system in (12) has a volume of between 1 and 2 mL, that of the µ-TAS reported here is about 90 µL. However, the amount of reagent required for an analysis drops by a factor of 270. This is due in part to the flow rates in the µ-TAS being one-tenth of the corresponding rates in the conventional system, as mentioned above. The additional decrease in reagent consumption is a result of the stop flow procedure applied, which saw addition of reagents over a period of only 7 seconds during an analysis. In contrast, the nature of the system in (12) dictated that reagent pumps ran continuously throughout an analysis. Thus, where 1.7 mL of ascorbic acid was needed in the conventional system (0.56 mL/min over 3 min), only 6.2 µL was required in the µ-TAS analysis. Similarly, only 3.0 µL of molybdate reagent were added in the µ-TAS, as opposed to 0.8 mL in the larger system. Use of piezoelectrically driven micro pumps allows reproducible addition of microlitre amounts of solution in small, sub-microlitre steps. This is difficult to impossible to achieve with most commercially available pumps, whose lowest volume resolutions lie in the microlitre range.

Conclusion

Initial experiments have demonstrated the feasibility of flow manifolds having a stacked configuration for chemical analysis systems. A substantial reduction in system size may be accomplished when micromachined silicon elements are employed. Implementation of the concept of merging zones of sample and reagents results in more efficient consumption of reagents. Incorporation of micro pumps and small volume detector cells into the stack, as well as elements having small channel dimensions, will yield a truly miniaturized total analysis system.

Acknowledgments

E. Verpoorte thanks M. Garn, A. Spielmann, and S. Haemmerli for several very helpful discussions. Part of this work has been financially supported by the Commission for the Promotion of Scientific Research (CERS) and the Swiss Foundation for Microtechnology Research (FSRM).

Literature Cited

(1) Manz, A.; Graber, N.; Widmer, H. M. *Sensors and Actuators* **1990,** *B1* 244.
(2) Ruzicka, J.; Hansen, E. H. *Flow Injection Analysis*; John Wiley and Sons: New York, 1988; 2nd Ed.
(3) Valcarcel, M.; Luque de Castro, M. D. *Flow Injection Analysis: Principles and Applications*; John Wiley and Sons: New York, 1987.
(4) Van der Linden, W. E. *Trends in Anal. Chem.* **1987,** *6*, 37-40.
(5) *Micro-Column High Performance Liquid Chromatography*; Kucera, P., Ed.; Elsevier: Amsterdam, 1984.
(6) 2nd International Symposium on High-Performance Capillary Electrophoresis *J. Chromatogr.* **1990,** *516*, 1.
(7) Ruzicka, J.; Hansen, E. H. *Anal. Chim. Acta* **1984,** *161*, 1-25.
(8) Van Lintel, T. G.; van der Pol, F. C. M.; Bouwstra, S. *Sensors and Actuators* **1988,** *15*, 153.
(9) Van der Schoot, B. H.; Jeanneret, S.; van den Berg, A.; De Rooij, N. F. *Sensors and Actuators* **1992,** *B6*, pp. 57-60.
(10) Van der Schoot, B. H.; Jeanneret, S.; Van den Berg, A.; De Rooij, N. F. *Anal. Methods & Instrumentation* **1993,** *1*, 38-42.
(11) Murphy, J.; Riley, J. P. *Anal. Chim. Acta* **1962,** *27*, 31-36.
(12) Spielmann, A.; Garn, M.; Haemmerli, S.; Manz, A.; Widmer, H. M. *Proceedings of the Seminar of the Swiss Committee of Analytical Chemistry*, Bern, Switzerland, October 16, 1992; p.An 49.
(13) Linares, P.; Luque de Castro, M. D.; Valcarcel, M. *Journal of Automatic Chemistry* **1992,** *14*, 173-175
(14) Bruno, A. E.; Gassmann, E.; Periclès, N.;Anton, K. *Anal. Chem.* **1989,** *61*, 876

RECEIVED May 2, 1994

Chapter 22

Multilayered Coatings of Perfluorinated Ionomer Membranes and Poly(phenylenediamine) for the Protection of Glucose Sensors In Vivo

D. Jed Harrison[1], Francis Moussy[1,3], Stephen Jakeway[1], Zhoughui Fan[1], and Ray V. Rajotte[2,3]

Departments of [1]Chemistry and [2]Surgery and Medicine, University of Alberta, Edmonton, Alberta T6G 2G2, Canada
[3]Surgical–Medical Research Institute, University of Alberta, Edmonton, Alberta T6G 2N8, Canada

A miniature, needle-type glucose sensor based on a tri-layer membrane configuration has been prepared and evaluated both *in vitro* and *in vivo*. The perfluorinated ionomer, Nafion, was used as a protective, biocompatible, outer coating, and poly(o-phenylenediamine) as an inner coating to reduce interference by small, electroactive compounds. Glucose oxidase immobilized in a bovine serum albumin matrix was sandwiched between these coatings. Heat curing of the assembled sensor at 120 °C improved the stability of the Nafion layer and extended the sensor lifetime. This also reduced interferences by ascorbic acid, uric acid and acetaminophen. Even after 10 days in dogs the sensor current closely followed the plasma glucose level during a glucose tolerance test in active dogs, with a delay of 3-5 min, corresponding to the known lag time for subcutaneous glucose levels. The sensors failed after 10 days of implantation due to breakage of the leads.

The potential uses of a glucose sensor for the treatment of diabetes include continuous glucose monitoring, application as an alarm device for detecting hypoglycemia and, ultimately, use as a component of a closed-loop insulin delivery system where an implanted sensor would monitor glucose levels in the body to control insulin delivery. The first demonstrated use of a glucose sensor for continuous monitoring of physiological glucose levels prompted considerable interest (*1*). However, the difficulties encountered, involving issues of biocompatibility, protein fouling and interferences, have meant a truly implantable sensor remains elusive (*2*). Despite this, there has been a continued effort (*2-14*) due to the expected importance of implantable glucose sensors as therapeutic tools for the treatment of diabetes.

The use of semipermeable polymeric overcoatings to protect electrodes in complex sample matrices, or to otherwise tailor their reactivity and selectivity

0097–6156/94/0561–0255$08.00/0

characteristics, is an attractive and successful approach to electrode design. We have used Nafion, the DuPont perfluorinated sulfonate ionomer, as a protective and selective coating material for a glucose sensor in blood (*12-14*). Nafion provides a relatively biocompatible interface with the biological matrix, invoking a tissue response that is either similar to, or less than the response to medical grade silastic, depending on the implantation site (*15*). More recently (*14*), we have used an additional membrane underneath the Nafion and enzyme layers to reduce electrochemical interferences at the Pt electrode. This membrane is based on the electrochemical polymerization of *o*-phenylenediamine to form poly(phenylenediamine) (PPD), which was suggested by Malitesta et al. (*16*) and Sasso et al. (*17*) to reduce interferences.

Initial *in vivo* studies with a needle-type glucose sensor indicated that the multi-layered films were not completely stable (*14,18*). Since Nafion is known to require curing at elevated temperatures to obtain good stability (*19,20*) this result was not unexpected. We report in this paper the effect of thermal curing of the assembled sensor on its performance and lifetime, as well as the transport properties of cured Nafion for neutral species such as glucose. In addition, some aspects of the selectivity of the PPD layer are also presented.

Experimental Section

High-purity glucose oxidase (Aspergillus Niger, Calbiochem, La Jolla, CA, USA), bovine serum albumin (Fraction V, 98-99 % albumin, Sigma), and glutaraldehyde (25 % aqueous solution, Sigma) were used as received. All other chemicals were reagent grade. Solutions were prepared from doubly distilled, deionized water.

A pH 7.4 phosphate buffer solution (PBS) was prepared from phosphate salts ($\mu = 0.05$ M) with sodium benzoate (5 mM) and ethylenediaminetetraacetic acid (1 mM) as preservatives and NaCl (0.1 M) as electrolyte. Glucose (0.1 M) was added and allowed to mutarotate overnight at room temperature, then stored at 4°C. Solutions of interfering species in pH 7.4 buffer were prepared just before use, as was 5 mM *o*-phenylenediamine (Aldrich) in an acetate buffer (pH 5.5). Nafion solutions (5 wt%, Solution Technology Inc., Mendenhall, PA, USA) of 0.5, 1 and 3 wt% were prepared by dilution with 1:1 2-propanol and water.

The electrochemical instrumentation used, fabrication of the needle-type sensor with an incorporated Ag/AgCl reference electrode, and electrodeposition of the PPD film have been described previously (14). Rotating disk studies of Nafion and PPD membranes were performed as described elsewhere (*19*). Nafion membrane thicknesses were determined using ellipsometry (*19*). Nafion and glucose oxidase membranes were cured for 1 h in an oven, or overnight at room temperature (*18*).

Results and Discussion

The assembled glucose sensor was 0.5 mm in diameter, comparable to the size of a 25 gauge needle, and could be inserted under the skin through an 18 gauge needle (*14*). The needle-type geometry was chosen for the sensor to minimize the biological response to the implant and to facilitate the subcutaneous implantation procedure (*2-5*). As suggested by the electron micrograph in Figure 1, coiled Pt and Ag/AgCl wires were wrapped around a Cu wire and served as the working and reference electrodes, respectively. The working electrode was coiled to increase the sensitivity of the sensor by increasing the area of the indicating electrode. The working electrode was coated with PPD and glucose oxidase (GOx) immobilized in bovine serum albumin. The entire sensor was then coated in Nafion (*14*).

The sensor assembly was used after curing at room temperature or at elevated temperatures. However, to evaluate the polymer properties simpler electrode geometries such as Pt disks were also used in the work described below, in order to allow measurement of diffusion coefficients (*19*).

The Nafion coating must protect the underlying sensor elements from compounds in the body, selectively transport O_2 relative to glucose, and should preferably inhibit interfering species such as ascorbic acid (*2,13-15*). The permeability of Nafion to several compounds was determined using a Nafion-coated rotating disk electrode (*19*). From plots of the inverse of the observed disk current as a function of the inverse of the square root of the rotation rate (a Koutecky-Levich plot) the permeability, P_m, was obtained. This value was determined for a series of Nafion thicknesses, measured by ellipsometry, and the effective diffusion coefficients were obtained, according to equation 1.

$$P_m = \alpha \, D_m / \delta_m \qquad (5)$$

where α is the extraction coefficient of solute between the solution and the film, D_m is the diffusion coefficient in the film, and δ_m is the film thickness. Figure 2 shows a plot of permeability versus the inverse of Nafion membrane thickness for hydroquinone oxidation at pH 7.4. Hydroquinone was selected as a model for glucose, due to its much simpler electrochemistry. It is neutral at pH 7.4 and has a structure and functional groups that are roughly similar to glucose, although there are obvious differences owing to its aromaticity.

Similar experiments with several compounds gave the data in Table I. The Nafion films were cured at room-temperature or 120 °C and were cast from the isopropanol:water mixture as supplied by the manufacturer (*19*). Also given in the table are values obtained by others for the diffusion of O_2 within Nafion films prepared by casting from solution (*21,22*). The high temperature cure clearly

Table I. Effective Diffusion Coefficients in Nafion

Compounds	αD_m $(x\ 10^8\ cm^2/s)$ [a]	
	22 °C cure (± 0.5)	120 °C cure (± 0.3)
O_2	400-600 [b]	600 [b]
Glucose	3.4	
hydroquinone (pH 7.4)	5.4	2.6
ferroceneTMA+	1.8	1.1
$Fe(CN)_6^{4-}$	< 0.5 ± 0.1	< 0.2 ± 0.1

[a]Values are from reference 19 except those superscripted with (b).

[b]Values are from references 21 and 22.

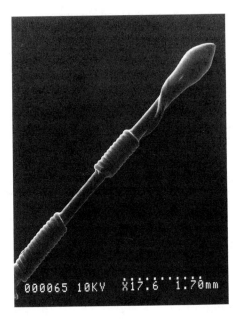

Figure 1.   Scanning electron micrograph of a needle-type glucose sensor coated
             with a tri-layer membrane. Using 10 kV acceleration no gold coating
             was needed to obtain the image. The coiled Pt electrode is located near
             the tip of the sensor.

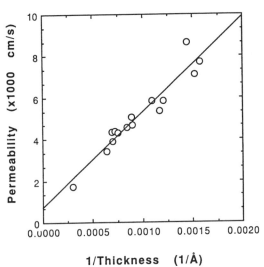

Figure 2.   Plot of Nafion permeability obtained from rotating disk measurements
             versus the inverse of Nafion thickness, determined by ellipsometry.

lowers the diffusion coefficients within Nafion for all the compounds studied except O_2. This is consistent with the increased crystallinity of the polymer observed by others following curing (*20*). The data for ferrocyanide reflects the presence of pinholes in the films, which were prepared by spin coating a single layer. (This was concluded on the basis of the much higher value measured compared to that reported by others, and compared to values we measured for multiple-dip coated electrodes of various geometries (*19*).) The currents for glucose, hydroquinone and [(trimethylammonio)methyl]-ferrocene (FcTMA$^+$) were significantly higher than for ferrocyanide, so that pinholes resulted in no more than a 10% error in the diffusion coefficients.

The data in Table I shows that Nafion is more permeable to the neutral species than to large cations or highly charged anions. This presumably results from the absence of electrostatic interactions for the neutral compounds. Comparison of the diffusion coefficients for O_2 and glucose shows that Nafion is 120 to 180-fold more permeable to O_2 when cured at room temperature. Assuming the ratio of effective diffusion coefficients for glucose and hydroquinone are the same for Nafion cured at any temperature, we estimate that O_2 has about 350 times greater permeability than glucose after curing at 120 °C. Whether cured or not the differences in permeability mean that Nafion selectively enhances the rate of O_2 transport relative to glucose. Consequently, the sensor response will be limited by the glucose supply rather than the O_2 tension in solution, which reduces the O_2 sensitivity of the sensor substantially. In addition, by limiting the rate of supply of glucose the linear response of the sensor is extended up to 20 mM (*14*), which is sufficient for blood glucose measurements for diabetic patients (*2*).

Curing Nafion at high temperatures proved to be necessary, as the Nafion coating deteriorated when sensors were implanted in animals after about 1 week if it was not cured. Sensors implanted for two weeks were found to still respond to glucose once removed from the animal, however, all of the AgCl on the sensor reference electrodes had been lost, due to direct contact with the surrounding tissue (*14,18*). High temperature curing of the Nafion before implantation prevented this effect from occurring (*18*). However, this made it necessary to test the thermal stability of sensors based on immobilized glucose oxidase (GOx), since the Nafion layer overcoats the enzyme layer in a fully assembled sensor.

The thermal stability of immobilized GOx based sensors was tested using small planar Pt electrodes coated with GOx immobilized in bovine serum albumin. These electrodes were heated at elevated temperatures 1 h. No Nafion coating was used, to ensure that the rate limiting step was not glucose diffusion through the Nafion. Figure 3 gives the electrode activity measured in a 5.6 mM glucose solution (at room temperature), pH 7.4, both before and after treatment at the indicated temperature. The activity is expressed as the ratio of the current after treatment to that obtained before curing. It can be seen that substantial activity remains even when using curing temperatures as high as 150 °C. It should be noted that these studies do not mean that the enzyme activity is unaffected at 120 °C. Rather, enough enzyme activity must be retained by the immobilized GOx layer that the sensor current is not enzyme activity limited, so that mass transport of glucose remains the limiting factor.

The above results show that curing of the assembled sensor was readily accomplished at 120 °C without seriously affecting performance. Planar electrodes coated with enzyme and Nafion showed a drop in activity consistent with that predicted by the data in Table I and Figure 3. The response times of completely assembled, needle type sensors remained essentially unchanged (33 ± 13 s), but the sensitivity to glucose decreased by a factor of about 7 from its value of 25 nA/mM before curing. This drop arises in part from the effect of curing on the effective diffusion coefficients, but must have another source as well. We believe it may indicate the presence of pinholes in the uncured needle-type sensor that were sealed by the curing process.

The effect of interferences on the glucose signal was also reduced by curing the assembled sensor. Table II gives data for sensors cured at room temperature or for 1 h at 120 °C. In Table II the percentage increase in sensor current when the interferent was added to a 5.6 mM glucose solution is expressed as the apparent error. The interferents were added at their maximum normal physiological levels in blood, so this is the maximum error that should be observed in blood for the compounds studied under normal conditions. Curing Nafion at 120 °C completely eliminated the interference of ascorbic acid and substantially reduced the interferences of uric acid and acetaminophen. However, in addition to curing the Nafion an additional coating was also utilized to enhance the selectivity of the sensor assembly. This coating was electropolymerized phenylenediamine (PPD), deposited directly on the Pt electrode before the GOx and Nafion layers were added. This layer eliminated ascorbic acid interference even without curing the Nafion, and also reduced the interference by uric acid. It had very little effect on the interference arising from acetaminophen, for which curing the Nafion layer was more beneficial.

Table II. Sensor Response to Various Interferences

Interferent	Concentration[a]	% error relative to 5.6 mM glucose[b]			
		no PPD RT cure	no PPD 120 °C	with PPD RT cure	with PPD 120 °C
Ascorbic acid	0.11 mM	5	0	0	/
Uric acid	0.48 mM	39	12 ± 6	5 ± 3	/
Urea	4.30 mM	/	0	0	/
Acetaminophen	0.17 mM	/	21 ± 8	57 ± 29	23 ± 4

[a]Interferent is added at the maximum physiological concentration in blood.
[b]Error is expressed as the apparent % increase in glucose concentration when the interferent is added to a 5.6 mM glucose, pH 7.4 buffer.

Implantation of assembled sensors in dogs was undertaken once the problem with Nafion stability was overcome by curing at elevated temperature. Sensors were implanted in the back of a dog anesthetized with halothane using a 2 cm diam., double velour, dacron phlange (Meadox, Oakland, NJ) placed around a modified Konigsberg skin button connector to which the sensor's leads were attached. A venous catheter was also implanted for blood sampling, and was inserted in the right, external jugular. The use of dacron phlanges promotes tissue growth and prevents infection and removal of the implants. The sensor was tested immediately after surgery (and stabilization) and then every few days. Between tests, no potential was applied. During testing the sensor was biased at +0.7 V vs the implanted Ag/AgCl electrode. After the sensor signal stabilized (about 40 min) a blood sample was taken. A bolus of glucose (0.5 g/kg body weight) was then injected through the indwelling venous catheter and blood was taken at varying intervals. Glycemia of the dog was measured using a Beckman Glucose analyser II and compared to the current measured for the sensor.

The implanted sensor performed for 10 days with very little change in sensitivity, although there was an improvement in the response time after a few days. We believe the latter may be related to the initial formation of a blood clot during the surgery performed to implant the sensor, as sensors implanted on an acute basis through an 18 gauge needle always exhibit better response times (*14*). (The needle implantation procedure causes much less bleeding that the incision needed to implant the skin button, but the button makes for a more stable chronic implant in terms of freedom of movement for the animal.) Figure 4 shows the response of the sensor on the 10th day of implantation, as well as the plasma glucose levels determined from blood samples taken from the indwelling catheter. The signal can be seen to accurately follow the glycemia of the dog following injection of a bolus of glucose. The 5 minute delay observed during the initial increase in glucose level corresponds to the known lag time between subcutaneous and blood glucose levels (*4*), and is not due to the sensor response time. The sensitivity remained essentially constant during the 10 day implant study. Histological examination of the subcutaneous tissue surrounding the sensor after 2 week implantation did not show any major inflammatory response or fibrous reaction. This confirms our previous results concerning the biocompatibility of Nafion. During these 10 days the sensor was polarized for about 2 hours per day. However, the enzyme was continuously reacting with glucose and O_2 to produce H_2O_2. Consequently, the histological study, and the stable sensitivity suggests that the production of H_2O_2 does not represent a problem to either the enzyme or the biological tissue surrounding the sensor.

The performance of the sensor on the 10th day suggests that it could have continued responding for a much longer period. Unfortunately, we have been hampered by repeated electrical failure of the leads and solder joints at the skin button connector, so that the sensor assemblies fail after about 10 days *in vivo*. Explanted sensors still function, so the failures are related to the lead insulation, rather than the sensor or the reference electrode. We are presently working on methods to solve this difficulty.

Conclusion

The requirement of the polymer coatings for satisfactory performance shows the necessity to design for the effect of the sample matrix on the transduction process, rather than focussing on the sensing or transducing events and elements alone. The choice of dialysis membranes used to enhance selectivity and durability of implanted glucose sensors is a key aspect in determining their performance and successful development. In this work we have shown that Nafion coated sensors function well when implanted subcutaneously. The combination of poly(phenylenediamine) undercoating with Nafion overcoating provides a sensor with better selectivity and improved performance in a biological matrix, relative to the use of either material alone. Curing the Nafion film at 120 °C or higher was shown to reduce the effect of interferences and enhance the lifetime of the sensor. Importantly, this curing procedure proved to be compatible with immobilized glucose oxidase based sensors. The assembled sensor appears to be promising, at least for acute implantations and short term chronic implantations. Longer term evaluation of the implanted sensors is in progress.

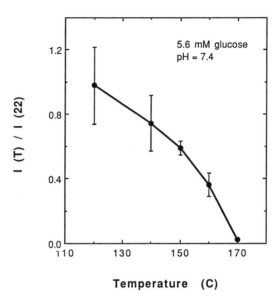

Temperature (C)

Figure 3.   Effect of high temperature curing on the activity of immobilized glucose oxidase. Electrode current was measured in 5.6 mM glucose before, I(22), and after, I(T), curing at the indicated temperature. The ratio of these currents is plotted versus the curing temperature.

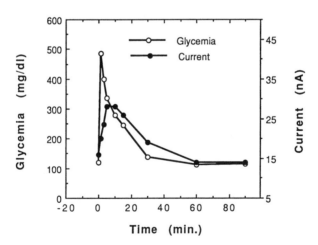

Time (min.)

Figure 4.   The glycemia after an intraveneous injection of glucose in a dog is shown, as determined with a Beckman analyzer from blood samples. The current response of a subcutaneously implanted sensor to the same injection is also shown. The sensor had been in place for 10 days at the time of this test, and was polarized at 0.7 V vs Ag/AgCl for several hours every 2 to 3 days.

Acknowledgments

The authors wish to acknowledge the financial support of the Alberta Heritage Foundation for Medical Research, the Natural Sciences and Engineering Research Council, the Medical Research Council of Canada and the Muttart Diabetes Treatment and Training Centre. F.M. thanks the University of Alberta for additional support, and Z.F thanks the Alberta Microelectronic Centre for a stipend.

Literature Cited

(1)   Albisser, A.M.; Leibel, B.S.; Ewart, T.G.; Davidovac, Z.; Botz, C.K.; Zingg, W.; Schipper, H.; and Gander, R. *Diabetes* **1974**, *3*, 397-404.
(2)   Reach, G.; Wilson, G.S. *Anal. Chem.* **1992**, *64*, 381A-386A.
(3)   Velho, G.D.; Reach, G.; Thevenot, D. In: Turner, A.P.F.; Karube, I.; Wilson, G.S. (Eds) *Biosensors: Fundamentals and applications.* Elsevier **1987**, pp 390-408.
(4)   Fisher, U.; Ertle, R.; Abel, P.; Rebrin, K.; Brunstein, E.; Hahn von Dorsche, H.; Freyse, E.J. *Diabetologia* **1987**, *30*, 940-945.
(5)   Pickup, J.C.; Shaw, G.W.; Claremont, D.J. *Diabetologia* **1989**, *32*, 213-217.
(6)   Schichiri, M.; Asakawa, N.; Yamasaki, Y.; Kawamori, R.; Abe, H. *Diabetes Care* **1986** , *9*, 298-301.
(7)   Bindra, D.S.; Zhang, Y.; Wilson, G.; Sternberg, R.; Thevenot, D.R.; Moatti, D.; Reach, G. *Anal. Chem.* **1991**, *63*, 1692-1696.
(8)   Moatti-Sirat, D.; Capron, F.; Poitout, V.; Reach, G.; Bindra, D.S.; Zhang, Y.; Wilson, G.S.; Thevenot, D.R. *Diabetologia* **1992**, *35*, 224-230.
(9)   Clark, L.C.; Duggan, C.A. *Diabetes Care* **1982**, *5*,174-180.
(10) Thevenot, D.R. *Diabetes Care* **1982**, *5*, 184-189.
(11) Gough, D.A.; Lucisano, J.Y.; Tse, P.H.S. *Anal. Chem.* **1985**, *57*, 2351-2357.
(12) Turner, R.F.B.; Harrison, D.J.; Rajotte, R.V.; Baltes, H.P. *Sensors and Actuators* **1990**, *B1,* 561-564.
(13) Harrison, D.J.; Turner, R.F.B.; Baltes, H.P. *Anal. Chem.* **1988**, *60*, 2002-2007.
(14) Moussy, F.; Harrison, D.J.; O'Brien, D.W.; Rajotte, R.V. *Anal .Chem.* **1993**, *65*, 2072-2077.
(15) Turner, R.F.B.; Harrison, D.J.; Rajotte, R.V. *Biomaterials* **1991**, *12*, 361-368.
(16) Malitesta, C.; Palmisano, F.; Torsi, L.; Zambonin, P.G. *Anal Chem.* **1990**, *62*, 2735-2740.
(17) Sasso, S.V.; Pierce, R.J.; Walla, R.; Yacynych, A.M. *Anal. Chem.* **1990**, *62*, 1111-1117.
(18) Moussy, F.; Harrison, D.J. *Anal. Chem.* submitted, **1993**.
(19) Fan, Z.; Harrison, D.J. *Anal. Chem.* **1992**, *64*, 1304-1311.
(20) Moore, R. B.; Martin, C. R. *Macromolecules* **1988**, *21* , 1334-1339.
(21) Gottesfeld, S.; Raistrick, I. D.; Srinivasan, S. *J. Electrochem. Soc.* **1987**, *134* , 1455-1462.
(22) Lawson, D. R.; Whiteley, L. D.; Martin, C. R. *J. Electrochem. Soc.* **1988**, *135*, 2247-2253.

RECEIVED April 12, 1994

Chapter 23

Chemically Sensitive Interfaces on Surface Acoustic Wave Devices

Antonio J. Ricco[1], Richard M. Crooks[2], Chuanjing Xu[2], and Ronald E. Allred[3]

[1]Microsensor Research and Development, Sandia National Laboratories, Albuquerque, NM 87185–0351
[2]Texas A&M University, College Station, TX 77843–3255
[3]Adherent Technologies, Inc., Albuquerque, NM 87123

Using surface acoustic wave (SAW) devices, three approaches to the effective use of chemically sensitive interfaces that are *not* highly chemically selective have been examined: (1) molecular identification from time-resolved permeation transients; (2) using multifrequency SAW devices to determine the frequency dependence of analyte/film interactions; (3) use of an array of SAW devices bearing diverse chemically sensitive interfaces to produce a distinct response pattern for each analyte. In addition to their well-known sensitivity to mass changes (0.0035 monolayer of N_2 can be measured), SAW devices respond to the mechanical and electronic properties of thin films, enhancing response information content but making a thorough understanding of the perturbation critical. Simultaneous measurement of changes in frequency and attenuation, which can provide the information necessary to determine the type of perturbation, are used as part of the above discrimination schemes.

Surface acoustic wave (SAW) devices have been studied in detail for chemical sensing applications (*1-12*). Nearly all this work has relied on some sort of chemically sensitive interface, many of which, however, are not particularly chemically *selective*. Table I summarizes the different classes of materials that have been examined for SAW-based chemical sensing applications, with a few examples in each category. Bearing in mind that SAW devices respond to changes in mass/area, none of the materials in Table I can be claimed to be entirely immune to interference from nonspecific adsorption, particularly for interferants with vapor pressures significantly below ambient pressure.

Approaches to Discrimination for Chemical Sensors. Fortunately, there are several reasons why perfect chemical specificity is not always a prerequisite to a useful chemical sensor. In some applications, interferants are simply not present and the

0097–6156/94/0561–0264$08.00/0
© 1994 American Chemical Society

Table I. Chemically Sensitive Films and Analytes for SAW-Based Sensors

Group	Chemically Sensitive Film	Analyte(s) [Limit of Detection]
Metals	Pd	H_2 [50 ppm]
	Pt	NH_3 [0.5%]
	Cu	H_2S [100 ppb]
Metal oxides	WO_3	H_2S [10 ppm]
	ZnO	Organic solvents
	t-$PtCl_2$(ethylene)(pyridine)	Styrene [5 ppm]; vinyl acetate [5 ppm]; butadiene
	CuPc	Cl_2, Br_2, I_2
Metal complexes	H_2Pc, CoPc, CuPc, FePc, MgPc, NiPc, PbPc	NO_2 [500 ppb]
	Cu^{2+}/mercaptoundecanoic acid self-assembled monolayer	DIMP [100 ppb], DMMP
	Co(II) complexes of isonitrilo-benzoylacetate and tetramethyl-ethylenediamine	DIMP
Organic compounds	Triethanolamine	SO_2 [10 ppb]
	Pyridinium tetracyanoquinodimethane	NO_2
	Aminopropyltriethoxysilane	Nitrobenzene and its derivatives
Organic polymers	Polybutadiene	O_3
	Polyethylene maleate	Cyclopentadiene [200 ppm-min]
	Ethyl cellulose, fluoropolyol, phenoxy resin, poly(amidoxime), poly-1-butadiene, poly(epichloro-hydrin), polyethylene, poly-(ethylene maleate), poly(iso-prene), poly(methylmethacrylate), polystyrene, poly(vinyl chloride), poly(vinyl stearate), 1,1,1-tri-fluoroisopropyl methyl siloxane	Benzene, 1-butanol, 2-buta-none, 1,2-dichloroethane, dichloropentane, *N,N*-di-methylacetamide, dimethyl-phosphite, dodecane, methane-sulfonyl fluoride, octane, α-pinene oxide, *i*-propyl-acetate, toluene, triamylphos-phite, tributylphosphate, water
	Ethyl cellulose, fluoropolyol, poly(ethylene maleate), poly-(vinylpyrrolidone)	Chemical warfare agents and their simulants [30 ppb typical for organophosphonates]

Pc = phthalocyanine; DIMP = diisopropylmethylphosphonate; DMMP = dimethyl-methylphosphonate identity of the analyte is known unequivocally. An example is monitoring the concentration of solvent vapors in the exhaust from a cleaning station (*13*): the solvent being used is known and the point of interest is how changes in procedure (e.g., the amount of air admixed to the sprayed solvent) affect solvent emissions, and whether or not the emission level is legal. A nonselective polymer film such as polyisobutylene makes an excellent sensing layer for such an application, since it rapidly and reversibly absorbs a wide range of organic solvents.

Response to a broad range of chemical species can make a sensor more useful. To monitor underground storage tanks containing gasoline and other fuels for leaks, a sensor that responds to any liquid fuel or its vapors, but does not respond to water, has the broadest applicability; plasma-polymerized tetrafluoroethylene has been shown to work well in this capacity (*14*). Another example is monitoring the "general corrosivity" of an industrial ambient: the SAW-measured rate of mass change of an ordinary copper film can be utilized (*15*), providing sensitivity to corrosion rates of less than one molecular monolayer per day.

When chemical specificity is called for, a highly selective thin film is one solution. For analytes of major concern—Hg, trichloroethylene, Cr^{VI}, and certain organophosphonates—it may be worthwhile to use a complex, multistep procedure to synthesize a material that responds to a single analyte or narrow chemical class. In a few instances, very simple materials are quite selective for particular analytes; examples include gold films to detect mercury (if sulfur compounds are not present to interfere) and palladium films for the detection of hydrogen (unsaturated hydrocarbons can interfere in this case). When such simple, obvious interfaces are unavailable or inadequate, scrutiny of the literature of interfacial chemistry, bulk-phase coordination chemistry, and catalysis (*16*) may point the way for the development of a tailored interface. But to utilize the one-analyte/one-(new)-film approach for general-purpose chemical detection systems, which could be called upon to recognize tens or hundreds of analytes in the presence of many interferants, is impractical, so alternatives must be examined.

Time resolution of analytes can provide excellent molecular discrimination. By adding gas- or liquid-phase chromatography as the "front end" of a chemical sensor system, many different analytes and, importantly, complex mixtures, can be identified and quantified using a single nonselective film, which might more accurately be called a detector in this context. Though this adds complexity to the sensing system, it provides versatility in return. A less complex (and somewhat less powerful) technique is to utilize the time response of the chemically sensitive film itself: the transient signal obtained when an aliquot of analyte is sorbed by the chemically sensitive interface is often a function of the analyte's physicochemical properties. In this case, an example of which is presented in the Results and Discussion section, the only added complexity is in the software and perhaps the computational hardware used to analyze the data.

A final means to circumvent the need for perfect chemical selectivity is to utilize the response from several sensors, applying the techniques of chemometrics or pattern recognition to identify and quantify the analyte(s). In this Chapter and within the context of SAW device-based sensor systems, we examine two variations on this theme. The first may be considered a form of spectroscopy: the utilization of a multifrequency SAW device to probe the interactions between various analytes and a single chemically sensitive interface at several frequencies (*17*).   As described in more detail below, this technique is useful because SAW devices respond with differing frequency dependencies to many different physical perturbations.

The second, more conventional array/pattern-recognition approach involves selecting an array of chemically sensitive interfaces that produces a unique response pattern for each analyte.   A number of research groups are active in this research area (*18-30*), several of them studying acoustic wave devices (*26-30*).   A key to obtaining a unique pattern for each of many different analytes is "chemical orthogonality" of the sensitive interfaces: if one member of the array responds very similarly to another, there is little added value in its response.   This requirement means that chemically sensitive interfaces must have a degree of selectivity, but the selectivity need not (and, for some pattern-recognition techniques, *must not* (*30*)) be as perfect as in the one-film/one-analyte approach.   For chemical detection using nonbiological interfaces, the sort of selectivity that works best with pattern recognition, namely a strong affinity for one or a few related analytes, together with weaker affinities for many other species, allows the consideration of many more candidate chemically sensitive interface materials than does the requirement for perfect selectivity.

Two Classes of Chemically Sensitive Interface.   It is our ultimate goal to select materials from widely different chemical classes—metals, metal oxides, semiconductors, organic polymers, coordination complexes, and organized thin films—to form a chemically disparate array.   Because of our fairly extensive experience with metals, oxides, and ordinary polymers, we are currently focusing our efforts on two newer classes of materials, self-assembled monolayers and plasma-grafted polymer films.

Self-assembled monolayer (SAMs) show promise for a diverse set of thin-film applications in areas such as chemical sensors, electronic and optical devices, lubrication, model biological membranes, electron-transfer barriers, catalysis, and separations (*31*).   Self assembly occurs when the appropriate substrate is exposed to a dilute gas stream or solution of the monolayer-forming molecule.   SAMs rely on a relatively strong (> 40 kcal/mol) (*32*) chemisorptive interaction between a head group and the substrate to orient all the molecules "head down"; van der Waals interactions between the bodies (typically long alkyl chains) are optimized when all the chains line up in the same relative orientation, producing an ordered monolayer. The most ordered layers result when the spacing between adjacent head-group adsorption sites on the substrate is close to the optimal van der Waals spacing of the molecular bodies.

Plasma-grafted films, a promising new class of polymeric materials, have been little explored to date (*33*).   They have the potential to incorporate a wide range of functional groups in a polymer matrix that is very open and permeable.   The process of plasma grafting, outlined in **Scheme 1**, begins with a brief exposure of the substrate to a plasma containing a gaseous species that forms relatively stable free radicals; we have used hydrogen-rich vinyl compounds.   The result is a thin, highly crosslinked (typical of plasma-polymerized films) "base" layer terminated by a considerable density of free radicals.   With the plasma "switched off," the molecule to be grafted is admitted to the chamber without breaking vacuum; this molecule must contain a vinyl or similar unit that readily undergoes free-radical polymerization. Polymerization is mainly straight-chain, resulting in the "kelp forest-like" grafted layer shown in Scheme 1. Polymerization ceases when the radicals have been quenched (e.g., by O_2 admitted when the chamber is opened to air or by recombination of radicals on adjacent chains to form loops). Such films *do not* have the ordered nature of SAMs: there is no epitaxy involved and as the chains become longer, they fold over on themselves. The point is that a high density of the desired functional group can be incorporated into a thin film while maintaining a highly permeable morphology.

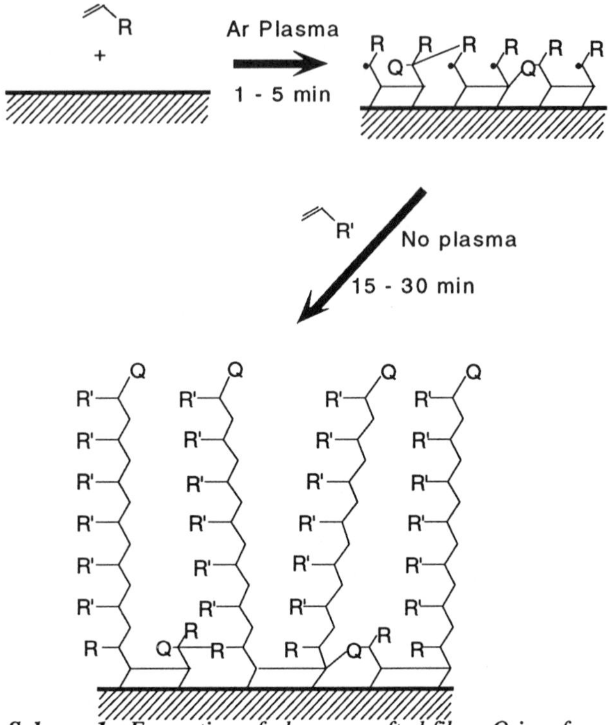

Scheme 1. Formation of plasma-grafted film. Q is a free-radical quencher.

SAW Devices. In principle, any change in a physical property of the SAW device surface or a thin film on the surface that affects either (or both) of the wave propagation parameters, velocity and attenuation, can be utilized in the construction of SAW-based chemical sensors; changes in pressure, temperature, mass, film (visco)elastic properties, electrical conductivity, and dielectric coefficient can all affect the SAW. In practice, most SAW-based chemical sensors have relied on the superior mass sensitivity of the SAW velocity to operate in a microbalance mode: mass resolution can exceed 100 pg/cm^2, about 10^{-3} monolayers of $C_7H_{15}SH$. Several reports from our laboratory have described the use of perturbations other than mass loading for SAW-based chemical sensing (*1,34-37*), and we believe use of these perturbations in combination with the SAW's mass-detection capabilities can lead to more versatile chemical sensors. To utilize all the information contained in SAW sensor response and to recognize non-mass-related perturbations, it is necessary to monitor the SAW attenuation in addition to its velocity (*37*).

To monitor changes in SAW propagation characteristics, an oscillator loop is often utilized (*10,38*). In this circuit, the SAW device serves as the frequency-control element and frequency changes are proportional to SAW velocity shifts; attenuation changes are proportional to changes in insertion loss (*38*). The devices used in this work cause loop oscillation at approximately 100 MHz; short-term (several seconds) frequency stability of 1 Hz is typical (*39*). Thus, a frequency change of 1 part in 10^8, and a similarly small velocity change, is measurable. This measurement accuracy, combined with the fact that SAW energy is distributed within one acoustic wavelength of the surface (*40*), is the reason for the SAW's exquisite sensitivity to surface perturbations.

Experimental Methods and Materials

ST-cut, *X*-propagating quartz SAW devices were designed and fabricated at Sandia National Laboratories (SNL) (*17,39*). Interdigital transducers (IDTs) have finger lengths of 50 Λ (Λ is the acoustic wavelength), finger width and separation of $\Lambda/4$, and 200 nm-thick Cr-on-Au metallization. Single-frequency devices have a center frequency (f_c) of 97 MHz, $\Lambda = 32$ µm, 50 finger-pairs/IDT, and a center-to-center IDT spacing of 230 Λ. Seven-frequency devices have $f_c = 25$ through 200 MHz in $\sqrt{2}$ steps, $\Lambda = 124$ through 15.5 µm in $1/\sqrt{2}$ steps, 25 finger-pairs/IDT, and center-to-center IDT spacings of 200 Λ (the highest and lowest frequency devices were not used for the measurements described in this Chapter). Single-frequency measurements utilized an oscillator loop (*38*), and multifrequency experiments utilized a computer-controlled phase-locked loop (*17*); in both cases, frequency and insertion loss were monitored simultaneously. ZnO-on-Si resonators, designed and fabricated at Purdue University (*41*) were operated in an oscillator loop configuration.

Metal films (Aesar, 99.999%) were thermally evaporated from tungsten baskets at 0.1 to 5 Å/s in a cryogenically pumped high-vacuum system with base pressure below 10^{-7} Torr; Au films utilized a 10 - 15 nm-thick Cr adhesion underlayer. SAW devices were masked to expose between 50 and 75% of the region between input and output IDTs for metal deposition.

Test gases/vapors were provided by a computer-controlled gas-flow system (*42*). Streams containing organophosphonates, water, and various organic compounds were produced by passing high-purity N_2 through a gas-washing bottle with a 50-mm diameter fritted disk; this vapor-saturated gas stream was diluted as required with additional N_2. Total flow rates were 1 l/min; the gas, organic liquid, and the SAW device were all thermostatted at 23° C. For time-resolved permeation experiments, molar volumes were calculated from the density of the bulk liquid.

Chemicals and solvents, reagent grade or better, were used as received (Alfa, Aldrich, or Fisher). Prior to monolayer self assembly, SAW devices were silanized using $ClSi(CH_3)_3$ (Petrarch) to prevent interactions between polar compounds (particularly the organophosphonates) and the quartz substrate from complicating interpretation of the response. SAMs of mercaptoundecanoic acid (MUA, $HS(CH_2)_{10}COOH$; Aldrich) were produced by soaking Au-coated SAW devices in a dilute solution of the thiol in ethanol overnight.

Polybutadiene/polystyrene (PB/PS, 28% PS by weight; Kraton D1102) ABA triblock copolymer films (A = PS) were spin cast from toluene onto the entire surface of the multifrequency device to give a film 200 nm thick by profilometry.

Acrylic acid was used as received (Polysciences, Inc., Warrington, PA). A flask containing this monomer was heated to 120 °C; a flow of 5 L/min of Ar entrained the acrylic acid, carrying it into the quartz plasma-deposition chamber, which was continuously pumped to maintain an indicated pressure of 1 Torr (thermocouple gauge). The RF plasma (13.56 MHz generator) was maintained at a forward power of approximately 50 W. After deposition of the base layer (5 min), the RF generator was switched off and grafting of acrylic acid onto the base layer allowed to continue for 15 - 30 min at an indicated pressure of 1 Torr.

Results and Discussion

Discrimination using Molecular Size-Dependent Response Transients. ZnO is one of a limited number of materials that have been sputter-deposited in thin-film form with retention of most of their piezoelectric activity: under the appropriate conditions, the majority of the sputter-deposited crystallites are oriented with their c axes normal to the substrate surface (*41*). Work by Martin *et al.* demonstrated the use of this technology to produce thin-film ZnO-on-Si-substrate SAW resonators that can be used as chemical sensors with no additional chemically sensitive interface (*2*).

Figure 1a shows the response of a 110-MHz ZnO-on-Si resonator to an ethanol-saturated stream of flowing N_2. Note the initial, rapid frequency decrease due to the physisorption of C_2H_5OH on readily accessible ZnO grain surfaces. The slower portion of the response transient represents percolation of ethanol into the film (presumably along grain boundaries); this portion of the transient can be fit to a simple exponential expression, allowing the extraction of a characteristic time constant. When the ethanol-saturated flow is replaced by dry N_2, rapid desorption of the surface-physisorbed molecules occurs, followed by much slower (but eventually complete) desorption from the interior of the film.

Repetition of the experiment of Figure 1a for a number of different organic solvents reveals an interesting trend: there is an inverse correlation between the time constant for the slow, permeation-related transient and the logarithm of the molar volume of the permeating species (Figure 1b) (*2*). This trend is similar to that observed in gel-permeation chromatography, and is explained by a size distribution of pores and cracks that provides a progressively more tortuous path to the adsorption sites as the sizes of the molecule decreases (*43*).

A similar but qualitatively opposite trend has been observed for the diffusion of small solvent molecules into a thin polyimide film: an inverse correlation was found between molar volume and the logarithm of diffusion coefficient, meaning that the time to attain equilibrium is longest for the largest species (*36*). In this case, larger molecules move more slowly because there is greater hindrance to their passage between/among polymer chains. For both the polyimide and ZnO examples, relatively nonselective films allow identification of chemical species according to the rate at which they move through the chemically sensitive interface.

Multifrequency SAW Devices for Chemical Discrimination (*17*). Each of the numerous physical perturbations to which SAW devices respond has a particular frequency (f) dependence for velocity shifts and, in general, a different frequency dependence for attenuation changes. Examining the frequency-dependent response of a chemically sensitive interface on a SAW device thus provides spectroscopic information. The data we present here are for a combination of mass-loading and viscoelastic perturbations, which involve energy transfer between the SAW device and a thin polymer film (*44*).

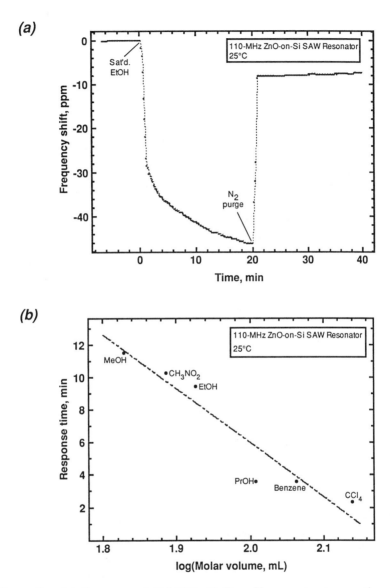

Figure 1. (*a*) Response of 110-MHz ZnO-on-Si resonator to an ethanol-saturated stream of flowing N_2. (*b*) Time constant for the permeation-related adsorption transient *vs.* the logarithm of the molar volume of the permeating species. (Adapted from ref. 2.)

Figure 2a shows schematically the multifrequency SAW device; the frequencies used range from 35 to 141 MHz. A single polybutadiene/polystyrene copolymer film was the chemically sensitive interface for all five delay lines. Figure 2b is a trichloroethylene (TCE) isotherm measured at 35 MHz, showing frequency shift (solid line) and attenuation change (dashed line) as a function of the fractional saturation pressure of TCE. In Figure 2c, the isotherm is replotted with attenuation against frequency shift on isometric axes, the variable parameter being TCE partial pressure, which increases as the curve is traversed from the origin to its distal end. The "loop" shape of this isotherm signals a combination of two SAW perturbations, namely a viscoelastic interaction as the polymer is plasticized by the TCE, and increased mass loading as the TCE concentration in the film increases (35). The isotherm is essentially a vectorial sum of a (somewhat distorted) semicircle, associated with polymer plasticization, and a (negative-going) horizontal line resulting from mass loading. In Figure 3a, attenuation vs. velocity TCE isotherms are shown for all five frequencies. Note that any particular TCE concentration yields a unique set of five points in the attenuation/velocity plane.

Figure 3b is a five-frequency set of isotherms for n-pentane. Note the marked difference in the shape of these curves compared to those for TCE (Figure 3a). The principal cause of the difference is pentane's lower density, less than one-half that of TCE: for pentane, the plasticization component of the response is dominant and the mass-loading component plays a minor part, so the isotherms are more nearly semicircular. A set of five isotherms for i-propanol (not shown) have their own distinctive set of (nearly linear) shapes, readily distinguishable from those of TCE and pentane. Thus, for a given concentration of a particular analyte, the set of five response points in the attenuation-frequency plane form a relatively unique signature using a *nonselective* chemically sensitive interface.

Single-Frequency SAW Arrays for Chemical Discrimination. An array of five identical 97-MHz ST-quartz SAW devices was utilized in combination with two general classes of film: self-assembled monolayers and plasma-grafted films. We are currently exploring these two classes of relatively new, general-purpose materials that are readily tailorable to provide moderate chemical selectivity for a particular species or chemical class.

Figure 4a is the frequency shift resulting from exposure to TCE at 10% of saturation for five SAW devices. Three were coated with self-assembled monolayers (n-hexadecane thiol, MUA, and MUA/Cu^{2+}) and two with plasma-grafted acrylic acid films (5 min plasma-polymerized base + 15 min graft; 5 min plasma-polymerized base + 30 min graft). In Figure 4b, the change in insertion loss (proportional to attenuation change) is shown for the same experiment. Thus, this five-device experiment provides a total of 10 data points for a single concentration of one analyte. Experiments similar to those depicted in Figure 4 have been carried out for a number of analytes at various concentrations, with the general result that the "pattern" of five frequency and five attenuation responses is not only distinct for each analyte, but is often concentration dependent as well.

To obtain responses to a wide range of concentrations for each analyte, isotherms were recorded. Frequency shifts and insertion loss changes were measured for each of the five coated SAW devices over a four-hour period as the concentration of the

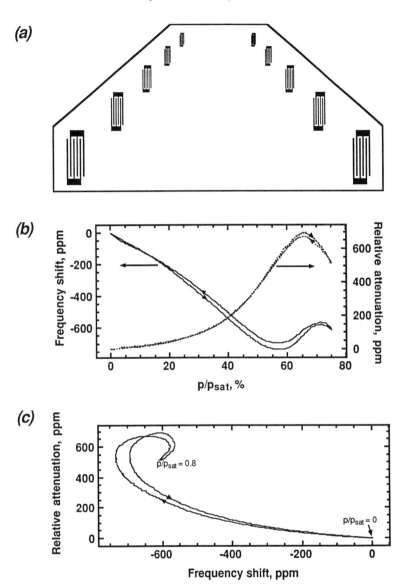

Figure 2. *(a)* Multifrequency SAW device. *(b)* Trichloroethylene (TCE) isotherm for PB/PS film measured at 35 MHz, showing frequency shift (solid line) and attenuation change (dashed line) as a function of the fractional saturation pressure of TCE. *(c)* Isotherm of Figure 2b replotted with attenuation against frequency shift on isometric axes; the variable parameter is TCE partial pressure.

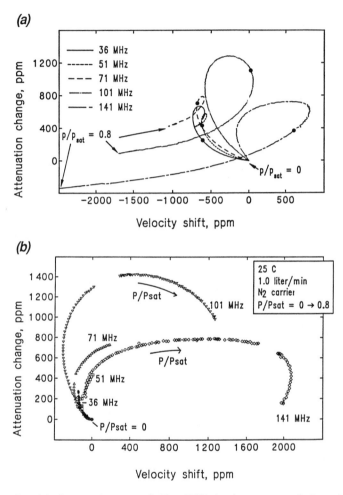

Figure 3. (a) Attenuation-vs.-velocity TCE isotherms recorded at the five indicated frequencies. The five filled circles are the response for a 5%-by-volume concentration of TCE in N_2. b) Five-frequency set of isotherms for n-pentane. (Reproduced with permission from ref. 17. Copyright 1993 Elsevier Sequoia.)

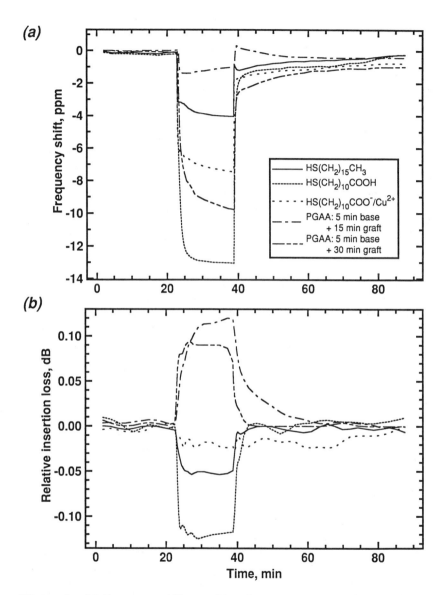

Figure 4. (*a*) Frequency shifts resulting from exposure to TCE at 10% of saturation for five SAW devices bearing the indicated thin films. (*b*) Change in insertion loss (proportional to attenuation change) for the same experiment and films as (a).

analyte was ramped from zero to 50% of saturation and back again to zero. Figure 5 contains isotherms for CCl_4, cyclohexane, and DMMP (dimethylmethylphosphonate), as indicated, with frequency shifts as the abscissa and insertion loss changes as the ordinate. For clarity, only the adsorption branches are shown; the desorption branches typically deviated by only a few percent. The films are the same as those described above for Figure 4. One isotherm typically provides frequency and insertion loss data points for several hundred concentrations of a particular analyte, which can be advantageous for many pattern-recognition techniques.

Comparison of the isotherms of Figure 5 shows that the three analytes can readily be distinguished from one another using this set of films. We have obtained isotherms for a range of organics, including DIMP (diisopropylmethylphosphonate), DMMP, CCl_4, trichloroethylene, acetone, methanol, n-propanol, pinacolyl alcohol, benzene, toluene, n-hexane, cyclohexane, i-octane, as well as water; every one of these analytes has a distinguishing response pattern.

Finally, the nonlinearities with concentration of some of the data of Figure 5 deserves mention. For example, the MUA-coated device isotherm for CCl_4 (Figure 5a) has a partial "loop" shape centered near 22 ppm on the frequency-shift axis. Such behavior is indicative of a local maximum in the energy transferred from the SAW device to the analyte-containing film (the same behavior shown in Figures 2b and 2c for TCE in PB/PS). This results either when the effective acoustic thickness of the film is an integral multiple of one-quarter acoustic wavelength (44) or when some intrinsic relaxation process in the film matches the period of the SAW. Note that the effective acoustic film thickness depends upon film modulus and density, and hence dissolved analyte concentration; the relaxation time(s) associated with polymer chain motion also depends, in general, upon the concentration of any dissolved species that affects chain movement. Such nonlinearities result in more distinctive isotherms and thus, in a general sense, are likely to increase the number of different analytes that can be identified with a single film; their potential drawback is that some traditional chemometrics approaches (such as multiple-regression analysis) only work with relatively linear systems. We are currently exploring multidimensional cluster analysis as a means to analyze our data, a method for which nonlinearities generally are not an obstacle (45,46).

Conclusions

Use of the full information content available in sensor responses allows identification of many analytes using chemically sensitive interfaces that are *not* perfectly chemically selective. Regardless of the sensor platform, the time dependence of the response to an analyte can depend on such factors as molecular size and shape, as well as the strength of the film/analyte interaction, providing one means of distinguishing various species. For SAW sensors, both the velocity and attenuation of the acoustic wave can be monitored simultaneously, providing information about the physical nature of the interface/analyte interaction that, again, can help distinguish one species from another (47). This technique is even more powerful when the frequency dependence of velocity and attenuation shifts are determined by making measurements at multiple frequencies.

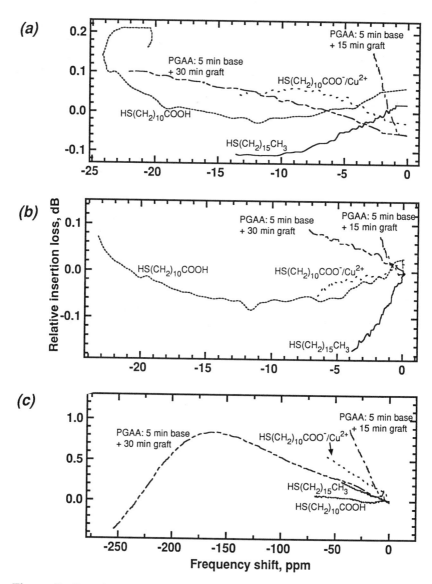

Figure 5. Insertion-loss-change-vs.-frequency-shift isotherms for five-SAW-device array. *(a)* CCl$_4$; *(b)* cyclohexane; *(c)* DMMP. In each case, the origin corresponds to p/p$_{sat}$ = 0 and the distal ends of the curves to p/p$_{sat}$ = 0.5. PGAA is plasma-grafted acrylic acid.

While the development of highly selective chemical interfaces for a few key analytes is important, the use of sensor arrays in combination with pattern recognition offers the promise of more general chemical analysis. Two SAW responses per chemically sensitive interface provide more information than single-parameter measurements. While pattern recognition does not require perfect chemical selectivity, it works best when the elements of the array are chemically diverse. The use of new materials such as self-assembled monolayers and plasma-grafted films, that offer relatively simple means to introduce a wide range of chemical functionalities, is one important means to provide such chemical diversity.

Acknowledgments

We gratefully acknowledge helpful technical discussions on pattern recognition with Gordon C. Osbourn of SNL, collaboration with AJR on the SAW array test fixture by Stephen J. Martin, and the excellent technical assistance of Mark Hill, Mary-Anne Mitchell, Alan Staton, and Barb Wampler of SNL for this work. Work at SNL is supported by the US DOE under contract DE-AC04-94AL85000.

Literature Cited

1. Ricco, A. J.; Martin, S. J.; Zipperian, T. E. *Sensors and Actuators* **1985**, *8*, 319.
2. Martin, S. J.; Schweizer, K. S.; Schwartz, S. S.; Gunshor, R. L. *Proc. 1984 IEEE Ultrasonics Symp.*; IEEE: New York, New York, 1984; pp. 207-212.
3. Wenzel, S. W.; White, R. M. *Proc. 1989 IEEE Ultrasonics Symp.*; IEEE: New York, New York, 1989; pp. 595-598.
4. Zellers, E. T.; Hassold, N.; White, R. M.; Rappaport, S. M. *Anal. Chem.* **1990**, *62*, 1227.
5. Ballantine, D. S., Jr.; Wohltjen, H. *Anal. Chem.* **1989**, *61*, 704A.
6. Rose-Pehrsson, S. L.; Grate, J. W.; Ballantine, D. S., Jr.; Jurs, P. C. *Anal. Chem.* **1988**, *60*, 2801.
7. Heckl, W. M.; Marassi, F. M.; Kallury, K. M. R.; Stone, D. C.; Thompson, M. *Anal. Chem.* **1990**, *62*, 32.
8. Nieuwenhuizen, M. S.; Venema, A. *Sensors and Materials* **1989**, *5*, 261.
9. Falconer, R. S.; Lec, R.; Vetelino, J. F.; Xu, Z. *Proc. 1989 IEEE Ultrasonics Symp.*; IEEE: New York, New York, 1989; pp. 585-590.
10. Wohltjen, H. *Sensors and Actuators* **1984**, *5*, 307.
11. D'Amico, A.; Verona, E. *Sensors and Actuators*, **1989**, *17*, 55.
12. Roberts, G. G.; Holcroft, B.; Barraud, A.; Richard, J. *Thin Solid Films* **1988**, *160*, 53.
13. Frye, G. C.; Martin, S. J.; Cernosek, R. W.; Pfeifer, K. B. *Int. J. Env. Conscious Manf.* **1992**, *1*, 37.
14. Butler, M. A.; Ricco, A. J.; Buss, R. *J. Electrochem. Soc.* **1990**, *137*, 1325.
15. Frye, G. C.; Martin, S. J. *Appl. Spectrosc. Rev.* **1991**, *1&2*, 73.
16. Kepley, L. J.; Crooks, R. M.; Ricco, A. J. *Anal. Chem.* **1992**, *64*, 3191.
17. Ricco, A. J.; Martin, S. J. *Sensors and Actuators* **1993**, *B10*, 123.
18. Stetter, J. R.; Findlay, M. W.; Maclay, G. J. *Sensors and Actuators B* **1990**, 1, 43.

19. Horner, G.; Hierold, Chr. *Sensors and Actuators B* **1990**, *2*, 173.
20. Müller, R. *Sensors and Actuators B* **1991**, *4*, 35.
21. Schierbaum, K. D.; Weimar, U.; Göpel, W. *Sensors and Actuators B* **1990**, *2*, 71.
22. Winquist, F.; Sundgren, H.; Hedborg, E.; Spetz, A.; Lundström, I. *Sensors and Actuators B* **1992**, *6*, 157.
23. Shurmer, H. V.; Gardner, J. W. *Sensors and Actuators B* **1992**, *8*, 1.
24. Bott, B.; Jones, T. A. *Sensors and Actuators* **1986**, *9*, 19.
25. Davide, F. A. M.; D'Amico, A. *Sensors and Actuators A* **1992**, *32*, 507.
26. Ballantine, D. S., Jr.; Rose, S. L.; Grate, J. W.; Wohltjen, H. *Anal. Chem.* **1986**, *58*, 3058.
27. Schmautz, A. *Sensors and Actuators B* **1992**, *6*, 38.
28. Carey, W. P.; Beebe, K. R.; Kowalski, B. R. *Anal. Chem.* **1987**, *59*, 1529.
29. Nakamoto, T.; Fukuda, A.; Moriizumi, T.; Asakura, Y. *Sensors and Actuators B* **1991**, *3*, 221.
30. Zellers, E. T.; Patrash, S. J. *Anal. Chem.* **1992**, *65*, 2055.
31. Swalen, J. D.; Allara, D. L.; Andrade, J. D.; Chandross, E. A; Garoff, S.; Israelachvili, J.; McCarthy, T. J.; Murray, R. W.; Pease, R. F. *Langmuir* **1987**, *3*, 932.
32. DuBois, L. H.; Nuzzo, R. G. *Ann. Rev. Phys. Chem.* **1992**, *43*, 437.
33. Hsieh, Y. L.; Wu, M. *J. Appl. Polymer Sci.* **1991**, *43*, 2067.
34. Martin S. J.; Ricco, A. J. *Sensors and Actuators*, **1990**, *A22*, 712.
35. Martin, S. J.; Frye, G. C. *Appl. Phys. Lett.* **1990**, *57*, 1867.
36. Frye, G. C.; Martin, S. J.; Ricco, A. J. *Sensors and Materials*, **1989**, *1*, 335.
37. Martin, S. J.; Ricco, A. J. *Proc. 1989 IEEE Ultrasonics Symp.*; IEEE: New York, New York, 1989; pp. 621-625.
38. Ricco, A. J.; Martin, S. J. *Thin Solid Films*, **1992**, *206*, 94.
39. Ricco, A. J.; Frye, G. C.; Martin, S. J. *Langmuir*, **1989**, *5*, 273.
40. Auld, B. A. *Acoustic Waves and Fields in Solids*; John Wiley & Sons: New York, New York, 1973; Vol. 2.
41. Gunshor, R. L.; Martin, S. J.; Pierret, R. F. *Jap. J. Appl. Phys.* **1983**, *Suppl. 22-1*, 37.
42. Martin, S. J.; Ricco, A. J.; Ginley, D. S.; Zipperian, T. E. *IEEE Trans. on UFFC* **1987**, *UFFC-34*, 142.
43. Altgelt, K. H.; Segal, L. *Gel Permeation Chromatography*; Marcel Dekker: New York, New York, 1971.
44. Martin, S. J.; Frye, G. C. *Proc. 1991 IEEE Ultrasonics Symp.*; IEEE: New York, New York, 1991; pp. 393-398.
45. Hand, D. J. *Discrimination and Classification*; John Wiley & Sons: New York, New York, 1981; Ch. 7.
46. Jain, A. K. In *Handbook of Pattern Recognition and Image Processing*; Young, T. Y.; Fu, K.-S., Eds.; Academic Press, Inc.: San Diego, California, 1986; Ch. 2.
47. Frye, G. C.; Martin, S. J. *Sensors and Materials* **1991**, *2*, 187.

RECEIVED March 24, 1994

Chapter 24

Surface and Interfacial Properties of Surface Acoustic Wave Gas Sensors

R. Andrew McGill[1], J. W. Grate[2,3], and Mark R. Anderson[2]

[1]Geo-Centers, Inc., 10903 Indian Head Highway, Fort Washington, MD 20744
[2]Naval Research Laboratory, Code 6170, Washington, DC 20375

Surface Acoustic Wave (SAW) devices coated with a thin layer of polymeric material provide highly sensitive microsensors for the detection and monitoring of vapors and gases. The properties of gas/solid and polymer/solid interfaces upon exposure to vapor are conveniently monitored by SAW devices. Interfacial adsorption at the polymer/solid interface is indicated by fast mass loading effects and slow anomalous SAW responses. Chemical modification of uncoated SAW device surfaces by silanisation techniques and cleaning treatments including plasma etching and solvent rinsing are shown to impact greatly on the interfacial properties of SAW devices. Silanisation with diphenyltetramethydisilazane is applied to minimize interfacial adsorption and maintain polymer wetability. For a non-polar polymer coating of polyisobutylene the interaction of water vapor with the polymer/SAW device interface is dramatically reduced by this method.

Interfacial adsorption of vapors and gases are common to most polymer coated surfaces and in many instances is undesirable. For example, in composite materials it can lead to delamination of polymer coating from substrate and, for painted surfaces which have been exposed to water or organic vapor, paint can crack and peel. In gas-liquid chromatography (GLC) it is common to coat a relatively inert support with a polymeric coating. However, if the coating is not complete or if vapor adsorption to the polymer/support interface is significant it can lead to peak tailing and affect the retention time of vapor, which will reflect the absorption processes in the polymer coating and the adsorption processes at the interface. This can lead to inaccurate thermodynamic measurements, such as the gas-liquid partition coefficient, commonly measured by GLC (1). Similarly when vapor is exposed to polymer coated SAW devices, both absorption into the polymer film and adsorption to the polymer/SAW device interface are possible sorption processes. Adsorption processes are undesirable and can lead to anomalous SAW responses which are relatively slowly responding and in the opposite direction to normal mass loading effects. The solubility properties of the interface and not the bulk polymer coating are reflected by these interactions.

[3]Current address: Pacific Northwest Laboratory, Richland, WA 99352

Surface Acoustic Wave devices operate by applying a time varying electric field to the surface of a piezoelectric material via an interdigital transducer (IDT) element laid down on its surface. The SAW resonator devices used in this work consist of two aluminium IDT arrays that convert electrical energy into mechanical energy (and vice versa), and a set of reflector arrays on each end of an ST-cut quartz crystal, Figure 1. When a time varying radio frequency potential is applied to one set of IDT's, a synchronous mechanical deformation of quartz with coincident generation of an acoustic wave in the surface of the quartz crystal results. This wave travels across the surface of the crystal and is received by the second set of IDT's which translates the wave back into an electrical signal. The wave continues across the crystal and is reflected at each end of the crystal surface, resulting in a standing wave developed by constructive interference of the reflected waves. The propagation of the surface acoustic wave is perturbed by the presence or accumulation of material on the surface of the SAW device, and the detection of chemical species is monitored by a change in SAW frequency.

Operation of a SAW device as a chemical sensor normally involves the deposition of some chemically selective material onto the surface of the SAW device (2-5). In this work thin polymer films are coated over the entire surface of the SAW device. When a vapor is exposed to the polymer-coated SAW device, it distributes between the gas, bulk polymer phase, and interfaces, and comes to rapid thermodynamic equilibrium. Changes in mass and modulus that result from vapor absorption perturb the propagation of the surface acoustic wave (6), and ordinarily result in a reduction of the observed resonant frequency. The ratio between the concentration of the vapor in the gas phase, C_v, and the concentration of vapor in the polymer phase, C_p, is known as the partition coefficient, K_p, given by equation 1. The portion of the observed response of a SAW device to vapor, Δf_v^{ab}, due to mass loading of the absorbent layer can also be related to the partition coefficient by equation 2, where the amount of polymer coating on the SAW device surface is expressed as a frequency shift, Δf_p, and δ_p is the density of the polymer coating (2). Note that equation 2 assumes that polymer viscoelastic effects upon vapor sorption are negligible.

$$K_p = C_p/C_v \qquad (1)$$

$$\Delta f_v^{ab} = \Delta f_p C_v K_p / \delta_p \qquad (2)$$

In addition to vapor partitioning into the polymer coating (absorption), vapor can also adsorb at the gas/polymer interface (gp) or diffuse and adsorb at the SAW device/polymer interface (sp), Figure 2. The extent of vapor adsorption at these interfaces should be viewed relative to the amount of polymer coating and the vapor concentration. The SAW chemical sensors used in this work are typically coated with 250KHz of polymer which corresponds to an average thickness of 529Å or about 75 molecular layers of polymer. This is a very thin coating and thus SAW responses from interfacial adsorption may contribute significantly and, in some instances can be the major component of the sensor response.

The frequency shift caused by adsorption onto an uncoated SAW device or to the SAW device/polymer interface, Δf_v^{ad}, can be described by equation 3 (2,6). Note that here we are dealing with mass-loading effects of adsorbed vapors and not absorbed vapors. k_1 and k_2 are material constants for quartz, F is the unperturbed resonant frequency of the SAW device, m_v is the mass of vapor adsorbed at the SAW device/polymer interface or the SAW device/gas interface (for coated and uncoated SAW

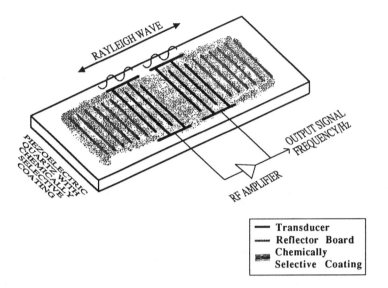

——	Transducer
⬝⬝⬝⬝⬝	Reflector Board
▓▓▓	Chemically Selective Coating

Figure 1. Surface acoustic wave resonator (SAW resonator) with chemically selective coating.

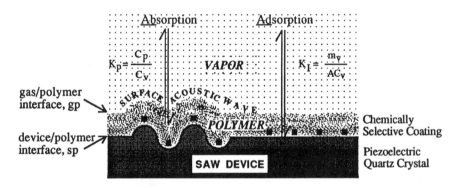

Figure 2. Oblique view of a polymer coated SAW device showing the reversible vapor sorption processes. The figure is not drawn to scale.

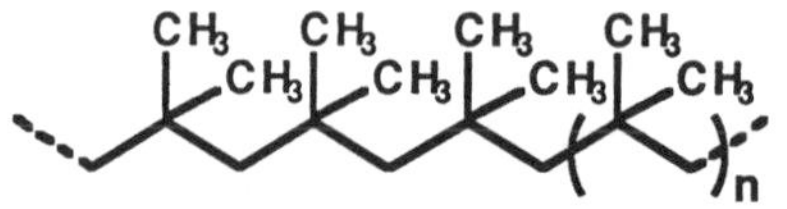

Figure 3. Polyisobutylene, a non-polar polymer coating for a SAW device.

$$\Delta f_V^{\underline{ad}} = (k_1 + k_2)F^2 m_V/A \qquad (3)$$

devices respectively), and A is the active surface area of the SAW device. The interfacial adsorption coefficient, K_I, is given by equation 4.

$$K_I = C_I/C_V = m_V/AC_V \qquad (4)$$

The variable C_I is the concentration of vapor present at the interface and has the units of mass per unit area. Rearranging equation 4 and substituting into equation 3 for m_V/A gives equation 5.

$$\Delta f_V^{\underline{ad}} = (k_1 + k_2)F^2 C_V K_I \qquad (5)$$

If we assume that significant adsorption occurs only at the SAW device/polymer interface, then equation 2 can be recast to equation 6 which reflects the additional adsorption process. Note that a similar additional term would be required to describe any response due to adsorption at the gas/polymer interface.

$$\Delta f_V^{\underline{ab+ad}} = \Delta f_p C_V K_p/\delta_p + (k_1 + k_2)F^2 C_V K_I \qquad (6)$$

$$\underline{Ab}\text{sorption} \qquad\qquad \underline{Ad}\text{sorption}$$

Both terms in equation 6 correspond to gravimetric processes and they may not be sufficient to accurately describe all processes that contribute to a SAW device response to vapor. Recently we have shown that viscoelastic changes in the polymer coating when exposed to vapor contribute to the overall SAW device signal (*6*) and to account for this an additional term is needed. The viscoelastic or modulus term (mod) will be proportional to the amount of polymer coating and the concentration of vapor in the polymer film, $C_V K_p$. In equation 7 a third term is included to describes the viscoelastic contribution in terms of polymer swelling (*6*). The variable α in equation 7 is the coefficient of thermal expansion of the polymer, and A_{SAW} represents the kilohertz change in frequency due to a 1°C change in temperature per kilohertz of coating on the SAW device surface, and δ_L is the density of the liquid solute vapor. Note that use of the word swelling here includes all effects not accounted for by gravimetric processes.

$$\Delta f_V^{\underline{ab+ad+mod}} = \Delta f_p C_V K_p/\delta_p + (k_1 + k_2)F^2 C_V K_I + (\Delta f_p C_V K_p/\delta_L)(A_{SAW}/\alpha) \qquad (7)$$

$$\underline{Ab}\text{sorption} \qquad \underline{Ad}\text{sorption} \qquad \underline{Mod}\text{ulus}$$

The polymeric coatings typically used as SAW device coatings are chosen to be selective towards a particular class of analyte vapor. For example, polyethyleneimine (PEI), a strong hydrogen-bond basic material, may be chosen as a SAW device coating to enhance the detection of hydrogen-bond acidic solute vapors such as water or alcohols (*7*). Polyisobutylene (PIB), Figure 3, a non-polar material, may be chosen to enhance the detection of non-polar solute vapors such as alkanes. SAW sensor systems have been designed around arrays of SAW devices, each device with a different polymeric coating. The array gives rise to a pattern of responses depending on the solubility properties of the coatings used (*8-13*) and the solubility properties of the vapor analyte of interest (*14*). Individual vapors can then be identified using pattern recognition techniques.

Consider the situation when the non-polar PIB polymer-coated SAW device is exposed to water vapor. This will be the normal situation when the SAW device is exposed to analyte vapor in ambient air. Note that the concentration of water vapor will vary over a large range, while the analyte of interest may be present at much lower concentration. Water molecules are dipolar, hydrogen-bond acidic and hydrogen-bond basic and do not readily dissolve in PIB, but can adsorb to the relatively polar interface between the SAW device surface and the polymer coating. The chemical nature of this interface depends upon the SAW fabrication or finishing techniques used, which will be more completely described later. For example, consider a SAW device fabricated such that the surface the polymer coats is bare quartz with aluminium metallisation for the IDT's. The water molecules will readily adsorb to surface silanol, silylether groups and aluminium oxide present at the interface, Figure 4a. The SAW device/polymer interfacial adsorption of water gives rise to a response which is in addition to, and may be in excess of the polymer coating absorption derived response. This is expected to be particularly marked at low vapor concentrations, where the adsorption sites are not saturated with adsorbate. In some instances interfacial adsorption gives rise to a relatively slow changing frequency response in the opposite direction to normal mass loading effects, and we refer to this as the anomalous effect (*15*). Interfacial adsorption processes are undesirable and could lead to misleading results and affect the pattern of responses developed by an array of SAW devices, especially when detecting trace amounts of analyte vapor. The interfacial response does not reflect the solubility properties of PIB which was chosen to enhance the SAW device selectivity and sensitivity towards non-polar solute vapors, and yet the SAW device manifests significant responses to water vapor, a strongly dipolar molecule.

In order to study surface responses of bare SAW devices, and interfacial responses of polymer coated SAW devices, a number of SAW devices were prepared with different surface finishing techniques to investigate SAW responses from different chemical compositions present at the interface between the SAW device and the polymer coating. The results of the study lead to recommendations for SAW device preparation prior to polymer coating, which minimize the effects of interfacial adsorption of vapors and gases on uncoated and polymer coated SAW devices.

Experimental

Materials. PIB was obtained from Aldrich (Cat. No. 18,145-5); diphenyltetramethyldisilazane was obtained from Pfaltz & Bauer, Inc. (Cat. No. D52405), which is referred to as DPTMDS; HPLC grade chloroform was obtained from Aldrich (Cat. No. 27,063-6). These reagents were used as received. The water used to generate the vapor stream was triply distilled, with the last two distillations in an all-quartz still.

SAW Devices and Electronics. 200MHz SAW resonator devices (*16*) and driver electronics were obtained from Femtometrics, Costa Mesa, CA. The SAW devices are fabricated on ST cut quartz with aluminum metallisation applied by vacuum deposition, for the reflector boards and signal transducers. The SAW devices were obtained with and without a thin protective overlay of silicon dioxide which was applied to a target thickness of 30 to 50nm. The devices are epoxy mounted on rectangular five-pin headers and gold leads attached by wire bonding, one device per header. The driver electronics were designed to provide both the absolute frequency of the test device and a difference frequency by subtraction from a sealed reference SAW device. In these experiments only the absolute frequency was monitored.

The SAW devices with and without the silica overlay were treated with three finishing techniques: solvent cleaning, plasma cleaning, and silanisation in that order. After each one of the finishing techniques was applied the bare SAW devices were exposed to water vapor at concentrations ranging from 50 to 12,000mg/m^3 and their responses recorded. These experiments were repeated with a PIB coating at an average polymer thickness of about 529Å for 250KHz of polymer, and this corresponds to 74 polymeric layers. In some additional experiments amounts of 50KHz, 100Khz, 150KHz, and 200KHz were used. The coating thickness, h in metres, was calculated using equation 8, where Δf_p is -250 x10^3 Hz, and F is 199 x 10^6 Hz, δ_p is 918 kg/m^3, and k_1 and k_2 are taken as -9 x 10^{-8} and-4 x 10^{-8} m^{2}skg^{-1}. The volume occupied by one polymer molecule was estimated by dividing the molecular weight (380,000) by the density of the polymer (0.918 g/ml). The length of a PIB polymer molecule was estimated at about 17000Å by assuming that for every C-C bond in the polymer backbone, the length of the chain is increased by 1.54Sin(109.3/2)Å. The length and angle between a C-C single bond was taken as 1.54Å and 109.3^o respectively. Assuming the linear polymer chain occupies the volume of a tube the molecule diameter is calculated at 7.17Å. Together with the coating thickness the number of polymeric layers is computed at 74.

$$\Delta f_p = (k_1 + k_2)F^2 h \delta_p \tag{8}$$

PIB was chosen because water vapor <u>ab</u>sorption is very poor into this apolar, hydrophobic material. Water was studied as the test vapor because from prior experience it had showed the worst anomalous responses and its polar and hydrogen-bond properties would favour interfacial adsorption at "polar" silica and alumina sites. The design of the experiments was specifically intended to be a worst case scenario for vapor <u>ab</u>sorption in the polymer film and a best case scenario for vapor <u>ad</u>sorption at the SAW device surface/polymer interface. This arrangement tests the SAW device in the most favourable environment for evaluating interfacial effects. The results and implications for SAW device fabrication are discussed.

SAW Device Finishing Techniques

1.   Solvent-Cleaning Method. As received SAW devices were dip rinsed with chloroform, methanol and acetone.

2.   Plasma-Cleaning Method. SAW devices were solvent cleaned as above and placed in a Harrick plasma cleaner (*17*). The chamber was evacuated until the pressure was approximately 100 millitorr and the air plasma developed by turning the r.f. power on. The air plasma produced a light sky blue color, and the pressure within the chamber was observed to increase as the SAW devices were cleaned. We interpret this to be the volatilisation of organic/inorganic material present on the surface of the SAW device as the cleaning procedure advanced. The return to the lowest pressure reading normally took no longer than 1 minute but the plasma cleaning process was typically extended to between 15 to 20 minutes.

3.   Silanisation Procedures. A wide variety of methods and reagents exist to silanise inorganic surfaces. At both silicon dioxide and aluminium oxide surfaces there exist hydroxy terminated chains (Si-OH, Al-OH) which can be functionalised with reactive organosilanes. From a literature search (*18-22*) and after testing a number of organosilanes under various conditions, diphenyltetramethyldisilazane (DPTMDS) was chosen to prepare PhMe2Si-silicon-dioxide and PhMe2Si-aluminium-oxide SAW surfaces, Figure 4b. One of the aims of the silane modification experiments was to

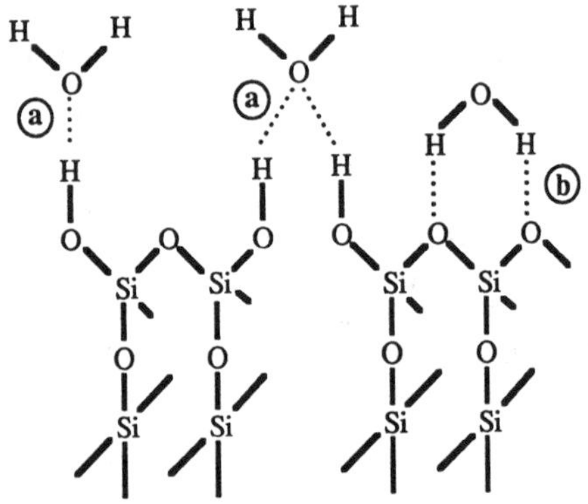

Figure 4a. Water adsorption on silica and quartz surfaces at silanol and silylether groups. Site "a" hydrogen-bond basic water...hydrogen-bond acidic silanol. Site "b" hydrogen-bond acidic water...hydrogen-bond basic silylether.

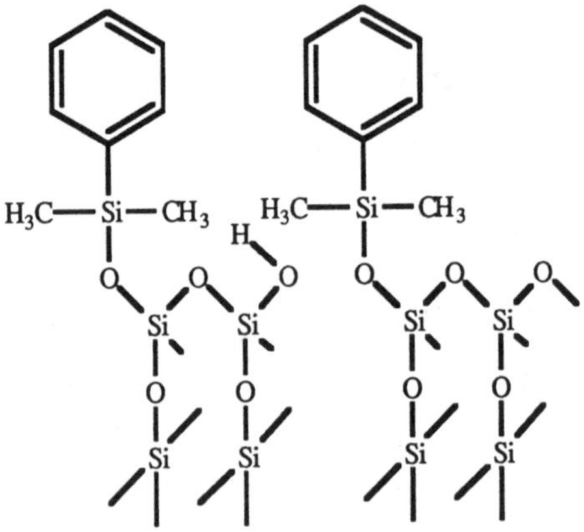

Figure 4b. Diphenyltetramethyldisilazane treated silica and quartz surfaces.

eliminate the polar sites at the surface of the SAW device that are targets for interfacial adsorption. After the preparation of non-polar surfaces, they can become difficult to coat with stable thin polymer coatings. To remedy this an organosilane capable of preparing surfaces with moderate polarity/polarisability and no significant hydrogen-bonding capability was chosen.

The experimental procedure involved immersing a plasma cleaned SAW device into liquid DPTMDS, pulling vacuum on the reaction chamber and heating to 170°C. The reaction was allowed to continue for 12 hours, and the silanised sensor was solvent cleaned with, toluene, methanol, and chloroform.

Polymer Coating and Vapor Testing

Polymer coating and vapor testing were as described in an earlier publication (6). In brief, polymer films were applied using an airbrush and a dilute solution of PIB in HPLC-grade chloroform, and the frequency of the SAW was monitored in real time during the coating process. The frequency of the SAW device was measured during coating and vapor testing using a Philips PM6674 Universal frequency counter with GPIB interface. The frequency versus time data was acquired and stored on an IBM compatible PC equipped with a National Instruments AT-GPIB interface card running software developed at the Naval Research Laboratory. The frequency data was obtained to 1Hz resolution.

The vapor source for testing was a Microsensors, Inc. (Bowling Green, KY) VG-7000 vapor generator. Water vapor was generated by bubbling dry N_2 through a thermostated teflon container maintained at 15°C, and diluted from saturation by the VG7000. The vapor concentrations were 50, 99, 335, 793, 1549, 3097, 6194, and 12388 mg/m^3 and tested in that order. Note that at 15°C, 12388 mg/m^3 corresponds to water vapor near saturation. The test cycle for each concentration was: a) 15 minutes of equilibration at the present dilution to allow the generator to stabilize, while pure carrier gas sweeps any residual vapor out of the SAW device from the previous experiment, b) 5 minutes of sensor exposure to the vapor stream, c) 10 minutes of sensor exposure to pure carrier gas to observe the rate of desorption and retention effects. The entire set of experiments was controlled by the computer using NRL developed software. The SAW device was thermostated at 25°C by clamping a copper heat sink attached to a circulating bath, to a nickel lid sealing the SAW device. Heat transfer was improved by applying a thin film of a thermally conductive paste. The test vapor was passed to the SAW device at 100 ml/min through a set of stainless steel tubes silver-brazed to the nickel lid.

Results and Discussion

The results from water isotherm experiments carried out on bare SAW devices are illustrated in Figure 5 and Figure 6. In Figure 5, SAW responses for uncoated devices fabricated with and without the protective overlay of silica are compared for solvent and plasma cleaning techniques. It is apparent that fast mass loading responses (decrease in frequency) are indicated for the plasma cleaned devices, Figures 5c and 5d, with relatively minor slowly developed anomalous responses, which are in the opposite direction to mass loading effects. The solvent rinsed device without the silica overlay, Figure 5a, showed particularly pronounced anomalous responses, while the solvent rinsed device with the silica overlay, Figure 5b, showed more normal mass loading responses. However baseline recovery after turning off the water vapor was not as rapid as the plasma cleaned device with the silica overlay, Figures 5b and 5d. From these experiments it is clear that the plasma technique is preferred over the solvent cleaning method.

In Figure 6, SAW responses for uncoated devices fabricated with and without silica are compared for plasma cleaned and silanised surfaces. The anomalous profiles

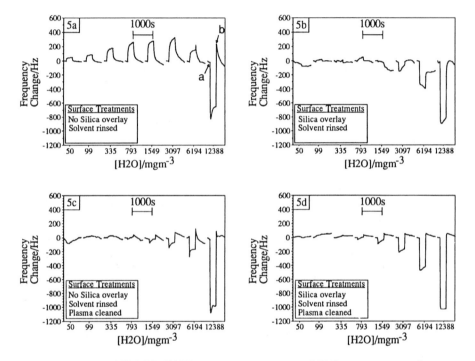

Figure 5. Bare 200 MHz SAW resonator responses at 25°C to water vapor from 50 - 12388mg/m³, for devices solvent rinsed and plasma cleaned, and manufactured with and without a protective overlay of silica. The vapor is on at "a" and off at "b".

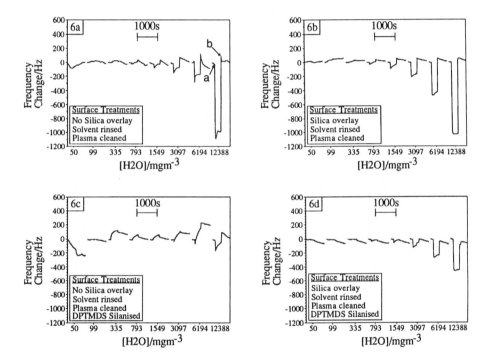

Figure 6. Bare 200 MHz SAW resonator responses at 25°C to water vapor from 50 - 12388mg/m³, for devices plasma cleaned and silanised with DPTMDS (see Figure 4b), and manufactured with and without a protective overlay of silica. The vapor is on at "a" and off at "b".

are more pronounced for the devices tested without the silica overlay, Figures 6a and 6c. The responses to water vapor for the silanised devices, Figures 6c and 6d are similar in shape to the plasma cleaned devices, Figures 6a and 6b, but the normal mass-loading responses are considerably smaller. This shows that the silanisation procedure has been successful in reducing the number of adsorption sites available on the surface of the SAW device.

In Figure 7, SAW responses for PIB-coated devices fabricated with and without silica are compared for plasma cleaned and silanised surfaces. The plasma cleaned devices, Figures 7a and 7b, showed both mass loading and anomalous responses and the plasma cleaned device without the silica overlay, Figure 7a shows similar response profiles and magnitude to the equivalent device without the polymer coating, Figure 6a. The rather large anomalous responses at the higher concentrations may be due in part to polymer/SAW device dewetting or delaminating processes brought on by the presence of water at the interface. The PIB coated, silanised device without silica overlay, Figure 7c shows relatively small mass loading effects but the anomalous responses are still present. By comparison, the PIB-coated, silanised device with the silica overlay, Figure 7d performed better than any of the other devices, showing the smallest mass responses to water vapor with little or no apparent anomalous contribution. Because the water vapor responses for PIB-coated, silanised SAW devices, Figures 7c and 7d are much smaller than identical experiments with SAW devices not treated with silane it is clear that the majority of SAW device response in this instance is due to interfacial adsorption of vapor and not polymer absorption.

In order to investigate this further, water vapor responses at 12388 mg/m^3 were determined at polymer thicknesses of 0, 50, 100, 150, 200, and 250KHz. The SAW frequency responses (Figure 8) were almost identical in shape and magnitude. If part or all of the SAW response was due to vapor absorption, the size of the frequency shift, on exposure to water vapor would have been expected to be significantly larger for thicker PIB coatings in accordance with equation 2. This is further evidence, that by far the majority of the SAW response in these experiments is due to interfacial effects. In this case, the adsorption term in equation 6 is very dominant and equation 6 can be simplified to equation 5. The similar responses for the uncoated SAW device and polymer coated device at various thicknesses is particularly significant and indicates that there is no significant contribution from a gas/polymer absorption process.

$$\Delta f_v^{ad} \quad = \quad (k_1 + k_2)F^2 C_v K_I \qquad\qquad (5)$$

Adsorption

Note that the SAW device response in Figure 7a is slightly larger than that shown in Figure 8 (250KHz coating) even though the sensor is the same device with the same amount of coating. The data collected for Figure 8 was taken 2 weeks after the initial coating on the plasma cleaned device. This may indicate that the "polar" nature of the surface of an uncoated plasma cleaned SAW device may change somewhat with time. One possible explanation for this may be the condensation of neighboring silanol groups over time (generated in the plasma by cleaving silylether chains) to silylether chains, which are less prone to water vapor adsorption.

The validity of equation 5 can be tested by examining the responses of uncoated SAW devices (Figures 6a and 6b) and PIB coated SAW devices (Figures 7a and 7d). In equation 5 k_1, k_2, and F are constants and K_I is expected to be a constant over some range of vapor concentration. A plot of frequency response to vapor, Δf_v, and the concentration of vapor, C_v should yield a straight line provided the surface or interface is not saturated with vapor, with the slope equal to $(k_1+k_2)F^2 K_I$. Figure 9 shows an example of this for a plasma cleaned, uncoated SAW device with a protective silica

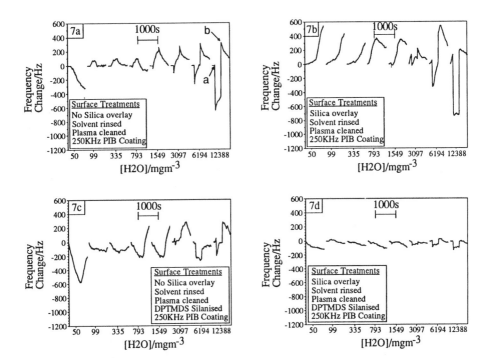

Figure 7. Polyisobutylene coated 200 MHz SAW resonator responses at 25°C to water vapor from 50 - 12388mg/m³, for devices plasma cleaned and silanised with DPTMDS (see figure 4b), and manufactured with and without a protective overlay of silica. The vapor is on at "a" and off at "b".

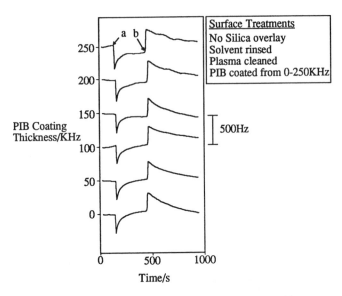

Figure 8. Polyisobutylene coated 200 MHz SAW resonator responses at 25°C to water vapor at 12388mg/m³, for different polymer coating thicknesses ranging from 0 - 250KHz (0 - 529 Å), for a plasma cleaned device manufactured without the protective overlay of silica. The vapor is on at "a" and off at "b".

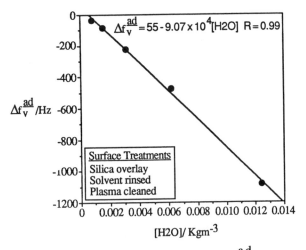

Figure 9. Mass loading water vapor responses, Δf_v^{ad}, at different vapor concentrations for a bare 200MHz SAW resonator prepared by plasma cleaning and manufactured with a protective overlay of silica. The data corresponds to the mass loading responses in Figure 6b. The adsorption coefficient K_I can be determined from the slope of the line.

overlay. The line obtained is linear, and from the slope it is a simple matter to calculate the SAW device/gas adsorption coefficient, K_I. k_1 and k_2 are taken as -9 x 10^{-8} and -4 x 10^{-8} m^2skg^{-1} (2,5,6) and F is 199 x 10^6 Hz, and K_I is computed for water as 1.76 x 10^{-5} m^{-1}. Note that the adsorption coefficient is determined from the normal mass loading component of the responses, the anomalous portions are not used. For a plasma cleaned, uncoated SAW device manufactured without the silica overlay K_I is calculated as 1.82 x 10^{-5} m^{-1}. The similarity of these two adsorption coefficients indicates that the adsorption processes occuring on the SAW devices with and without silica are similar. For a PIB coated SAW device (with no silica overlay) K_I is calculated as 1.33 x 10^{-5} m^{-1}, which is smaller than the equivalent uncoated SAW device. This may indicate that polar polymer end groups adsorb to some adsorption sites on the quartz surface of the SAW device, resulting in a lower interfacial energy, the water vapor has to compete with the polymer end groups for adsorption sites and consequently there is less water adsorption. For the PIB coated SAW device (with silica overlay), which was silane treated prior to polymer coating, the K_I is calculated as 0.27 x 10^{-5} m^{-1}. This is the lowest adsorption coefficient determined for the various surfaces studied and indicates that approximately 85% of the adsorption sites have been eliminated by silanisation.

Concluding Remarks

The use of silica as an overlay treatment of the SAW device is to protect the aluminum electrodes from corrosion and electrical shorting problems from dust particles. The surface chemistry of silica is very similar to the quartz crystal it is coated upon and both are "polar" in nature, which encourages interfacial adsorption of "polar" vapors. This is undesirable and ideally, the protective overlay would be relatively non-polar in nature. However it is important that the SAW device surface is wettable by thin polymer films, so a very non-polar surface could be disadvantageous if polymers of a "polar" nature are used. Adsorption of vapor at the SAW device/polymer coating interface is due to "polar" sites ubiquitous to the surfaces of inorganic oxides. These "polar" sites, especially silanol and aluminium hydroxide, can be controlled by covalent attachment of a phenyl silane to eliminate hydroxyl species. At high surface coverage the phenyldimethyl surface may also act to sterically hinder adsorption of vapor at any residual "polar sites" remaining after silane treatment (23). The determination of adsorption coefficients is a useful method for comparing the properties of different surfaces. A general method to determine the adsorption coefficient, K_I, and partition coefficient, K_p, for different polymer coatings and vapors which exhibit absorption, swelling and adsorption processes would use equation 7. A plot of vapor response versus coating thickness for one vapor concentration would give a slope and an intercept from which K_p and K_I can be determined respectively. Note that, even if all the parameters in the modulus term are not available, K_I will still be calculable from the intercept of the line.

The anomalous SAW device responses in this work are characterized by a slow increase in SAW device frequency that decays over a period of about 15 minutes. This in direct contrast to mass loading effects which are characterized by rapid decreases in frequency over a period of seconds. The rise in frequency for anomalous responses suggests a viscoelastic effect with some stiffening process at the air-SAW device or polymer-SAW device interface. The relatively slow development of the anomalous effect suggests some surface reorganization effect. Water is capable of amphihydrogen-bonding and can form two hydrogen-bonds simultaneously through the two lone pairs on the oxygen or the two hydrogen atoms. For water vapor interacting with a silica surface, Figure 4a, the stiffening process might be accounted for by the structure making processes of water with silanol and silylether groups. The slow development of the anomalous response is not clearly understood, but it does not appear to be diffusion related. SAW devices without the silica overlay expose quartz and aluminium oxide to water vapor. This surface is not expected to be permeable, but anomalous responses are still observed. Anomalous responses in this work appear to be more marked for SAW devices manufactured without the silica overlay, however the exact mechanism of the anomalous response remains to be resolved.

Acknowledgement

This work was supported by the Office of Naval Research, Technology Directorate (block funds administered by Joe Brumfield, Naval Surface Warfare Center, Dahlgren, VA) and performed at the Naval Research Laboratory (NRL 6170).

Literature Cited

(1)   Conder, J. R.; Young, C. L. *Physicochemical Measurements by Gas Chromatography*; John Wiley & Sons: New York, NY, 1979.
(2   Wohltjen, H. *Sens. Actuators* 1984, 5, 307-325.
(3)   Snow, A.; Wohltjen, H. *Anal. Chem.* 1984, 56, 1411-1416.
(4)   Ballantine, D. S., Jr.; Rose, S. L.; Grate, J. W.; Wohltjen, H. *Anal. Chem.* 1986, 58, 3058-3066.
(5)   Grate, J. W.; Snow, A.; Ballantine, D. S., Jr.; Wohltjen, H.; Abraham, M. H.; McGill, R. A.; Sasson, P. *Anal. Chem.* 1988, 60, 869-875.
(6)   Grate, J. W.; Klusty, M.; McGill, R. A.; Abraham, M. H.; Whiting, G. S.; Andonian-Haftvan, J. *Anal. Chem.*, 1992, 64, 610-624.
(7)   Grate, J. W.; Rose-Pehrsson, S. L.; Venezky, D. L.; Klusty, M.; Wohltjen, H. *Anal. Chem.* 1993, 65, 1868-1881.
(8)   McGill, R. A. *Ph.D. thesis*, University of Surrey, England, 1988.
(9)   Abraham, M.H.; Grellier, P.L.; McGill, R. A.; Doherty, R. M.; Kamlet, M. J.; Hall, T. M.; Taft, R. W.; Carr, P.W.; Koros, W. J. *Polymer* 1987, 28, 1363-1369.
(10)   McGill, R. A.; Paley, M. S.; Harris, J. M. *Macromol.*, 1992, 25, 3015-19.
(11)   Paley, M. S; McGill, R. A.; Howard, S. C.; Wallace, S. E.; Harris, J. M. *Macromol.*, 1990, 23, 4557-64.
(12)   McGill, R. A.; Paley, M. S.; Souresrafil, N; Harris, J. M. *Polymer Preprints.* 1990, 31, 578-79.
(13)   Abraham, M. H.; Andovian-Haftvan, J.; Du, M. D.; Diart, V.; Whiting, G. S.; Grate, J. W.; McGill, R. A. Submitted to *Sensors and Actuators* 1993.
(14)   Abraham, M. H. *Chem. Soc Rev.*1993, 73-83.
(15)   McGill, R. A.; Grate, J. W. *Electrochem. Soc. Proc.* 1992, 92-2, 1014.
(16)   Bowers, W. D.; Duong, R.; Chuan, R. L. *Rev. Sci. Instrum.* 1991, 62, 1624-1629.
(17)   Goldfinger, G. (Ed). *Clean Surfaces: Their preparation & characterization for interfacial studies*, Marcel Dekker, Inc., New York, 1970.
(18). Grob,K.;Grob, G.J.*High Res.Chrom.&Chrom. Comm.,*1980,3, 197-198.
(19). McMurtrey, K. D. *J. Liq. Chrom.* 1988, 11, 3375-3384.
(20). Welsch,T;Frank,H.*J.High Res. Chrom.& Chrom. Comm.,* 1985,8,709-714.
(21). Rutten, G.; van de Ven, A.; de Haan, J.; van de Ven, L.; Rijks, J. *J. High Res. Chrom. & Chrom. Comm.*, 1984, 7, 607-614.
(22)   Harris, J. M.; Dust, J. M.; McGill, R. A.; Harris, P. A.; Edgell, M. in *Water Soluble Polymers* Editors. S. W. Shalaby, C. L. McCormick, G. B. Butler ACS Symposium Series 467: Washington, D. C., 1991, 418-429.; Harris, J. M.; .Dust, J. M.;McGill, R. A.; Harris, P. A.; Edgell, M. *Polymer Preprints* 1989, 30(2), 356-357.
(23)   Kirkland, J. J.; J. L. Glajch, Farlee, R. D. *Anal. Chem.*, 1989, 61, 2-11.

RECEIVED March 25, 1994

Chapter 25

Electropolymerized Films in the Development of Biosensors

Allan Witkowski, Seung-Tae Yang, Sylvia Daunert[1], and Leonidas G. Bachas[1]

Department of Chemistry and Center of Membrane Sciences, University of Kentucky, Lexington, KY 40506-0055

The principles of chemical recognition have been widely used in analytical chemistry for the development of selective sensing devices. The recognition properties of several ionophores have been exploited to enable the design of advanced materials suitable for the preparation of fiber optic and potentiometric sensors. Specifically, sensors were prepared by electrochemically growing poly[Co(II)tetra(o-aminophenyl)porphyrin] and poly[Co(II)tetra(p-hydroxyphenyl)porphyrin] on glassy carbon electrodes and indium(tin) oxide (ITO) slides. Further, a novel polymeric material was synthesized by the electrochemically mediated attachment of biotin to a surface. These films should allow for the formation of well-defined biotinylated interfaces, which can be utilized for numerous analytical purposes using the biotin-avidin interaction.

Electrochemical polymerization (electropolymerization) has been recognized as a useful technique for the preparation of pinhole-free membranes (1, 2) that have found numerous applications in electrocatalysis, amperometric biosensors, etc. These membranes are prepared by the electrodeposition of a polymeric film on the surface of an electrode that is immersed in a solution of an appropriate monomer. In order for polymerization to occur, groups that can be electrochemically oxidized or reduced to form a polymer need to be present in the monomer.

This chapter reviews recent developments in our group regarding sensors based on electropolymerized porphyrins and biotin. An inherent advantage of these sensors, compared to conventional ones, is the retention of the active components (i.e., ionophore, metal-selective indicator, affinity ligand, etc.) in the polymer membrane. Indeed, leaching of the plasticizer or the ionophore from the polymer membrane to the aqueous sample solution reduces the lifetime of sensors (3-6). The existence of such leaching has been established previously by a variety of techniques (for a review of this subject see reference 7). This leaching usually worsens the detection limits of the sensors and results in a gradual deterioration of the response (8). Potentiometric and fiber optic sensors based on electropolymerized films present

[1]Corresponding authors

0097-6156/94/0561-0295$08.00/0

296   INTERFACIAL DESIGN AND CHEMICAL SENSING

the advantage that they do not need any plasticizer and that the ionophore is firmly attached to the electrode surface.

Potentiometric sensors based on electropolymerized porphyrins

The response of ion-selective electrodes (ISEs) based on ionophore-impregnated polymer membranes (typically, plasticized poly(vinyl chloride) (PVC)) is mainly controlled by selective transport of ions across the aqueous solution/membrane interface (*3, 7, 9-13*). This results in a separation of charges, and therefore, in an electrical potential. Thus, the activity of the ion in a sample can be determined by the respective change in the potentiometric signal across the membrane. Most of the anion-responsive electrodes initially developed have been based on long-chain tetraalkylammonium salts (*14*). These electrodes respond in decreasing order to the following anions: lipophilic anions > perchlorate > thiocyanate ~ iodide > nitrate ~ bromide > nitrite > chloride > bicarbonate > phosphate > sulfate. This is also the order of decreasing hydrophobicity (known as the Hofmeister series (*15*)), and therefore, the anions for which the electrodes have the greatest selectivity are the large hydrophobic ions. Indeed, there is a direct relation between the Hofmeister selectivity series and the free energies of hydration of the respective anions (*14*). Therefore, ISEs prepared with quaternary ammonium salts are essentially "nonselective".

In order to develop electrodes with selectivity patterns that deviate from the Hofmeister series, a specific interaction must occur between the ionophore and the anion (*12*). Metallomacrocycles satisfy this requirement and have been used as ionophores in the design of "truly" anion-selective electrodes. These ionophores include porphyrins, corrins, and phthalocyanines coordinated with different metals (*12, 16*). When metallomacrocyclic complexes based on these ring systems are used as ionophores in anion-selective electrodes, one or two axial ligands are necessary to complete the coordination sphere of the metal ion. The proposed mechanisms for the selective anion-carrying ability of metallomacrocycles in membranes usually involve the exchange of the axial ligand by the anion of interest (*17-19*). The axial-ligand exchange properties derive from the nature of the coordination between the metal ion and the ligand, and can be influenced by changing either the metal center or some of the ring constituents (*18-29*). Therefore, by manipulating the structure of the complexes, it should be possible to control the selectivity characteristics of the ionophores. Our approach toward achieving this goal is to use electropolymerization to control the polymer structure around the ionophore.

Electropolymerized metalloporphyrins have been used previously in voltammetric and amperometric studies (*30-33*). A mechanism for the polymerization reaction has been proposed which, in the case of aminotetraphenylporphyrin, implicates a radical cation of the porphyrin that is produced in a manner similar to that of the oxidative electropolymerization of aniline (*34*). It has also been reported that during the polymerization the essential porphyrinic monomer structure is preserved (*30, 31, 35, 36*).

Co(II)tetra(*o*-aminophenyl)porphyrin, [Co(*o*-NH$_2$)TPP] (Figure 1), was electropolymerized by cyclic voltammetry on a glassy-carbon surface (*37*). The prepared electrode presented near-Nernstian response to thiocyanate, its preferred anion, and had good detection limits (5 x 10^{-7} M thiocyanate). The typical response time of the electrode was less than 25 s. The selectivity pattern presented by this electrode toward a series of anions is illustrated in Figure 2. The deviation of this anion-selectivity pattern from that of the Hofmeister series (*15*) suggests a selective interaction of the immobilized porphyrin with thiocyanate. Furthermore, the electrodes still had essentially the same slopes and detection limits after two months.

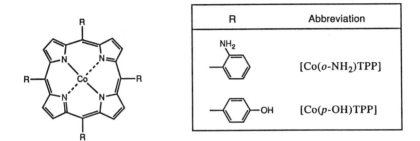

Figure 1. Structures of cobalt(II)tetraphenylporphyrin derivatives.

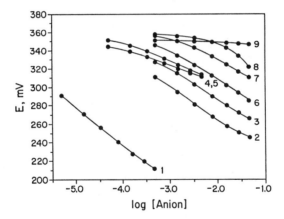

Figure 2. Selectivity pattern of an ion-selective electrode based on electropolymerized [Co(*o*-NH₂)TPP]. The electrode was exposed to (1) thiocyanate, (2) perchlorate, (3) iodide, (4) nitrite, (5) salicylate, (6) bromide, (7) chloride, (8) bicarbonate, and (9) phosphate. (Reproduced with permission from ref. *37*. Copyright 1991 American Chemical Society.)

Several investigators have developed PVC-based electrodes by using lipophilic Co(III)tetraphenylporphyrins as ionophores (20, 22, 23). These electrodes were also selective for thiocyanate but had short lifetimes, as demonstrated by the fact that their slopes deteriorated in just a few weeks (22, 23). The shorter lifetime of PVC-based ISEs has been attributed to leaching of the ionophore from the sensor matrix (3-6). Hodinar and Jyo have also proposed that decreased slopes may be due to crystallization of the ionophore in the membrane (22). Both the PVC electrodes and the electropolymerized films that utilize cobalt porphyrins as ionophores respond primarily to thiocyanate, but their selectivity properties are quite different. In fact, based on data obtained from references 20 and 22, the $\log k^{pot}_{SCN^-,I^-}$ (i.e., logarithm of the selectivity coefficient) calculated for the soluble ionophores was -0.11 and -0.89, respectively. This indicates that in both cases there is only a slight discrimination for thiocyanate over iodide. On the other hand, the electropolymerized [Co(o-NH₂)TPP] films display a much larger discrimination between these two anions, as shown by a value of the $\log k^{pot}_{SCN^-,I^-}$ of -3.3. This was attributed to differences in the ion-recognition properties of the porphyrin resulting from its immobilization during electropolymerization. In such a case, the porphyrin units are oriented in the film in a three-dimensional cross-linked structure. This arrangement allows an additional control on selectivity by forming an ion "sieve". The molecular sieving property of electropolymerized porphyrin films has been demonstrated by Murray and coworkers by obtaining permeability data in ultrathin (15-4600 Å) poly[Co(o-NH₂)TPP] films (38). As a consequence, in the potentiometric sensors reported here, bulkier anions may be precluded from an effective interaction with the ionophore.

The proposed electrodes resemble the coated-wire type of electrodes. Because the latter electrodes do not have a well-defined membrane-wire interface, they usually suffer from drifting potentials that limit their applicability (39). On the contrary, the studied ISEs based on poly[Co(o-NH₂)TPP] had reproducible starting potentials over a period of two months (the standard deviation of the starting potential was 6.2 mV for 11 experiments). This is probably a result of the Co^{2+}/Co^{3+} couple defining the potential across the membrane-glassy carbon interface. It should be noted, however, that the membrane potential of these sensors is affected by the presence of strong redox species, such as ferricyanide, in the sample solution.

Another porphyrin derivative that was used for sensor development is Co(II)tetra(p-hydroxyphenyl)porphyrin [Co(p-OH)TPP] (Figure 1) (40). Using cyclic voltammetry to electropolymerize the monomer on glassy carbon electrodes, a pH-sensitive poly[Co(p-OH)TPP] film was prepared. The pH response of these electrodes has been characterized, with a typical pH calibration curve shown in Figure 3. The poly[Co(p-OH)TPP] electrodes have near-Nernstian slopes (-52 to -54 mV/pH) and linear response from pH 2 to 12.

Unlike poly[Co(o-NH₂)TPP] electrodes, sensors based on poly[Co(p-OH)TPP] do not respond to anions such as perchlorate, iodide, salicylate, nitrate, or thiocyanate, even at concentrations as high as 5×10^{-3} M (experiments performed at pH 6.0). This lack of response to anions has been attributed to Donnan exclusion due to the polyanionic character of the poly[Co(p-OH)TPP] film. The polyanionic nature of these films was verified experimentally based on the incorporation of the electroactive cation $Ru(NH_3)_6^{3+}$ (40).

The pH response of this sensor has been compared to that of bare glassy carbon and polyphenol-coated electrodes (see Figure 3), both of which show a non-linear change in potential with pH over the pH 2 to 12 range. The response of such films has been attributed to proton diffusion to the underlying electrode surface, as evidenced by their slow response times. Since the poly[Co(p-OH)TPP] electrodes respond linearly to pH in a much faster manner (response times less than 10 s), a selective interaction between the protons and porphyrin is suspected. In such a case, the response is determined by the protonation/deprotonation equilibria of hydroxyl groups coordinated to the central cobalt metal.

Fiber optic sensors based on electropolymerized porphyrins

Fiber optic chemical sensors (FOCS) have been reported for several different types of analytes such as ions, metals, pH, gases, enzymes, etc. These optical sensors can often be used as alternatives to electrochemical sensing devices. However, with a few exceptions (e.g., halides, cyanide, etc.), there is a lack of anion-selective FOCS.
In general, fiber optic chemical sensors take advantage of the change in the optical properties of an analyte-selective reagent phase. Typically, the basis of fiber optic sensors is the interaction of an immobilized reagent phase with the analyte of interest in a sample solution. In these devices, optical fibers are used as optical waveguides that carry light to and from an analyte-selective reagent phase at the tip of the sensor, where the chemical interaction takes place. In the presence of the analyte of interest, a change in the optical signal is observed that can be related to the analyte concentration in the sample solution.
Immobilization of reagents for optical sensors can be achieved in a number of ways including entrapping of the reagent within polymer matrices. In that respect, a number of polymer membrane films have been employed in optical sensors. Examples of these membranes include acryloylfluorescein copolymerized with acrylamide or 2-hydroxyethyl methacrylate in the development of both pH-sensitive films (*41*) and CO_2 sensors (*42*), fluorescent polymers (i.e., poly(phenylquinoline), poly(biphenylquinoline), and poly(phenylquinoxaline)) to develop fiber optic sensors for high acidities (*43*), etc. There are also numerous reports of optical sensors based on the immobilization of reagents on or in polymer films (*44, 45*). Given the advantages of electropolymerization in preparing ISEs with improved characteristics, the feasibility of using electropolymerization in the development of fiber optic sensors for ions was investigated as well.
Optical sensors based on poly[Co(*p*-OH)TPP] were constructed by exploiting the fact that the visible absorption spectrum of Co(*p*-OH)TPP displays significant differences as a function of the solution pH (*40*). Therefore, Co(*p*-OH)TPP was electropolymerized by cyclic voltammetry onto transparent ITO glass slides. Two hundred cycles between 0.0 and 1.3 V yielded films that adhered well to the slides and displayed a change in absorbance with a corresponding change in pH. Fiber optic sensors were constructed by placing a piece of the poly[Co(*p*-OH)TPP] coated ITO slide at the common end of a bifurcated fiber optic bundle (Figure 4). One leg of the bundle carries monochromatic light to the sensing layer, while the other leg transmits the collected light to a PMT detector interfaced with a strip-chart recorder. The light measured by the detector was monitored as the analyte solution composition was varied.
Figure 5 shows the pH response of the poly[Co(*p*-OH)TPP] optical sensor from pH 8-12. The relative standard deviation of the pH measured by the sensor was less than 5%. Although a change in signal was observed as a function of pH, the device did not respond to anions such as acetate, nitrate, nitrite, iodide, salicylate, phosphate, thiocyanate, or sulfate at concentrations up to 0.1 M. It is believed that the absence of response to anions is due to Donnan exclusion, as described above.
Our data on fiber optic and ISE sensors based on poly[Co(*p*-OH)TPP] also provide several interesting comparisons. First, approximately 5 min were required to achieve the steady-state response of the poly[Co(*p*-OH)TPP] fiber optic sensor. Although this is longer than the corresponding period for the ISE, the additional time is expected because response of the optical sensor represents a change in a bulk property of the polymer membrane as opposed to merely an interfacial phenomenon in the case of ISEs. Unlike the ISEs, the optical sensors also have a more limited working range. This is because no change in absorbance of the film can be observed

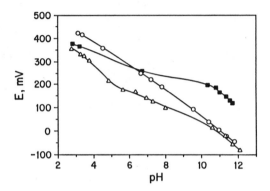

Figure 3. Potentiometric pH response of a bare glassy-carbon electrode (Δ), and electrodes based on electropolymerized [Co(p-OH)TPP] (O) and electropolymerized phenol (■). (Reproduced with permission from ref. *40*. Copyright 1993 American Chemical Society.)

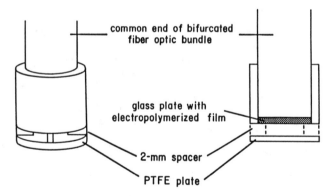

Figure 4. Schematic of the construction of the fiber optic sensor. (Reproduced with permission from ref. *40*. Copyright 1993 American Chemical Society.)

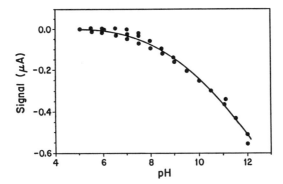

Figure 5. Response to pH of the fiber optic sensor based on electropolymerized poly[Co(*p*-OH)TPP] on indium(tin) oxide glass slides. The y-axis refers to the difference between the PMT current (in microamperes) at the specified pH and that at pH 5.0. (Reproduced with permission from ref. *40*. Copyright 1993 American Chemical Society.)

until a significant proportion of the porphyrin has hydroxide coordinated, which occurs only in a solution of high pH.

   Similarly, an anion-selective fiber optic sensor was prepared using electropolymerized Co(*o*-NH₂)TPP. The polymeric layer was grown by cycling an ITO slide between -0.1 V and +1.2 (or 1.3 V) for a total of 100 cycles at 200 mV/s. The actual sensor was constructed as described above (Figure 4). The absorption spectrum of a poly[Co(*o*-NH₂)TPP] slide changes in the presence of various concentrations of nitrite. As the nitrite concentration increases, a hyperchromic effect is seen due to the replacement of axial water ligands of the cobalt center with nitrite from the sample solution. These sensors were influenced by the ionic strength of the buffer solution in which they were used, and the solution pH also affected sensor response. In fact, a hydroxide interference was observed at pH > 6.5.

   The response of these optical sensors toward nitrite is shown in Figure 6. The sensor was selective for nitrite, yielding a detection limit of 10^{-8} M. The poly[Co(*o*-NH₂)TPP] sensors had slightly longer response times compared to those based on poly[Co(*p*-OH)TPP] (15 min versus 5 min, respectively). The response of the poly[Co(*o*-NH₂)TPP] sensor was stable for at least two months, with an enhancement in lifetime resulting from the covalent immobilization of the monomer during electropolymerization. It should be noted that the poly[Co(*o*-NH₂)TPP] sensor still coordinates with thiocyanate, but there is no detectable change in film absorbance in the presence of this anion. Thus, thiocyanate can not be directly measured with this sensor, but it will act as an interferent for nitrite determinations. Chloride does not interfere with the fiber optic sensor until concentrations higher than 10^{-2} M.

Design of a Novel Biotinylated Interface

The use of biotinylated molecules is widespread in numerous branches of chemistry because of the well-characterized biotin-avidin interaction. In order to address the need for the controlled production of biotinylated surfaces, we have envisioned the development of a covalently attached layer of electropolymerized biotin molecules to provide a well-defined surface capable of a variety of applications. We have synthesized a biotin tyramine derivative (Figure 7) and successfully electropolymerized a poly[biotin tyramine] film on both glassy carbon electrodes and

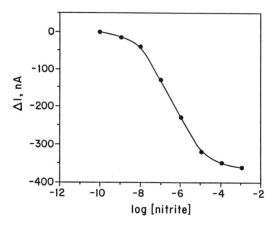

Figure 6. Nitrite response of a fiber optic sensor based on electropolymerized poly[Co(*o*-NH₂)TPP] on indium(tin) oxide glass slides. The y-axis refers to the detector current (in nanoamperes).

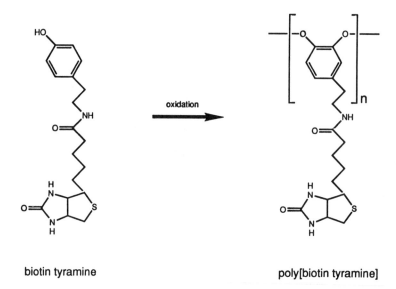

biotin tyramine   poly[biotin tyramine]

Figure 7. A proposed model for the electropolymerization of biotin tyramine.

ITO slides. Experiments are currently underway to characterize these surfaces and to optimize the growth conditions. These films should lead to highly ordered biotinylated surfaces which are covalently attached to the underlying support. These materials should have numerous potential applications in a variety of areas, including potentiometric and optical sensors, immunoassays, solid phase extraction, hybridization assays, and affinity chromatography.

Acknowledgments

The research from the authors' laboratories was supported by grants from the National Science Foundation (DMR-9000782 and EHR-9108764). The authors would also like to thank John R. Allen, Timothy L. Blair, Antonio Florido, and Shelley Wallace Young for their contributions to this work which is described in references *37* and *40*.

Literature Cited

1. Heinze, J. *Top. Curr. Chem.* **1990**, *152*, 1-47.
2. Ivaska, A. *Electroanalysis* **1991**, *3*, 247-254.
3. Moody, G. J.; Saad, B. B.; Thomas, J. D. R. *Sel. Electrode Rev.* **1988**, *10*, 71-106.
4. Davies, J. E. W.; Moody, G. J.; Price, W. M.; Thomas, J. D. R. *Lab. Practice* **1973**, *22*, 20-25.
5. Oesch, U.; Simon, W. *Anal. Chem.* **1980**, *52*, 692-700.
6. Martin, C. R.; Freiser, H. *Anal. Chem.* **1981**, *53*, 902-904.
7. Moody, G. J. *J. Biomed. Eng.* **1985**, *7*, 183-195.
8. Lawton, R. S.; Yacynych, A. M. *Anal. Chim. Acta* **1984**, *160*, 149-158.
9. Arnold, M. A.; Meyerhoff, M. E. *CRC Crit. Rev. Anal. Chem.* **1988**, *20*, 149-196.
10. Meyerhoff, M. E.; Opdycke, W. N. *Adv. Clin. Chem.* **1986**, *25*, 1-47.
11. Kuan, S. S.; Guilbault, G. G. In *Biosensors; Fundamentals and Applications* ; Turner, A. P. F.; Karube, I.; Wilson, G. S., Eds.; Oxford: New York, NY, 1987; pp 135-152.
12. Wotring, V. J.; Johnson, D. M.; Daunert, S.; Bachas, L. G. In *Immunochemical Assays and Biosensor Technology for the 1990's*; Nakamura, R. M.; Kasahara, Y.; Rechnitz, G. A., Eds.; American Society of Microbiology: Washington, D.C., 1992; pp 355-376.
13. Morf, W. E.; Simon, W. *Helv. Chim. Acta* **1986**, *69*, 1120-1131.
14. Yu, R.-Q. *Ion-Selective Electrode Rev.* **1986**, *8*, 153-172.
15. Hofmeister, F. *Arch. Exp. Pathol. Pharmakol.* **1888**, *24*, 247-260.
16. Chang, Q.; Park, S. B.; Kliza, D.; Cha, G. S.; Yim, H.; Meyerhoff, M. E. *Am. Biotechnol. Lab.* **1990**, *8*, 10-21.
17. Ma, S. C.; Chaniotakis, N. A.; Meyerhoff, M. E. *Anal. Chem.* **1988**, *60*, 2293-2299.
18. Chaniotakis, N. A. ; Park, S. B.; Meyerhoff, M. E. *Anal. Chem.* **1989**, *61*, 566-570.
19. Schulthess, P.; Amman, D.; Kräutler, B.; Caderas, C.; Stepánek, R.; Simon , W. *Anal. Chem.* **1985**, *57*, 1397-1401.
20. Ammann, D.; Huser, M.; Krautler, B.; Rusterholz, B.; Schulthess, P.; Lindemann, B.; Halder, E.; Simon, W. *Helv. Chim. Acta.* **1986**, *69*, 849-854.
21. Chaniotakis, N. A.; Chasser, A. M.; Meyerhoff, M. E.; Groves, J. T. *Anal. Chem.* **1988**, *60*, 185-188.
22. Hodinar, A.; Jyo, A. *Chemistry Lett.* **1988**, 993-996.
23. Hodinar, A.; Jyo, A. *Anal. Chem.* **1989**, *61*, 1169-1171.
24. Chang, Q.; Meyerhoff. M. E. *Anal. Chim. Acta* **1986**, *186*, 81-90.

25. Abe, H.; Kokufuta, E. *Bull. Chem. Soc. Jpn.* **1990**, *63*, 1360-1364.
26. Stepánek, R.; Kräutler, B.; Schulthess, P.; Lindemann, B.; Amman, D.; Simon, W. *Anal. Chim. Acta* **1986**, *182*, 83-90.
27. Daunert, S.; Witkowski, A.; Bachas, L. G. *Prog. Clin. Biol. Res.* **1989**, *292*, 215-225.
28. Florido, A.; Daunert, S.; Bachas, L. G. *Electroanalysis* **1991**, *3*, 177-182.
29. Daunert, S.; Bachas, L. G. *Anal. Chem.* **1989**, *61*, 499-503.
30. White, B. A.; Raybuck, S. A.; Bettelheim, A.; Pressprich, K.; Murray, R. W. In *Inorganic and Organometallic Polymers*; Zeldin, M.; Wynne, K. J.; Allcock, H. R., Eds.; American Chemical Society: Washington, D.C., 1988; pp 408-419.
31. White, B. A.; Murray, R. W. *J. Electroanal. Chem.* **1985**, *189*, 345-352.
32. Macor, K. A.; Spiro, T. G. *J. Electroanal. Chem.* **1984**, *163*, 223-236.
33. Malinski, T.; Ciszewski, A.; Fish, J. R.; Czuchajowski, L. *Anal. Chem.* **1990**, *62*, 909-914.
34. Ohnuki, Y.; Matsuda, H.; Ohsaka, T.; Oyama, N. *J. Electroanal. Chem.* **1983**, *158*, 55-67.
35. Bettelheim, A.; White, B. A.; Raybuck, S. A.; Murray, R. *Inorg. Chem.* **1987**, *26*, 1009-1017.
36. Bettelheim, A.; White, B. A.; Murray, R. W. *J. Electroanal. Chem.* **1987**, *217*, 271-286.
37. Daunert, S.; Wallace, S.; Florido, A.; Bachas, L. G. *Anal. Chem.* **1991**, *63*, 1676-1679.
38. Pressprich, K. A.; Maybury, S. G.; Thomas, R. E.; Linton, R. W.; Irene, E. A.; Murray, R. W. *J. Phys. Chem.* **1989**, *93*, 5568-5574.
39. Wang, J. *Electroanalytical Techniques in Clinical Chemistry and Laboratory Medicine*; VCH: New York, NY, 1988.
40. Blair, T. L.; Allen, J. R.; Daunert, S.; Bachas, L. G. *Anal. Chem.* **1993**, *65*, 2155-2158.
41. Munkholm, C.; Walt, D. R.; Milanovich, F. P.; Klainer, S. M. *Anal. Chem.* **1986**, *58*, 1427-1430.
42. Munkholm, C.; Walt, D. R.; Milanovich, F. P. *Talanta* **1988**, *35*, 109-112.
43. Carey, W. P.; Jorgensen, B. S. *Appl. Spectrosc.* **1991**, *45*, 834-838.
44. Wolfbeis, O. S. *Anal. Chim. Acta* **1991**, *250*, 181-201.
45. Arnold, M. A. *Anal. Chem.* **1992**, *64*, 1015A-1025A.

RECEIVED April 5, 1994

Chapter 26

Molecular Interfacing of Enzymes on the Electrode Surface

M. Aizawa, G. F. Khan, E. Kobatake, T. Haruyama, and Y. Ikariyama

Department of Bioengineering, Tokyo Institute of Technology, Nagatsuta, Midori-ku, Yokohama 227, Japan

Few redox enxymes undergo reversible electron transfer on the electrode surface primarily due to steric hindrance of the redox centers of these enzymes. Several molecular inter- faces have been designed to promote electron transfer of electrode-bound redox enzymes. The authors have successfully interfaced fructose dehydrogenase and alcohol dehy- drogenase on the electrode surface with conducting polymer of polypyrrole, which could cause these enzymes to make an electron transfer with retaining their enzymatic activity. The molecular-interfaced redox enzymes will find their application in fabricating biosensors that respond specifi- cally to the corresponding substrates in current.

Biomolecular devices, which are composed of function- ally active biomolecules, have long been explored for achieving such novel information processing systems as biosensors or biocomputers. Although there are many supramolecules in the biological systems that could work as biomolecular devices, few functionally active proteins have successfully been used to realize a biomolecular device, primarily due to difficulty in interfacing them with electronic materials. A new technology has emerged to utilize conducting polymer as a molecular wire and molecular interface to facilitate electron transfer of proteins to the electronic materi- al.
　　　Protein molecules are designed to make excellent intermolecular communication through a specific site of each molecule, most of them situated within a concave structure. Such sites do not allow good electronic

0097–6156/94/0561–0305$08.00/0

communication with electronic materials such as semiconductors and metals. The authors have proposed a
molecular interface for these protein molecules to
promote electronic communication (1,2).

Considering the difficulties in direct electron
transfer, several strategies have been proposed to
overcome these problems. Degani et al. (3) proposed
the incorporation of an electron relay for electronic
communication between glucose oxidase and a Pt electrode. The authors introduced a molecular interface of
conductive polymer which could work as a molecular wire
between an enzyme and an electrode (4,5). Some of the
reports also included electron mediators. Most of
these studies have been concerned with glucose oxidase.
Recently, our group has developed a new electronic
communication between fructose dehydrogenase (FDH) as
well as alcohol dehydrogenase (ADH) and a Pt electrode
through the polypyrrole interface (6 - 8).

In this study, we present a novel approach for
reversible electron transfer between the prosthetic
group of pyrrolo quinoline quinone (PQQ) enzyme (fructose dehydrogenase) and an electrode through a molecular interface. Fructose dehydrogenase (EC 1.1.99.11)
catalyzes the following reaction ; D-Fructose + acceptor $\rightleftharpoons$ 5-keto-D-fructose + reduced acceptor. The PQQ
moieties of randomly oriented fructose dehydrogenase
(FDH) which are very close to the transducer electrode
can easily transfer their electrons to the electrode
(7,8). However, the prosthetic groups of FDH located
far from the electrode can not provide their electrons,
as the distance from the electrode exceeds the electron
transfer distance. Therefore, to make the FDH molecules on the electrode surface electrochemically active, it is essential to introduce a kind of wiring to
assist the electron transfer from PQQ to the electrode.
In this study, an ultrathin conductive polypyrrole (PP)
membrane is employed as a molecular interface material.
We expect that due to the electronic communication
through the conductive interface the FDH molecules on
the electrode surface will become electrochemically
active.

In addition, we demonstrate that the enzyme activity of the membrane-bound FDH is dependent on an applied potential. The enzyme activity is electrochemically controllable without the loss of enzyme activity
in a certain range. This new finding will lead us to
design a novel bioelectronic molecular device.

Electrochemical synthesis of molecular-interfaced FDH
on Pt electrode : The molecular-interfaced FDH was
electrochemically prepared on the Pt electrode surface
by the following two steps ; potential-controlled
adsorption of FDH, and (2) electrochemical polymerization of polypyrrole.Electrochemical adsorption of FDH
(EC 1.1.99.11, MW : 141,000) from *Gluconobacter sp.* on

a Pt electrode surface was extensively studied in our previous paper (7). In this study, the optimum conditions clarified there for adsorption of an FDH monolayer were employed ; i.e., the enzyme (5 mg) was dissolved in 1 ml of 0.1 M phosphate buffer of pH 6.0, which was followed by voltage-assisted adsorption by applying a positive potential of 0.5 V for 10 min. A Pt plate having a surface of 1 sq. cm was used as a working electrode. The FDH-adsorbed electrode was then thoroughly washed with distilled water, and kept in a buffer solution of pH 4.5 until further experiments. All the electrochemical measurements were carried out by a conventional three-electrode system. The electrode potential is referred to the Ag/AgCl electrode.

Electrooxidative polymerization of pyrrole was performed on the FDH-adsorbed electrode in a solution containing 0.1 M pyrrole and 0.1 M KCl under anaerobic conditions at a potential of 0.7 V. The thickness of polypyrrole membrane was controlled by coulometry. After washing with distilled water the PP/FDH/Pt electrode was kept in McIlvaine buffer of pH 4.5 at 4°C for several hours until further experiments commenced.

The enzyme activity of the FDH/Pt and the PP/FDH/Pt electrode was determined by the following method : upon oxidation of D-fructose, the electron acceptor ferricyanide is reduced to ferrocyanide, which forms prussian blue with $Fe_2(SO_4)_3$. The appearance of prussian blue was measured spectrophotometrically at 660 nm. One unit of FDH catalyzes the oxidation of one micromole D-fructose (the formation of two micromole prussian blue) per min.

Reversible electron transfer between FDH and the electrode through PP interface : The differential pulse voltammograms of the FDH/Pt and the PP/FDH/Pt electrodes gave a pair of anodic and cathodic peaks (shown in Fig.1). These were attributed to the electrochemical oxidation and reduction of the quinoprotein at redox potentials of 0.08 V and 0.07 V (vs.Ag/AgCl), respectively. The anodic and cathodic peak shapes and peak currents of the PP/FDH/Pt electrode were identical, which suggested reversibility of the electron transfer process. One further point requires emphasis: the peak current of the PP/FDH/Pt electrode is about 8 times greater than that of the FDH/Pt electrode. The significant increase in redox peaks strongly supports our concept of the conductive biomolecular interface. Due to the incorporation of FDH molecules in the conductive polymer matrix, the prosthetic PQQ electrically communicates with the base electrode through the molecular interface. These results clearly indicate that conductive PP works as an ideal molecular interface.

The cyclic voltammogram of the PP/PDH/Pt electrode is presented in Fig.2. A pair of symmetric and revers-

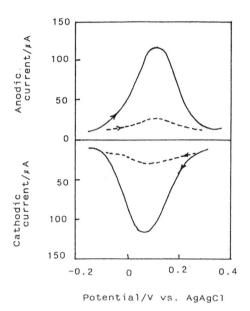

Potential/V vs. AgAgCl

Fig. 1 Differential pulse voltammograms of the FDH/Pt
(--------) and PP/FDH/Pt (————) electrodes in
McIlvaine buffer solution of pH 4.5.
Conditions : pulse amplitude, 50 mV; pulse
interval, 1000 ms; scan rate, 10 mV·s^{-1}.

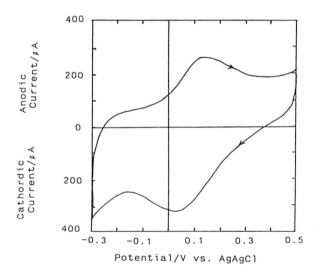

Potential/V vs. AgAgCl

Fig. 2 Cyclic voltammogram of the PP/FDH/Pt electrode
in McIlvaine buffer solution of pH 4.5.   Scan
rate : 500 mV·s^{-1}.

ible anodic and cathodic peaks are obtained. Since the peak current increases linearly with the increase of the potential scan rate, the amount of FDH molecule immobilized on the electrode surface was calculated to be 6.2 $\mu g \cdot cm^{-2}$ from the gradient of the relationship by assuming that the electrochemical reaction proceeds in a two-electron transfer process. This value was a slightly higher than the actual amount of FDH adsorbed on the electrode surface (6.0 $\mu g \cdot cm^{-2}$). These results also support our concept that the molecular interfaced FDH can reversibly and effectively transfer their electrons with the aid of the molecular interface.

Continuous current generation in the presence of D-fructose : If the potentials of the FDH/Pt and the PP/FDH/Pt electrodes are more positive than 0.1 V, the FDH at the electrode surface can be retained in the oxidized form because the redox potential of FDH is less than 0.1 V. As long as the applied potential is higher than the redox potential, the oxidized form of FDH oxidizes D-fructose by transferring its two electrons to the electrode, thus a continuous flow of anodic current is observed. At a lower potential at around 0.1 V the residual current of both electrodes was about 0.38 μA and 0.35 μA for the FDH/Pt and the PP/FDH/Pt electrodes, respectively. To make the residual current as small as possible and to observe the anodic current, the determination of D-fructose was performed at 0.4 V. The response curves of these electrodes upon the addition of substrate are presented in Fig. 3. To ascertain the selectivity of the enzyme electrode, we injected a series of saccharides. A typical example is shown in the figure. Satisfactory selectivity was obtained in the case of the PP/FDH/Pt electrode. However, the FDH/Pt electrode suffered from nonspecific response to saccharides such as glucose, probably due to the simultaneous surface oxide formation and saccharide adsorption at the noble metal surface. The response time was as rapid as 2-3 sec at the FDH/Pt and 3-5 sec at the PP/FDH/Pt electrode. The response current of the PP/FDH/Pt electrode to D-fructose was about 4 times larger than that of the FDH/Pt electrode, which suggested effective electron shuttling in the interface.

Potential controlled activity of FDH in the conductive interface : The potential dependence of the biocatalytic activity of the PP/FDH/Pt electrode is shown in Fig. 4. The dependence was investigated by adding 5mM fructose to a magnetically stirred solution, and the resulting current response was compared. The applied potential-current response curve was divided into three parts. First potential range was less than the redox potential of FDH (0.07 V). Negligible re-

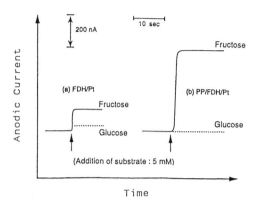

Fig. 3 The response curves of (a)FDH/Pt and
        (b)PP/FDH/Pt electrode upon the addition of 5
        mM substrates.   The potential was controlled at
        0.4 V.

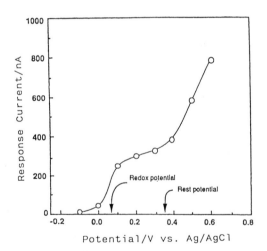

Fig. 4 Dependency of the electrochemical activity of
        the PP/FDH/Pt electrode on applied potential.
        Different potentials were applied to the same
        PP/FDH/Pt electrode. When the residual current
        became steady state, fructose solution was
        injected to a final concentration of 5 mM, and
        the resulting increase in anodic current was
        measured as the response current.

sponse current was generated in this potential range because less oxidation of fructose occurred, since very few FDH was at oxidized form. The second potential range extended from the redox potential of FDH to the rest potential of the PP/FDH/Pt electrode (0.35 V), in which a sharp increase of response current at round 0.1 V and then a gradual increase were observed. The third potential range was at a potential higher than the rest potential. The sharp increase in response current was observed up to 0.6 V. At a potential higher than this potential the response current fell sharply, probably due to the irreversible deactivation of FDH. The possible reason for the deactivation at higher potential may be : (1) higher electrical field causes conformational change of the FDH in such a way that the enzyme looses its prosthetic group, or it may change the structure of the enzyme into an inactive form, (2) higher potential drastically changes the pH inside the membrane interface in a manner that the enzyme loses its activity. However, enzyme activity at the PP/FDH/Pt electrode can be controlled reversibly in the potential range from 0.1 V to 0.6 V.

The increase in anodic current upon the addition of fructose can be explained by considering the following three processes : 1) Diffusion of substrate into the PP/solution interface, 2) FDH–PQQ is reduced to FDH–PQQH$_2$ by the added fructose, 3) Reoxidation of reduced FDH to oxidized FDH–PQQ by releasing electron to the electrode through the PP interface. All of these steps seem to be accelerated by the application of higher potential.

The performance of the PP-interfaced electrode was controllable as far as the applied potential was in the range from 0.1 V to 0.6 V without any deterioration of the enzyme. Since PP is doped state in this potential range, the π-conjugated polymer works as a molecular interface which carries electron between a protein molecule and an electrode.

Molecular-interfaced alcohol dehydrogenase with NAD and its response to ethanol. : It has been difficult to incorporate dehydrogenases that are coupled with NAD(P) into amperometric enzyme sensors owing to the irreversible electrochemical reaction of NAD. We have developed a molecular-interfaced alcohol dehydrogenase on the surface of an electrode on which NAD is electrochemically regenerated within a membrane matrix.

Alcohol dehydrogenase (ADH), Meldla's blue (MB) and NAD are immobilized within a conductive polypyrrole membrane on the platinum electrode surface. ADH (EC 1.1.1.1) catalyzes the following reaction ; Alcohol + NAD $\rightleftharpoons$ aldehyde or ketone + NADH.

A platinum disk electrode (0.2 cm^2) was electrolytically platinized in a platinum chloride solution to increase the surface area and enhance the adsorption

power. The platinized platinum electrode was then
immersed in a solution containing 10 mg ml^{-1} ADH, 0.75
mM MB and 6.2 mM NAD. We found that ADH could make a
very stable complex with MB at a molar ratio of 1 to 2.
After sufficient adsorption of these molecules on the
electrode surface, the electrode was transferred into
a solution containing 0.1 M pyrrole and 1 M KCl.
Electrochemical polymerization of pyrrole was conducted
at +0.7 V vs. Ag/AgCl. The electrolysis was stopped at
a total charge of 1 C cm^{-2}. An enzyme-entrapped polypyrrole membrane was deposited on the electrode surface.
    Cyclic voltammetry was performed with the ADH–NAD–
MB/polypyrrole electrode in 0.1 M phosphate buffer (pH
8.5) at a scan rate of 5 mV s^{-1}. The cyclic voltammogram is presented in Fig.5. A pair or redox peaks was
attributed to the oxidation and reduction ones of MB.
The substrate ethanol of ADH caused the anodic current
at +0.10 V vs. Ag/AgCl to increase. These results suggest a possible electron transfer from membrane–bound
ADH to the electrode through membrane–bound NAD and MB
with the help of the conductive polymer of polypyrrole.
    The electrode gave a steady–state current when the
electrode potential was maintained at +0.10 V vs.
Ag/AgCl in 0.1 M phosphate buffer. Addition of ethanol
to the buffer solution resulted in an increase in the
anodic current, which was attributed to the oxidation
of membrane–bound NADH. A steady response was obtained
within 40 sec. The increase in the anodic current was
linearly correlated with the concentration of ethanol.
    In a further study continuous electrolytic oxida-
tion of ethanol was performed with the molecular-
interfaced ADH/MB/NAD. The electrode potential was
controlled at 0.35 V vs. Ag/AgCl. Ethanol in a solution was enzymatically oxidized to acetaldehyde with
resulting in the reduction of NAD to NADH. The turnover
number of the NAD/NADH cycle was calculated from the
total conversion of ethanol and the coulomb during the
electrolysis. The results clearly showed that NAD was
electrochemically regenerated within the polypyrrole
membrane through the electron transfer from ADH, NAD
and MB to the electrode.

Conclusion
    Electronic communication of fructose dehydrogenase
(FDH) with a Pt electrode was accomplished through the
conducting polymer molecular interface. Electrons were
reversibly transferred between the active center of FDH
and the electrode surface when the electrode potential
was properly controlled. The enzyme activity of the
molecular-interfaced FDH was found to be modulated in
the presence of D-fructose by the electrode potential.
Electronic communication of alcohol dehydrogenase (ADH)
with a Pt electrode was also accomplished in the

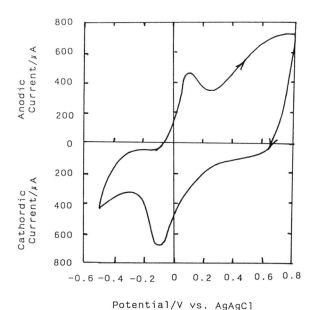

Potential/V vs. AgAgCl

Fig. 5   Cyclic voltammogram of the PP/ADH/NAD/MB/Pt
electrode at pH 8.0.   Scan rate : 5 mV•s^{-1}.

presence of NAD and Meldola's blue through the conducting polymer molecular interface.

Literature Cited

1. Aizawa, M., "Molecular Electronics-Science and
   Technology" (A.Aviram, ed.), Eng.Found., New York
   (1989)
2. Aizawa, M.; Yabuki, S.; Shinohara, H.   *"Molecular
   Electronics"*, (F. T. Hong., ed.), Plenum Press, New
   York 1989, p.269.
3. Degani, Y.; Heller, A., J.Phys.Chem., *1987* 91, 1288-
   1289.
4. Aizawa, M.; Yabuki, S.; Shinohara, H.; Chiba, T.
   Proc.Int.Symp. Electroorganic Synthesis (S.Torii,
   ed.) Elsevier, Amsterdam, 1987, pp.353-360.
5. Yabuki, S.; Shinohara, H.; Aizawa, M.
   J.Chem.Soc.Chem.Commun. 1989, 14, 945.
6. Yabuki, S.; Shinohara, H.; Ikariyama, Y.; Aizawa, M.
   J.Electroanal.Chem. 1990,277,179.
7. Shinohara, H.; Khan, G.F.; Ikariyama, Y.; Aizawa. M.
   *J.Electroanal.Chem.*1991, 304, 75.
8. Khan, G.F., Shinohara, H., Ikariyama, Y., and
   Aizawa, M.; *J.Electroanal.Chem.* 1991, 315, 263.

RECEIVED April 15, 1994

Chapter 27

Fluorescent Chemosensors of Ion and Molecule Recognition

Recent Applications to Pyrophosphate and to Dopamine Sensing

Anthony W. Czarnik

Department of Chemistry, Ohio State University, Columbus, OH 43210

1,8-Bis(TRPN)methylanthracene, available from 1,8-dibromomethylanthracene, binds phosphate and pyrophosphate anions in aqueous solution, pH 7, with signalling in the form of chelation enhanced fluorescence. The affinity of this convergent chemosensor for pyrophosphate (K_d= 2.9 µM) is over 2000-times that for phosphate (K_d= 6.3 mM); a reference nonconvergent chemosensor [1-mono(TRPN)methylanthracene] displays only a 100-fold difference in binding affinities. The ion discrimination thus realized permits real-time analysis in the hydrolysis of pyrophosphate by inorganic pyrophosphatase.
It has been known for almost 35 years that catechol complexes reversibly to boronic acids. We observe that 2-anthrylboronic acid complexes catechol in water with K_d 330 µM and concomitant 20-fold reduction in fluorescence intensity. L-DOPA and dopamine behave similarly, suggesting a mechanism for the development of real-time sensing schemes.

Over the past several decades, chemists have devoted their collective attention to the design of molecules that bind other molecules reversibly. A primary goal of such work is the desire to engineer synthetic catalysts from first principles. However, an equally appealing goal, and one with significant practical potential, is the design of chemosensors with selectivity for analytes such as anions (e.g., phosphate), zwitterions (e.g., amino acids), and neutral species (e.g., O_2). Insufficient attention has been paid to this application of ion and molecule recognition, given the potential utility of chemosensors as essential components in the construction of real-time monitoring devices. Because many sensing applications will take place in an aqueous environment, aqueous recognition is prerequisite. Considerable recent effort has focused on fluorescence as a signal transduction mechanism because of its sensitivity, its potential application to fiber

0097–6156/94/0561–0314$08.00/0

optic-based remote sensing schemes, and because the multiplicity of mechanisms for modulating fluorescence provides for the incorporation of many known binding mechanisms.

Anion Chemosensing: Pyrophosphate

While an ability to monitor the concentrations of simple anions in real-time is surely as desirable as that for metal ions, there is almost no literature on the chemosensing of monomeric, inorganic anions in water.   Batch methods have been described for fluorimetric chemodosimetry of several halides,[1] but *sensing* methods (i.e., nondestructive and reversible) reported to date all rely on the ability of the analyte (e.g., iodide) to quench collisionally the fluorescence of compounds such as rhodamine 6G.[2] Thus, it would seem useful to apply the substantial literature on anion recognition to this purpose.[3] As a signal type, fluorescence is both more sensitive than absorption spectroscopies and more amenable to the conception of signal transduction mechanisms on the molecular level.   In 1989, we reported previously that the aqueous complexation of anions such as phosphate and sulfate to anthrylpolyamines (e.g., **1**) result in chelation-enhanced fluorescence (CHEF) signalling.[4] Compound **1** proved to be the first fluorescent chemosensor for an inorganic anion that does not rely on an inherent quenching property of the anion.   Even so, it is a conception far from usable form.   For example, chemosensor **1** binds phosphate at pH 6 with K_d=150 mM; in other words, the midpoint of the titration occurs at a phosphate concentration of 150 mM, which is very high for many applications.   Likewise, selectivity between a variety of simple anions (sulfate, acetate) is nonexistent.   In an effort to improve upon the conceptual design of anion sensors, we synthesized a convergent chemosensor demonstrating anion binding in a more useful concentration range (i.e., μM K_d) and designed to discriminate between phosphate ('P$_i$') and pyrophosphate ('PP$_i$') ions based on size.

Chemosensor **3**, prepared by the reaction of 1,8-dibromomethylanthracene[5] with excess tris(3-aminopropyl)amine (TRPN), yields a pH-fluorescence titration with pK_a^{obsd}= 6.7; thus, anion binding could be measured at pH 7 even though that using **1** (pK_a^{obsd}= 5.4) could not.[6] Besides possessing a high charge density at neutral pH, the anion receptor groups of **3** are geometrically disposed to bind both sides of pyrophosphate simultaneously. Our previously reported rationale for CHEF signalling of phosphate[4] applies similarly to the current situation, as shown in Scheme I. We anticipated that the binding of pyrophosphate (non-fluorescent) to chemosensor **3** (low fluorescence at pH 7 due to photoinduced-electron transfer) would yield enhanced fluorescence as a result of intracomplex amine protonation. Anion titrations were thus carried out in pH 7 solution containing 4 μM **3** and excess cyclen (1 mM), which does not itself bind to **3** but does chelate adventitious transition metal ions that would otherwise bind to TRPN ligands in **3**. We were gratified to observe (Figure 1) that the binding of pyrophosphate to **3** occurs with fluorescence enhancement (λ_{ex} 368 nm, λ_{em} 414 nm), and that the 1:1 complexation occurs with K_d= 2.9 μM under these conditions. By contrast, the association of **3** to phosphate occurs with a K_d of 6.3 mM, which is only 13-times tighter binding than that of phosphate with 1-monoTRPNanthracene (not shown; K_d= 82 mM). Thus, chemosensor **3** displays a pyrophosphate/phosphate discrimination of 2200-fold at pH 7.

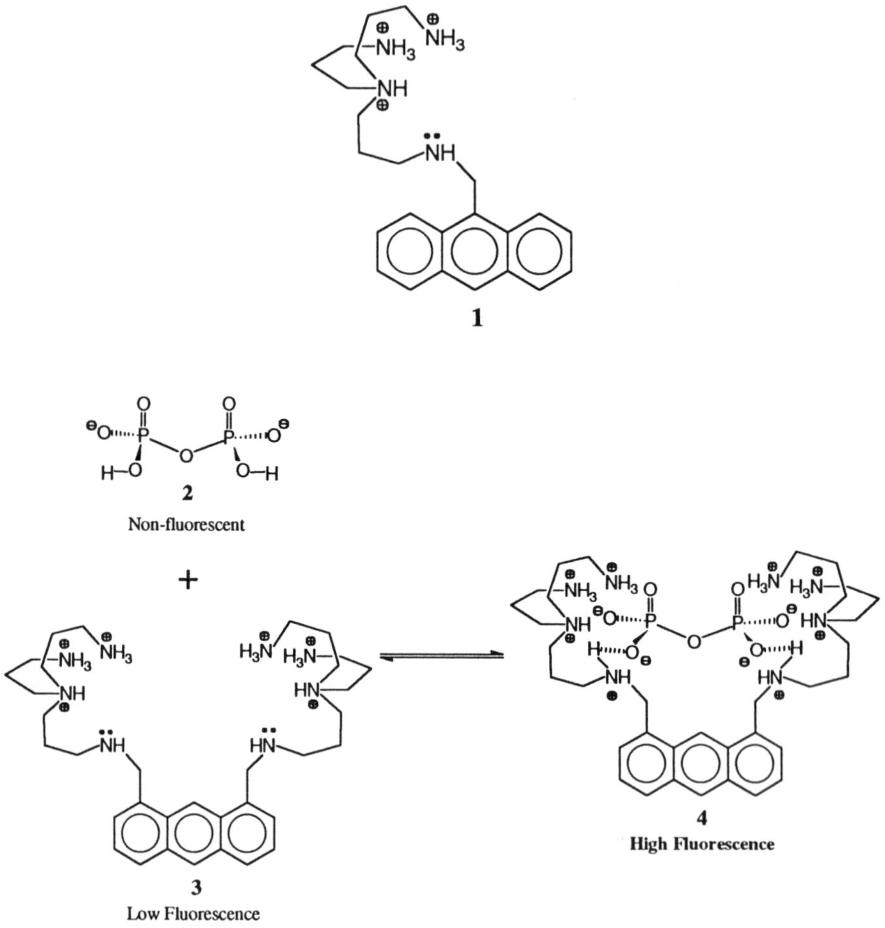

1

2

Non-fluorescent

+

3

Low Fluorescence

4

High Fluorescence

Scheme I. Reaction scheme rationalizing the observed fluorescence increase upon the binding of pyrophosphate to chemosensor **3**.

The ion selectivity, which is also readily apparent in Figure 1, allows for real-time assay of pyrophosphate hydrolysis. For example, the fluorescence of a reaction containing 20 µM pyrophosphate would be expected to decrease (by 54%) on conversion at constant pH to 40 µM phosphate (Scheme II). This reaction is catalyzed by inorganic pyrophosphatase, a Mg(II)-dependent hydrolase that serves to drive reactions dependent upon PP_i release by further hydrolysis to P_i.[7] Previously reported assay methods for this enzyme include; phosphomolybdate analysis, paper electrophoresis, starch gel electrophoresis, isotopic labelling, pH stat analysis and enyzmic-UV coupling.[8] To test the use of chemosensor 3 for this purpose, we prepared solutions containing 50 mM HEPES, 1 mM cyclen, 20 µM $Na_4P_2O_7$, 20 µM $Mg(ClO_4)_2$ and 4 µM 3. Inorganic PPase from baker's yeast was added, and after a brief mixing period (30 seconds) fluorescence intensity (λ_{ex} 368 nm; λ_{em} 414 nm) was monitored as a function of time. As shown in Figure 2, the hydrolysis of pyrophosphate generates the predicted fluorescence response, with the rate of hydrolysis increasing as does the enzyme concentration.

Neutral Chemosensing: Dopamine and Other Catechols

There is a substantial classical literature describing the fluorimetric indication of ions in water,[9] which has been augmented by more recent studies.[10] We have been studying additionally the synthesis of fluorescent chemosensors for *uncharged* organic analytes in water. The design of any fluorescent chemosensor requires knowledge of three topics: (1) how can one bind a specie with selectivity from water, (2) how can one generate signals from such binding events that are easy to measure, and (3) what mechanisms for binding and signal transduction intersect. Mechanisms have been elucidated by which ion binding events can be transduced to fluorescent events, but these is less literature available on how to design fluorescent chemosensors of neutral analytes.[11]

The complexation of carbohydrates with phenylboronic acid has been known for at least 40 years,[12] and the reversibility of that interaction serves as a basis for the chromatographic separation of sugars.[13] In 1959, Lorand and Edwards[14] reported association constants (listed here as K_d's) for aqueous associations of phenylboronic acid with many saturated polyols; binding interactions range from very weak (e.g., ethylene glycol, K_d=360 mM) to moderately strong (e.g., glucose, K_d=9.1 mM). We have utilized these results to evaluate a mechanism for fluorescent chemosensing based on polyol-boronic acid association.[15] This abiotic sensing scheme yields fluorescence quenching of 40% upon polyol chelation, owing to a modulation of the boronic acid pK_a.

Lorand and Edwards also described the binding of phenylboronic acid to catechol, with significantly stronger association (K_d=57 µM) than that exhibited by carbohydrates.[14] As efficient intramolecular quenching has been observed previously by other electron-rich organic compounds,[14] it seemed reasonable that the catechol complexes of anthrylboronic acids might similarly demonstrate chelation-enhanced quenching (**CHEQ**). Indeed, we found that anthrylboronic acids bind catechol and catecholamines from water, K_d=330 µM (for the 2-isomer), with concomitant 20-fold changes in fluorescence intensity.[16,17]

Fluorescence titrations of 6 and 9 with catechol (1,2-dihydroxybenzene; 7) were carried out using an 0.75 µ*M* solution of chemosensor at room temperature and p*H* 7.4 (Scheme III), and the results are shown in Figure 3. (Excitation was

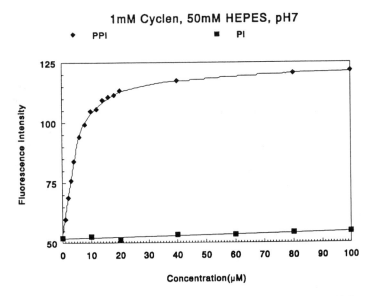

Figure 1. Fluorescence titrations of chemosensor **3** with P_i and PP_i.

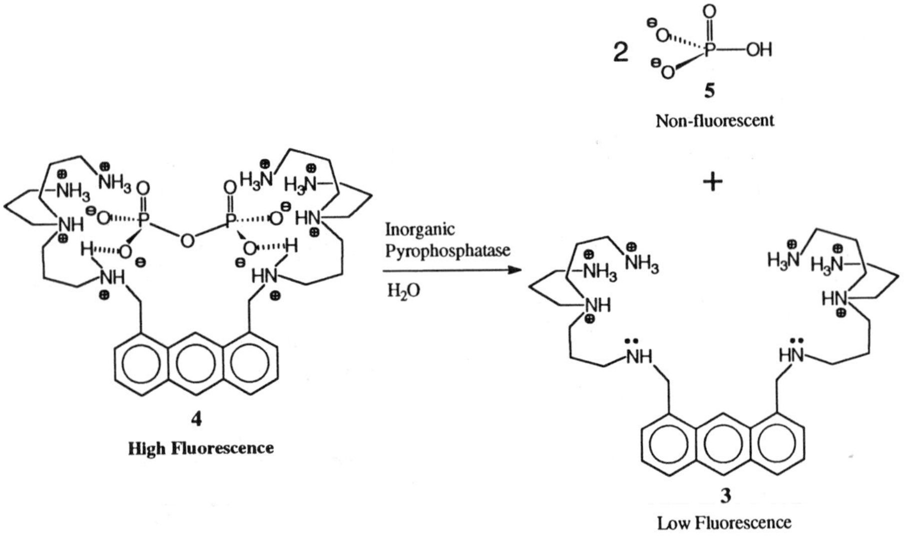

Scheme II. Reaction scheme rationalizing the observed fluorescence assay of inorganic pyrophosphatase activity using chemosensor **3**.

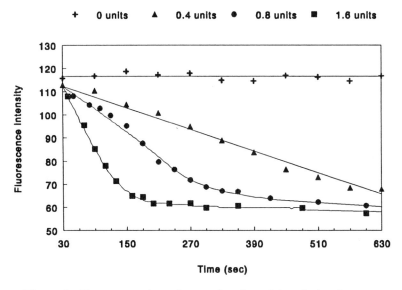

Figure 2. Fluorescence intensity as a function of time during the pyrophosphatase-catalyzed hydrolysis of PP_i to $2P_i$, as monitored using chemosensor **3**.

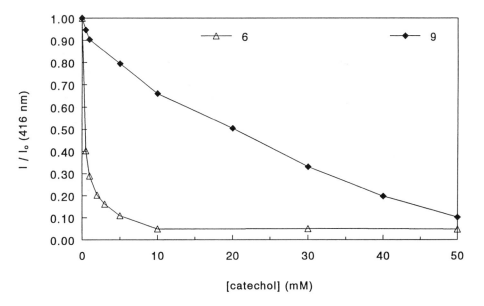

Figure 3. Fluorescence titrations of 2- and 9-anthrylboronic acids with catechol (-Δ- **6**; -◆- **9**). (Reproduced with permission from reference 16. Copyright 1993 Pergamon Press.)

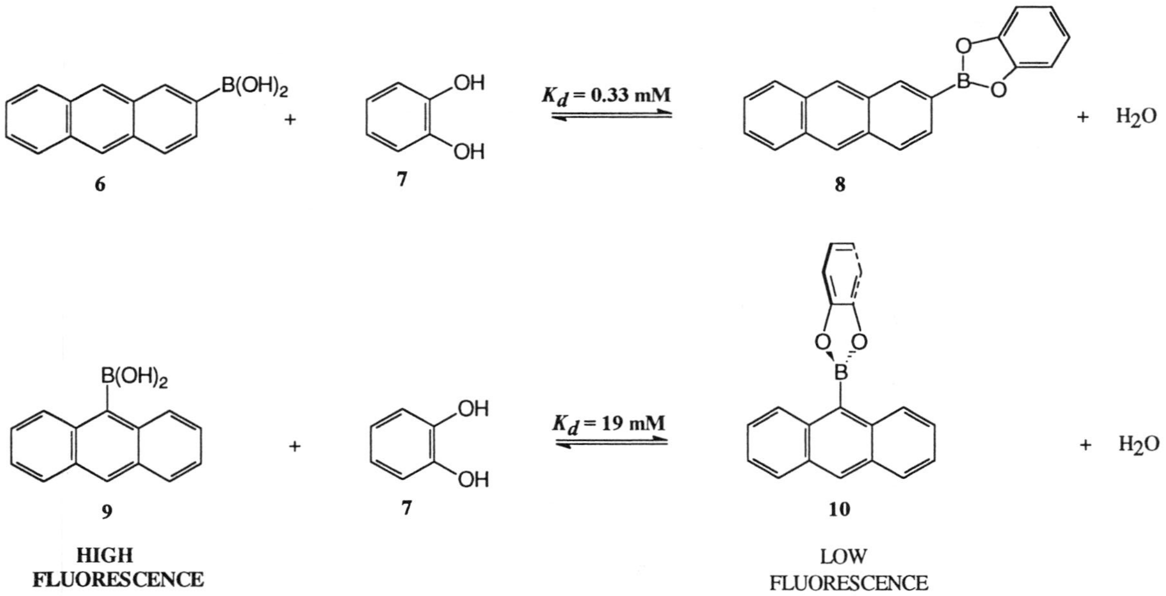

Scheme III. Reversible complexation of catechol by 2- and 9-anthrylboronic acids. The 2-isomer (**6**) displays higher affinity for catechol. (Reproduced with permission from reference 16. Copyright 1993 Pergamon Press.)

at 348 nm, and emission was observed at 416 nm; because catechol itself displays weak fluorescence (centered at 396 nm), the data shown include subtraction of anthracene-free reference solutions.) Several observations are notable. First, the reversible complexation of catechol to each anthrylboronic acid results in an approximately 20-fold **CHEQ** on saturation. Association of **6** or **9** with catechol (**7**) yields boronic esters **8** and **10**, respectively (Scheme III). The 57-fold weaker interaction with **9** may be attributable to peri-interactions, although this is speculation only. As a working hypothesis, we speculate that boronic esters **8** and **9** are of lower fluorescence due to photoinduced electron transfer (**PET**) from catechol into the excited S_1 state of anthracene (Scheme IV).

To evaluate further both the selectivity and intensity range for this mechanism of signal transduction, fluorescence titrations with a variety of potential ligands were compiled (Figure 4). Guaicol (**11**) and phenol (**12**), each structurally related to catechol, also complex with CHEQ, although the association is predictably weaker. Glucose (**13**), whose association with **3** has been described by us previously, also yields CHEQ; however, because the mechanism for fluorescence quenching is different, the magnitude is smaller. Additionally, the binding of catechol is considerably stronger than that of glucose.

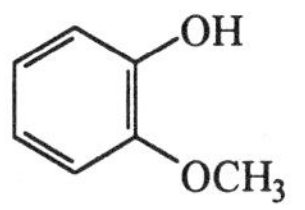

11 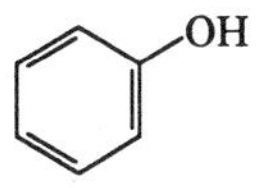 **12** 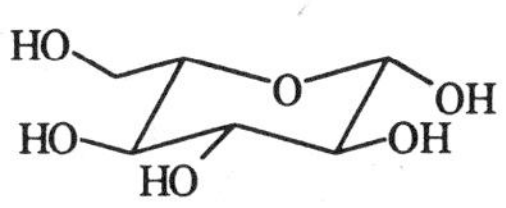 **13**

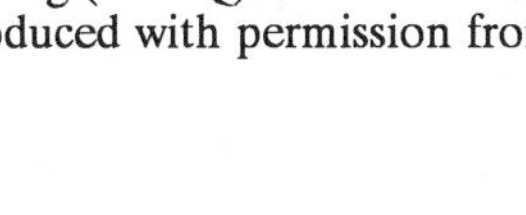

14  **15**

Scheme IV. A working hypothesis for the observed chelation-enhanced quenching (CHEQ) effect observed in the equilibrium shown in Scheme III. (Reproduced with permission from reference 16. Copyright 1993 Pergamon Press.)

Figure 4 also documents the chemosensing of the catecholamine dopamine (**14**) and its metabolic pregenitor, L-Dopa (**15**). We observe that each compound complexes to anthrylboronic acid **6** with roughly equal binding affinities and fluorescence quenching as does catechol (K_d 330 µM); this is again predictable, as both compounds posess the catechol group in their structures. Because dopamine is an important neurotransmitter that regulates functions in the brain, heart, kidney, vasculature, and gut, advances in the ability to visualize it in real-time may prove of interest.

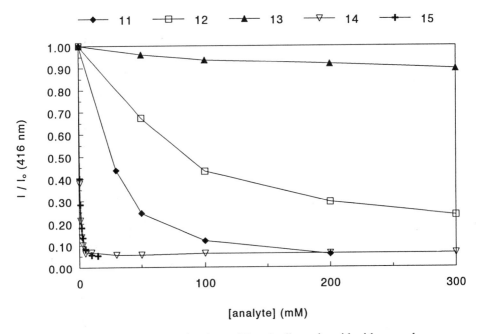

Figure 4. Fluorescence titrations of 2-anthrylboronic acid with several analytes (-◆- **11**; -□- **12**; -▲- **13**; -▽- **14**; -+- **15**). (Reproduced with permission from reference 16. Copyright 1993 Pergamon Press.)

References and Notes

1) Guilbault, G. G., in *Practical Fluorescence*, Guilbault, G. G., Ed., Marcel Dekker, Inc.: New York, 1990, Chapter 5.

2) Wyatt, W. A.; Bright, F. V.; Hieftje, G. M. *Anyl. Chem.* **1987**, *59*, 2272.

3) Lehn, J.-M. *J. Inclusion Phenom.* **1988**, *6*, 351. For an excellent overview of the great variation achievable in the design of anion receptors, see: Schmidtchen, F. P. *Nachr. Chem., Tech. Lab.* **1988**, *8*, 10.

4) Huston, M. E.; Akkaya, E. U.; Czarnik, A. W. *J. Am. Chem. Soc.* **1989**, *111*, 8735.

5) Akiyama, S.; Nakagawa, M. *Bull. Chem. Soc. Jap.* **1971**, *44*, 3158-3160.

6) While the fluorimetrically determined pK_a^{obsd} for **1** describes its +3→+4 ionization, pK_a^{obsd} for **3** likely is an aggregate of two +3→+4 ionizations, one at each TRPN group.

7) (a) Josse, J. and Wong, S. C. K., "Enzymes," 3rd ed., Vol. 4, pp 499, Boyer, P. D., Ed., Academic Press, New York, New York, 1971; (b) Butler, L. G. *ibid*, 529.

8) (a) Josse, J. *J. Biol. Chem.* **1966**, *241*, 1938; (b) Moe, O. A.; Butler, L. G. *J. Biol. Chem.* **1972**, *247*, 7308; (c) Cooperman, B. S.; Chiu, N. Y.; Bruckman, R. H.; Bunick, G. J.; McKenna, G. P. *Biochem.* **1973**, *12*, 1665; (d) Ryan, L. M.; Koxin, F.; McCarty, D. J.; *Arthritis Rheum.* **1979**, *22*, 892; (e) Baltscheffsky, M.; Nyrén, P. *Methods Enzymol.* **1988**, *126*, 538. Only the pH

stat-method is non-destructive with real-time signalling, but it is not useful in the μM concentration range.

9) (a) Schwarzenbach, G. & Flaschka, H. *Complexometric Titrations*, translation by H. Irving (Metheun, London, 1969); (b) West, T. S. *Complexometry with EDTA and Related Reagents* (BDH, London, 1969); (c) Bishop, E. ed., *Indicators* (Pergamon, NY, 1972); (d) Guilbault, G. G. *Practical Fluorescence* (Marcel Dekker, NY, 1973), chapter 6.

10) For lead references to recent work by several authors, see: *Fluorescent Chemosensors for Ion and Molecule Recognition*, Czarnik, A. W., Ed., ACS Books, Washington, 1993.

11) A notable exception is found in recent work by Ueno: Ueno, A., Kuwabara, T., Nakamura, A.; Toda, F. *Nature* **356**, 136 (1992).

12) Kuivila, H. G. , Keough, A. H.; Soboczenki, E. J. *J. Org. Chem.* **19**, 780 (1954).

13) (a) Weith, H. L., Wiebers, J. L.; Gilham, P. T. *Biochemistry* **9**, 4396 (1970); (b) Schott, H. *Angew. Chem. Int. Ed. Eng.* **11**, 824 (1972); (c) Barker, S. A., Hatt, B. W., Somers, P. J.; Woodbury, R. R. *Carbohyd. Res.* **26**, 55 (1973); (d). Yurkevich, A. M, Kolodkina, I. I., Ivanova, E. A.; Pichuzhkina, E. I *Carbohyd. Res.* **43**, 215 (1975).

14) Lorand, J. P.; Edwards, J. O. *J. Org. Chem.* **24**, 769 (1959).

15) Yoon, J.-Y.; Czarnik, A. W. *J. Am. Chem. Soc.* **114**, 5874 (1992).

16) Yoon, J.-Y.; Czarnik, A. W. *Bioorg. Med. Chem.* **1**, 267 (1993).

17) The current method is to be distinguished clearly from the well-known fluorimetric trihydroxyindol assay for catecholamines, which consumes the analyte (i.e., it is a batch analysis) and is not intended to afford real-time response: von Euler, U. S.; Lishajko, F. *Acta Physiol. Scand.* **51**, 348 (1961).

RECEIVED April 27, 1994

INDEXES

Author Index

Affiliation Index

Subject Index

Production: Meg Marshall
Indexing: Deborah H. Steiner
Acquisition: Barbara Pralle
Cover design: Alan Kahan

Printed and bound by Maple Press, York, PA

Highlights from ACS Books

Good Laboratory Practice Standards: Applications for Field and Laboratory Studies
Edited by Willa Y. Garner, Maureen S. Barge, and James P. Ussary
ACS Professional Reference Book; 572 pp; clothbound ISBN 0–8412–2192–8

Silent Spring Revisited
Edited by Gino J. Marco, Robert M. Hollingworth, and William Durham
214 pp; clothbound ISBN 0–8412–0980–4; paperback ISBN 0–8412–0981–2

The Microkinetics of Heterogeneous Catalysis
By James A. Dumesic, Dale F. Rudd, Luis M. Aparicio, James E. Rekoske,
and Andrés A. Treviño
ACS Professional Reference Book; 316 pp; clothbound ISBN 0–8412–2214–2

Helping Your Child Learn Science
By Nancy Paulu with Margery Martin; Illustrated by Margaret Scott
58 pp; paperback ISBN 0–8412–2626–1

Handbook of Chemical Property Estimation Methods
By Warren J. Lyman, William F. Reehl, and David H. Rosenblatt
960 pp; clothbound ISBN 0–8412–1761–0

Understanding Chemical Patents: A Guide for the Inventor
By John T. Maynard and Howard M. Peters
184 pp; clothbound ISBN 0–8412–1997–4; paperback ISBN 0–8412–1998–2

Spectroscopy of Polymers
By Jack L. Koenig
ACS Professional Reference Book; 328 pp;
clothbound ISBN 0–8412–1904–4; paperback ISBN 0–8412–1924–9

Harnessing Biotechnology for the 21st Century
Edited by Michael R. Ladisch and Arindam Bose
Conference Proceedings Series; 612 pp;
clothbound ISBN 0–8412–2477–3

From Caveman to Chemist: Circumstances and Achievements
By Hugh W. Salzberg
300 pp; clothbound ISBN 0–8412–1786–6; paperback ISBN 0–8412–1787–4

The Green Flame: Surviving Government Secrecy
By Andrew Dequasie
300 pp; clothbound ISBN 0–8412–1857–9

For further information and a free catalog of ACS books, contact:
American Chemical Society
Distribution Office, Department 225
1155 16th Street, NW, Washington, DC 20036
Telephone 800–227–5558

Bestsellers from ACS Books

The ACS Style Guide: A Manual for Authors and Editors
Edited by Janet S. Dodd
264 pp; clothbound ISBN 0–8412–0917–0; paperback ISBN 0–8412–0943–X

The Basics of Technical Communicating
By B. Edward Cain
ACS Professional Reference Book; 198 pp;
clothbound ISBN 0–8412–1451–4; paperback ISBN 0–8412–1452–2

Chemical Activities (student and teacher editions)
By Christie L. Borgford and Lee R. Summerlin
330 pp; spiralbound ISBN 0–8412–1417–4; teacher ed. ISBN 0–8412–1416–6

Chemical Demonstrations: A Sourcebook for Teachers,
Volumes 1 and 2, Second Edition
Volume 1 by Lee R. Summerlin and James L. Ealy, Jr.;
Vol. 1, 198 pp; spiralbound ISBN 0–8412–1481–6;
Volume 2 by Lee R. Summerlin, Christie L. Borgford, and Julie B. Ealy
Vol. 2, 234 pp; spiralbound ISBN 0–8412–1535–9

Chemistry and Crime: From Sherlock Holmes to Today's Courtroom
Edited by Samuel M. Gerber
135 pp; clothbound ISBN 0–8412–0784–4; paperback ISBN 0–8412–0785–2

Writing the Laboratory Notebook
By Howard M. Kanare
145 pp; clothbound ISBN 0–8412–0906–5; paperback ISBN 0–8412–0933–2

Developing a Chemical Hygiene Plan
By Jay A. Young, Warren K. Kingsley, and George H. Wahl, Jr.
paperback ISBN 0–8412–1876–5

Introduction to Microwave Sample Preparation: Theory and Practice
Edited by H. M. Kingston and Lois B. Jassie
263 pp; clothbound ISBN 0–8412–1450–6

Principles of Environmental Sampling
Edited by Lawrence H. Keith
ACS Professional Reference Book; 458 pp;
clothbound ISBN 0–8412–1173–6; paperback ISBN 0–8412–1437–9

Biotechnology and Materials Science: Chemistry for the Future
Edited by Mary L. Good (Jacqueline K. Barton, Associate Editor)
135 pp; clothbound ISBN 0–8412–1472–7; paperback ISBN 0–8412–1473–5

For further information and a free catalog of ACS books, contact:
American Chemical Society
Distribution Office, Department 225
1155 16th Street, NW, Washington, DC 20036
Telephone 800–227–5558